Advanced Electromagnetic Models for Materials Characterization and Nondestructive Evaluation

Scientific Computation

More information about this series at http://www.springer.com/series/718

Harold A. Sabbagh • R. Kim Murphy
Elias H. Sabbagh • Liming Zhou
Russell Wincheski

Advanced Electromagnetic Models for Materials Characterization and Nondestructive Evaluation

Springer

Harold A. Sabbagh
Victor Technologies, LLC
Bloomington, IN, USA

R. Kim Murphy
Victor Technologies, LLC
Bloomington, IN, USA

Elias H. Sabbagh
Victor Technologies, LLC
Bloomington, IN, USA

Liming Zhou
Hermitage, TN, USA

Russell Wincheski
NASA Langley Research Center
Hampton, VA, USA

ISSN 1434-8322 ISSN 2198-2589 (electronic)
Scientific Computation
ISBN 978-3-030-67954-5 ISBN 978-3-030-67956-9 (eBook)
https://doi.org/10.1007/978-3-030-67956-9

This Springer imprint is published by the registered company Springer Nature Switzerland AG
The registered company address is: Gewerbestrasse 11, 6330 Cham, Switzerland

Preface

This book is a continuation of a study begun in [111] in which we developed and applied computational electromagnetics and inverse techniques to quantitative non-destructive evaluation and materials characterization. While model-based inverse methods using a nonlinear least-squares algorithm (NLSE) were developed in [111], we develop two new 'voxel-based' algorithms in this book: the bilinear conjugate-gradient and set-theoretic algorithms.

NLSE is suitable when there are only a few parameters that are required to define a model. There are, of course, any number of practical problems that cannot be simply defined in terms of a few variables. For example, an actual fatigue crack may not be well-represented by a simple slot or EDM notch that could be parameterized for NLSE. Actual cracks may emanate from a number of sites ('multisite damage') or may simply meander through the structure, so that they are not so easily parametrized. This suggests that we need to reconstruct the anomaly voxel-by-voxel, which is what the set-theoretic and bilinear conjugate-gradient algorithms are designed to do.

By 'voxel-based', we mean that each voxel in the grid of the anomalous region is to be reconstructed in order to determine the anomaly, rather than to model the anomaly at the outset as a canonical structure that is defined by a few parameters that are to be reconstructed. In this sense, voxel-based algorithms are an attempt to eliminate the 'curse of dimensionality' malady that is present in NLSE due to the huge interpolation grid that is required for large problems [111].

We have studied two such voxel-based algorithms: the bilinear conjugate-gradient algorithm and the set-theoretic algorithm. The conjugate-gradient algorithm is rather well-known in the literature and reconstructs the grid in the usual manner of considering the mutual interaction of each voxel on the other, much as in solving the forward problem with **VIC-3D®**. On the contrary, the set-theoretic algorithm reconstructs each voxel independently of the other (after a first stage of linear data processing), i.e., it is non-Markovian and yields estimation-theoretic quality estimates of each voxel as well. Furthermore, the set-theoretic algorithm introduces some interesting statistical notions that are relevant to the research theme of this project. Both of these algorithms are well-suited to the volume-integral approach that is the basis of **VIC-3D®**.[1]

One interesting feature of set-theoretic estimation is that it is not an optimization algorithm, such as NLSE or bilinear conjugate-gradients. Such algorithms require derivatives, while set-theoretic estimation does not. In fact, we do not really invoke the notion of 'optimization' until the very end of the algorithm, and then we do it on a voxel-by-voxel basis using the robust estimators that are defined in [111, Chapter 13]. For this reason, we use the set-theoretic algorithm throughout the text.

Part II is the major portion of the book and deals with developing electromagnetic models for characterizing a variety of materials, from carbon-fiber reinforced polymer (cfrp) advanced composites to atherosclerotic lesions. This includes developing rigorous models for handling anisotropies and then applying the set-theoretic inversion algorithm to these materials. This plays a very significant role in such matters as microstructure quantification problems, which are important in the manufacture of cfrp prepregs, as well as in the analysis of completed structures. In addition, we introduce such matters as studying stochastic models for anisotropic materials that allow us to develop computational models for microstructure characterization, among other things. Part II ends with a discussion of electromagnetic models for biological tissue, which allows us to do such things as noninvasively detect and characterize atherosclerotic lesions.

The last two chapters of the book deal with the application of quantum effects on materials characterization. Chapter 11 deals with such things as paramagnetic spin dynamics and the spin Hamiltonian. These things, of course, are fundamental to the study of such practical systems as masers. Finally, in the last chapter, we begin the development of electromagnetic models for carbon nanotube-reinforced polymers through the use of quantum principles. These materials are finding application not only in structures but also in electromagnetic devices. In both of these chapters, we point out that the quantum calculations that are required to yield the electromagnetic properties that **VIC-3D®** requires are done off-line, and then these properties are imported into **VIC-3D®** to model the electromagnetic response of a device or structure.

Bloomington, IN, USA — Harold A. Sabbagh
Bloomington, IN, USA — R. Kim Murphy
Bloomington, IN, USA — Elias H. Sabbagh
Hermitage, TN, USA — Liming Zhou
Hampton, VA, USA — Russell Wincheski

[1] www.sabbagh.com.

Acknowledgments

Much of the work reported in this book was supported by the Air Force Research Laboratory, Dayton, Ohio, and the NASA Langley Research Center, Hampton, VA. Our colleagues, John Aldrin of Computational Tools, Gurnee, IL, and Jeremy Knopp of AFRL, made significant contributions as well.

Our thanks to Christoph Bauman, Dinesh Vinayagam, and Radha Lakshmanan of Springer for their excellent handling of the production of this book.

Contents

Part I
Voxel-Based Inversion Algorithms

Chapter 1
A Bilinear Conjugate-Gradient Inversion Algorithm

1.1 Optimization via Nonlinear Least-Squares

Standard methods for minimizing a real-valued function of several variables can be divided into two general classes: those that require second derivative information, usually referred to as Newton-type methods, and those requiring only first derivative information, referred to as gradient methods. There are several excellent texts which, in addition to discussing many of these methods in detail, also give suggestions on when to use certain techniques. See, for example, the texts by Fletcher[36], Hestenes[52] or Luenberger[64].

In this chapter, we concentrate on gradient techniques for minimizing Φ, the norm of the residuals, for basically two reasons. First, the Töplitz-Hankel structure of the operators in the original volume-integral equation allow us to use fast Fourier transform techniques when doing matrix multiplications in solving the forward problem, and secondly, the bilinearity of the entire system allows us to find the gradient of Φ in closed form, as well as performing exact line searches when minimizing Φ in a particular direction. We first presented this method in [102–104]; it is known in the recent literature as the 'contrast source inversion method' [1, 10, 131].

1.2 A Bilinear Conjugate-Gradient Inversion Algorithm Using Volume-Integrals

Bilinear Inversion Algorithm Consider a T/R configuration, in which a fixed transmitting coil excites the anomaly, which is assumed to have a conductivity vector, $\boldsymbol{\sigma}$, and a receive coil scans the anomaly at positions, $i = 1, \ldots, N_v$, where N_v is the number of 'views'. Conversely, we could assume a single transmitting coil,

H. A. Sabbagh et al., *Advanced Electromagnetic Models for Materials Characterization and Nondestructive Evaluation*, Scientific Computation,
https://doi.org/10.1007/978-3-030-67956-9_1

and an array of N_v receive coils occupying the positions of the original scanned receive coil. Furthermore, we could actually scan fewer than N_v positions, and then interpolate to complete the N_v data points. (See [107] and [14] for further discussions of multiview and multifrequency reconstruction methods.)

In any case, the transmitter generates an anomalous scattering current, $\mathbf{J}^{(x)}$, $\mathbf{J}^{(y)}$, $\mathbf{J}^{(z)}$, which is computed by **VIC-3D®** and the receive coil(s) produces an incident field, $\mathbf{E}_{0x}^{(R)}(i)$, $\mathbf{E}_{0y}^{(R)}(i)$, $\mathbf{E}_{0z}^{(R)}(i)$. These combine to produce the change-in-transfer impedance due to the anomaly,

$$Z^{(i)}(\mathbf{J}) = \mathbf{E}_{0x}^{(R)}(i) \cdot \mathbf{J}^{(x)} + \mathbf{E}_{0y}^{(R)}(i) \cdot \mathbf{J}^{(y)} + \mathbf{E}_{0z}^{(R)}(i) \cdot \mathbf{J}^{(z)} , \tag{1.1}$$

that is also computed directly by **VIC-3D®**.[1] Note that, unlike our NLSE algorithm, we do not consider $\mathbf{J}^{(x)}$, $\mathbf{J}^{(y)}$, $\mathbf{J}^{(z)}$ to be secondary variables that are dependent upon the primary unknown, $\boldsymbol{\sigma}$. Rather, the current-vectors along with the cell *resistivities*, $\boldsymbol{\rho}$, are the primary unknowns.

To determine the model equations for the inversion algorithm, return to the fundamental volume-integral electric equation that **VIC-3D®** solves for the anomalous currents:

$$\begin{bmatrix} \mathbf{E}^{(0x)} \\ \mathbf{E}^{(0y)} \\ \mathbf{E}^{(0z)} \end{bmatrix} = \begin{bmatrix} \mathbf{Q}^{(x)}(\boldsymbol{\rho}) & 0 & 0 \\ 0 & \mathbf{Q}^{(y)}(\boldsymbol{\rho}) & 0 \\ 0 & 0 & \mathbf{Q}^{(z)}(\boldsymbol{\rho}) \end{bmatrix} \begin{bmatrix} \mathbf{J}^{(x)} \\ \mathbf{J}^{(y)} \\ \mathbf{J}^{(z)} \end{bmatrix} + \begin{bmatrix} \mathbf{G}^{(xx)} & \mathbf{G}^{(xy)} & \mathbf{G}^{(xz)} \\ \mathbf{G}^{(yx)} & \mathbf{G}^{(yy)} & \mathbf{G}^{(yz)} \\ \mathbf{G}^{(zx)} & \mathbf{G}^{(zy)} & \mathbf{G}^{(zz)} \end{bmatrix} \begin{bmatrix} \mathbf{J}^{(x)} \\ \mathbf{J}^{(y)} \\ \mathbf{J}^{(z)} \end{bmatrix} . \tag{1.2}$$

The incident field on the left-hand side of (1.2) is due to the transmitting coil. The total electric-field moments are given by

$$\begin{aligned} \mathbf{E}^{(x)}(\boldsymbol{\rho}, \mathbf{J}^{(x)}) &= \mathbf{Q}^{(x)}(\boldsymbol{\rho}) \cdot \mathbf{J}^{(x)} \\ \mathbf{E}^{(y)}(\boldsymbol{\rho}, \mathbf{J}^{(y)}) &= \mathbf{Q}^{(y)}(\boldsymbol{\rho}) \cdot \mathbf{J}^{(y)} \\ \mathbf{E}^{(z)}(\boldsymbol{\rho}, \mathbf{J}^{(z)}) &= \mathbf{Q}^{(z)}(\boldsymbol{\rho}) \cdot \mathbf{J}^{(z)} , \end{aligned} \tag{1.3}$$

so that when this is substituted back into (1.2), we get the basic constraint equation between the primary variables, $\boldsymbol{\rho}$ and $\mathbf{J}$:

[1] We are considering only 'electric-electric' interactions in this chapter. We assume that all hosts and flaws are nonmagnetic.

$$\begin{bmatrix} \mathbf{E}^{(0x)} \\ \mathbf{E}^{(0y)} \\ \mathbf{E}^{(0z)} \end{bmatrix} = \begin{bmatrix} \mathbf{E}^{(x)}(\boldsymbol{\rho}, \mathbf{J}^{(x)}) \\ \mathbf{E}^{(y)}(\boldsymbol{\rho}, \mathbf{J}^{(y)}) \\ \mathbf{E}^{(z)}(\boldsymbol{\rho}, \mathbf{J}^{(z)}) \end{bmatrix} + \begin{bmatrix} \mathbf{G}^{(xx)} & \mathbf{G}^{(xy)} & \mathbf{G}^{(xz)} \\ \mathbf{G}^{(yx)} & \mathbf{G}^{(yy)} & \mathbf{G}^{(yz)} \\ \mathbf{G}^{(zx)} & \mathbf{G}^{(zy)} & \mathbf{G}^{(zz)} \end{bmatrix} \begin{bmatrix} \mathbf{J}^{(x)} \\ \mathbf{J}^{(y)} \\ \mathbf{J}^{(z)} \end{bmatrix} . \tag{1.4}$$

Thus, the two constraint equations that define the inverse model are the linear system, (1.1), and *bilinear* system, (1.4). A bilinear system is linear in each variable, when the other is held constant. In this case, the variables are $\boldsymbol{\rho}$ and $\mathbf{J}^{(x)}, \mathbf{J}^{(y)}, \mathbf{J}^{(z)}$.

Recall that the tri-diagonal matrices are symmetric, and have the following non-zero entries:

$$Q_{kK}^{(x)}(\boldsymbol{\rho}) = \frac{\delta x \delta y \delta z}{6} \begin{cases} \rho_{klm} & \text{if } K = k-1 \\ 2(\rho_{klm} + \rho_{k+1,lm}) & \text{if } K = k \\ \rho_{k+1,lm} & \text{if } K = k+1 \end{cases}$$

$$Q_{lL}^{(y)}(\boldsymbol{\rho}) = \frac{\delta x \delta y \delta z}{6} \begin{cases} \rho_{klm} & \text{if } L = l-1 \\ 2(\rho_{klm} + \rho_{k,l+1,m}) & \text{if } L = l \\ \rho_{k,l+1,m} & \text{if } L = l+1 \end{cases} \tag{1.5}$$

$$Q_{mM}^{(z)}(\boldsymbol{\rho}) = \frac{\delta x \delta y \delta z}{6} \begin{cases} \rho_{klm} & \text{if } M = m-1 \\ 2(\rho_{klm} + \rho_{klm+1}) & \text{if } M = m \\ \rho_{klm+1} & \text{if } M = m+1 \end{cases}$$

The first entry in each of these matrices is the lower diagonal, the second the main diagonal, and the third the upper diagonal. Recall that the resistivity of the klmth cell is the reciprocal of the conductivity, $\rho_{klm} = 1/\sigma_{klm}$. Note that if cell klm is filled with host material, then $\rho_{klm} = \infty$. This fact could be useful in allowing ρ_{klm} to act like a penalty term, forcing the associated current expansion coefficient, J_{klm}, to zero.

Now,

$$E_{klm}^{(x)} = \sum_K Q_{kK}^{(x)} J_{Klm}^{(x)}$$

$$E_{klm}^{(y)} = \sum_L Q_{lL}^{(y)} J_{kLm}^{(y)}$$

$$E_{klm}^{(z)} = \sum_M Q_{mM}^{(z)} J_{klM}^{(z)} , \tag{1.6}$$

is the form that is useful when singling out the currents at each cell. When we want to single out the resistivity at each cell, we simply rearrange (1.6):

$$E_{klm}^{(x)} = \sum_K P_{kK}^{(x)} \rho_{Klm}$$

$$E^{(y)}_{klm} = \sum_{L} P^{(y)}_{lL} \rho_{kLm}$$

$$E^{(z)}_{klm} = \sum_{M} P^{(z)}_{mM} \rho_{klM} , \tag{1.7}$$

where the matrix elements follow from (1.5) and (1.6)

$$P^{(x)}_{kK}(\mathbf{J}^{(x)}) = \frac{\delta x \delta y \delta z}{6} \begin{cases} J^{(x)}_{k-1lm} + 2J^{(x)}_{klm} & \text{if } K = k \\ 2J^{(x)}_{klm} + J^{(x)}_{k+1lm} & \text{if } K = k+1 \end{cases}$$

$$P^{(y)}_{lL}(\mathbf{J}^{(y)}) = \frac{\delta x \delta y \delta z}{6} \begin{cases} J^{(y)}_{kl-1m} + 2J^{(y)}_{klm} & \text{if } L = l \\ 2J^{(y)}_{klm} + J^{(y)}_{kl+1m} & \text{if } L = l+1 \end{cases} \tag{1.8}$$

$$P^{(z)}_{mM}(\mathbf{J}^{(z)}) = \frac{\delta x \delta y \delta z}{6} \begin{cases} J^{(z)}_{klm-1} + 2J^{(z)}_{klm} & \text{if } M = m \\ 2J^{(z)}_{klm} + J^{(z)}_{klm+1} & \text{if } M = m+1 \end{cases}$$

One approach to solving the system, (1.1) and (1.4), has been to use a linear ('Born') approximation, which means that the electric field within the flawed region is assumed to be the same as the electric field if the flaw were not present, namely the incident electric field. Thus, one would write $\mathbf{J} = \sigma \mathbf{E} \approx \sigma \mathbf{E}^0$, and then take moments of this equation. Substituting the resulting expressions for $\mathbf{J}^{(x,y,z)}_{klm}$ into (1.1) will produce a linear system for the unknown conductivities, σ_{klm}, that can be solved in a number of ways. See [105] for an example of this approach, in which the linear system is solved using the *algebraic reconstruction technique* (ART), and [24, 50, 51], as well as [111, pp. 282–285], for details on ART.

This approach, therefore, has the physical significance of ignoring the secondary sources produced by multiple scattering within the flaw. From a mathematical viewpoint, the second term of (1.4) (with the **G** matrices) is ignored. While linearization is occasionally accurate, it is not as general as the main subject of this chapter, bilinear conjugate-gradients. See [147] for a further discussion of the Born approximation, and [14] for an application of the Born approximation to solve a scattering problem of a three-dimensional flaw embedded in anisotropic composite materials.

Bilinear Conjugate-Gradients We use nonlinear conjugate-gradients to minimize the functional formed by the sum of squares of the residuals comprising (1.1) and (1.4):

$$\begin{aligned} \Phi(\boldsymbol{\rho}, \mathbf{J}) = \frac{1}{2} \Big\{ & \|\mathbf{E}^{(R)}_{0x} \cdot \mathbf{J}^{(x)} + \mathbf{E}^{(R)}_{0y} \cdot \mathbf{J}^{(y)} + \mathbf{E}^{(R)}_{0z} \cdot \mathbf{J}^{(z)} - \mathbf{Z}_{\text{meas}}\|^2 \\ & + \|\mathbf{E}^{(x)}(\boldsymbol{\rho}, \mathbf{J}^{(x)}) + \mathbf{G}^{(xx)} \cdot \mathbf{J}^{(x)} + \mathbf{G}^{(xy)} \cdot \mathbf{J}^{(y)} + \mathbf{G}^{(xz)} \cdot \mathbf{J}^{(z)} - \mathbf{E}^{(0x)}\|^2 \\ & + \|\mathbf{E}^{(y)}(\boldsymbol{\rho}, \mathbf{J}^{(y)}) + \mathbf{G}^{(yx)} \cdot \mathbf{J}^{(x)} + \mathbf{G}^{(yy)} \cdot \mathbf{J}^{(y)} + \mathbf{G}^{(yz)} \cdot \mathbf{J}^{(z)} - \mathbf{E}^{(0y)}\|^2 \end{aligned}$$

$$+\|\mathbf{E}^{(z)}(\boldsymbol{\rho},\mathbf{J}^{(z)})+\mathbf{G}^{(zx)}\cdot\mathbf{J}^{(x)}+\mathbf{G}^{(zy)}\cdot\mathbf{J}^{(y)}+\mathbf{G}^{(zz)}\cdot\mathbf{J}^{(z)}-\mathbf{E}^{(0z)}\|^2\Big\}$$

$$=\frac{1}{2}\Big\{\|\mathscr{A}^{(R)}(\mathbf{J})-\mathbf{Z}_{\text{meas}}\|^2$$

$$+\|\mathscr{A}^{(x)}(\boldsymbol{\rho},\mathbf{J})-\mathbf{E}^{(0x)}\|^2+\|\mathscr{A}^{(y)}(\boldsymbol{\rho},\mathbf{J})-\mathbf{E}^{(0y)}\|^2+\|\mathscr{A}^{(z)}(\boldsymbol{\rho},\mathbf{J})-\mathbf{E}^{(0z)}\|^2\Big\}$$

$$=\Phi^{(R)}(\mathbf{J})+\Phi^{(x)}(\boldsymbol{\rho},\mathbf{J})+\Phi^{(y)}(\boldsymbol{\rho},\mathbf{J})+\Phi^{(z)}(\boldsymbol{\rho},\mathbf{J}). \tag{1.9}$$

Now let's say a few words about the vector-matrix structure of these functionals. As shown in (1.1), $\mathscr{A}^{(R)}-\mathbf{Z}_{\text{meas}}$ is an $N_v\times 1$-vector, whose components are given by (1.1). This suggests that we should think of $\mathbf{E}_{0x}^{(R)}$, $\mathbf{E}_{0y}^{(R)}$, $\mathbf{E}_{0z}^{(R)}$, as $N_v\times N_c$-dimensional matrices, $E_{0x(i,klm)}^{(R)}$, $E_{0y(i,klm)}^{(R)}$, $E_{0z(i,klm)}^{(R)}$, $i=1,\ldots,N_v$, $klm=1,\ldots,N_c$, where N_c is the number of cells in the problem. Then, we have

$$\begin{aligned}
\Phi^{(R)}(\mathbf{J}) &= \frac{1}{2}\|\mathscr{A}^{(R)}(\mathbf{J})-\mathbf{Z}_{\text{meas}}\|^2\\
&= \frac{1}{2}\sum_{i=1}^{N_v}|\mathscr{A}_i^{(R)}(\mathbf{J})-Z_{\text{meas}}(i)|^2\\
&= \frac{1}{2}\sum_{i=1}^{N_v}\left(\mathscr{A}_i^{(R)}(\mathbf{J})-Z_{\text{meas}}(i)\right)\left(\mathscr{A}_i^{(R)}(\mathbf{J})-Z_{\text{meas}}(i)\right)^*\\
&= \frac{1}{2}\sum_{i=1}^{N_v}\left(\mathbf{E}_{0x}^{(R)}(i)\cdot\mathbf{J}^{(x)}+\mathbf{E}_{0y}^{(R)}(i)\cdot\mathbf{J}^{(y)}+\mathbf{E}_{0z}^{(R)}(i)\cdot\mathbf{J}^{(z)}-Z_{\text{meas}}(i)\right)\\
&\quad\left(\mathbf{E}_{0x}^{(R)}(i)\cdot\mathbf{J}^{(x)}+\mathbf{E}_{0y}^{(R)}(i)\cdot\mathbf{J}^{(y)}+\mathbf{E}_{0z}^{(R)}(i)\cdot\mathbf{J}^{(z)}-Z_{\text{meas}}(i)\right)^*,
\end{aligned} \tag{1.10}$$

where * denotes the complex-conjugate, and the vector dot-product in the last expression implies a sum over the cell indices, klm.

It is a straightforward calculation, using the final form of (1.10), to show that the gradient of $\Phi^{(R)}(\mathbf{J})$ with respect to its primary variables, $J_{klm}^{(x)}$, $J_{klm}^{(y)}$, and $J_{klm}^{(z)}$ is

$$\begin{aligned}
\frac{\partial\Phi^{(R)}}{\partial J_{klm}^{(x)}} &= \sum_i E_{0x(klm,i)}^{(R)H}\left(\mathbf{E}_{0x}^{(R)}(i)\cdot\mathbf{J}^{(x)}+\mathbf{E}_{0y}^{(R)}(i)\cdot\mathbf{J}^{(y)}+\mathbf{E}_{0z}^{(R)}(i)\cdot\mathbf{J}^{(z)}-Z_{\text{meas}}(i)\right)\\
\frac{\partial\Phi^{(R)}}{\partial J_{klm}^{(y)}} &= \sum_i E_{0y(klm,i)}^{(R)H}\left(\mathbf{E}_{0x}^{(R)}(i)\cdot\mathbf{J}^{(x)}+\mathbf{E}_{0y}^{(R)}(i)\cdot\mathbf{J}^{(y)}+\mathbf{E}_{0z}^{(R)}(i)\cdot\mathbf{J}^{(z)}-Z_{\text{meas}}(i)\right)\\
\frac{\partial\Phi^{(R)}}{\partial J_{klm}^{(z)}} &= \sum_i E_{0z(klm,i)}^{(R)H}\left(\mathbf{E}_{0x}^{(R)}(i)\cdot\mathbf{J}^{(x)}+\mathbf{E}_{0y}^{(R)}(i)\cdot\mathbf{J}^{(y)}+\mathbf{E}_{0z}^{(R)}(i)\cdot\mathbf{J}^{(z)}-Z_{\text{meas}}(i)\right),
\end{aligned} \tag{1.11}$$

where the superscript, H, on a matrix denotes the Hermitian of that matrix, namely the complex-conjugate of the transpose of the matrix. The term in parentheses is the ith component of the residual-vector, $\mathbf{RESID}(R)$.

The gradients of $\Phi^{(x)}$, $\Phi^{(y)}$, and $\Phi^{(z)}$ with respect to $\mathbf{J}^{(x)}$, $\mathbf{J}^{(y)}$, and $\mathbf{J}^{(z)}$ are given by:

$$\begin{aligned}
\frac{\partial \Phi^{(x)}}{\partial J_{klm}^{(x)}} &= \sum_{KLM} \left(Q_{klm,KLM}^{(x)T} + G_{klm,KLM}^{(xx)H} \right) (\mathbf{RESID}(X))_{KLM} \\
\frac{\partial \Phi^{(x)}}{\partial J_{klm}^{(y)}} &= \sum_{KLM} G_{klm,KLM}^{(xy)H} (\mathbf{RESID}(X))_{KLM} \\
\frac{\partial \Phi^{(x)}}{\partial J_{klm}^{(z)}} &= \sum_{KLM} G_{klm,KLM}^{(xz)H} (\mathbf{RESID}(X))_{KLM} \\
\frac{\partial \Phi^{(y)}}{\partial J_{klm}^{(x)}} &= \sum_{KLM} G_{klm,KLM}^{(yx)H} (\mathbf{RESID}(Y))_{KLM} \\
\frac{\partial \Phi^{(y)}}{\partial J_{klm}^{(y)}} &= \sum_{KLM} \left(Q_{klm,KLM}^{(y)T} + G_{klm,KLM}^{(yy)H} \right) (\mathbf{RESID}(Y))_{KLM} \\
\frac{\partial \Phi^{(y)}}{\partial J_{klm}^{(z)}} &= \sum_{KLM} G_{klm,KLM}^{(yz)H} (\mathbf{RESID}(Y))_{KLM} \\
\frac{\partial \Phi^{(z)}}{\partial J_{klm}^{(x)}} &= \sum_{KLM} G_{klm,KLM}^{(zx)H} (\mathbf{RESID}(Z))_{KLM} \\
\frac{\partial \Phi^{(z)}}{\partial J_{klm}^{(y)}} &= \sum_{KLM} G_{klm,KLM}^{(zy)H} (\mathbf{RESID}(Z))_{KLM} \\
\frac{\partial \Phi^{(z)}}{\partial J_{klm}^{(z)}} &= \sum_{KLM} \left(Q_{klm,KLM}^{(z)T} + G_{klm,KLM}^{(zz)H} \right) (\mathbf{RESID}(Z))_{KLM} \ , \qquad (1.12)
\end{aligned}$$

where

$$\begin{aligned}
\mathbf{RESID}(X) &= \mathbf{E}^{(x)}(\boldsymbol{\rho}, \mathbf{J}^{(x)}) + \mathbf{G}^{xx} \cdot \mathbf{J}^{(x)} + \mathbf{G}^{(xy)} \cdot \mathbf{J}^{(y)} + \mathbf{G}^{(xz)} \cdot \mathbf{J}^{(z)} - \mathbf{E}^{(0x)} \\
\mathbf{RESID}(Y) &= \mathbf{E}^{(y)}(\boldsymbol{\rho}, \mathbf{J}^{(y)}) + \mathbf{G}^{yx} \cdot \mathbf{J}^{(x)} + \mathbf{G}^{(yy)} \cdot \mathbf{J}^{(y)} + \mathbf{G}^{(yz)} \cdot \mathbf{J}^{(z)} - \mathbf{E}^{(0y)} \\
\mathbf{RESID}(Z) &= \mathbf{E}^{(z)}(\boldsymbol{\rho}, \mathbf{J}^{(z)}) + \mathbf{G}^{zx} \cdot \mathbf{J}^{(x)} + \mathbf{G}^{(zy)} \cdot \mathbf{J}^{(y)} + \mathbf{G}^{(zz)} \cdot \mathbf{J}^{(z)} - \mathbf{E}^{(0z)} \ . \qquad (1.13)
\end{aligned}$$

Note We use the notation for the gradient of a real function with respect to a complex variable to mean: $\frac{\partial \Phi}{\partial J} = \frac{\partial \Phi}{\partial R} + j\frac{\partial \Phi}{\partial I}$, where $R = \Re J$ and $I = \Im J$.

The final gradients are those of $\Phi^{(x)}$, $\Phi^{(y)}$, and $\Phi^{(z)}$ with respect to $\boldsymbol{\rho}$:

$$\frac{\partial \Phi^{(x)}}{\partial \rho_{klm}} = \Re \sum_{KLM} P^{(x)H}_{klm,KLM} (\mathbf{RESID}(X))_{KLM}$$
$$\frac{\partial \Phi^{(y)}}{\partial \rho_{klm}} = \Re \sum_{KLM} P^{(y)H}_{klm,KLM} (\mathbf{RESID}(Y))_{KLM}$$
$$\frac{\partial \Phi^{(z)}}{\partial \rho_{klm}} = \Re \sum_{KLM} P^{(z)H}_{klm,KLM} (\mathbf{RESID}(Z))_{KLM} \,. \tag{1.14}$$

Remember that the P-matrices that are defined in (1.7) and (1.8) are complex, because they are explicit functions of the complex currents.

The gradient of $\Phi(\boldsymbol{\rho}, \mathbf{J})$ is the sum of the various sub-gradients:

$$\frac{\partial \Phi}{\partial J^{(x)}_{klm}} = \frac{\partial \Phi^{(R)}}{\partial J^{(x)}_{klm}} + \frac{\partial \Phi^{(x)}}{\partial J^{(x)}_{klm}} + \frac{\partial \Phi^{(y)}}{\partial J^{(x)}_{klm}} + \frac{\partial \Phi^{(z)}}{\partial J^{(x)}_{klm}}$$
$$\frac{\partial \Phi}{\partial J^{(y)}_{klm}} = \frac{\partial \Phi^{(R)}}{\partial J^{(y)}_{klm}} + \frac{\partial \Phi^{(x)}}{\partial J^{(y)}_{klm}} + \frac{\partial \Phi^{(y)}}{\partial J^{(y)}_{klm}} + \frac{\partial \Phi^{(z)}}{\partial J^{(y)}_{klm}}$$
$$\frac{\partial \Phi}{\partial J^{(z)}_{klm}} = \frac{\partial \Phi^{(R)}}{\partial J^{(z)}_{klm}} + \frac{\partial \Phi^{(x)}}{\partial J^{(z)}_{klm}} + \frac{\partial \Phi^{(y)}}{\partial J^{(z)}_{klm}} + \frac{\partial \Phi^{(z)}}{\partial J^{(z)}_{klm}}$$
$$\frac{\partial \Phi}{\partial \rho_{klm}} = \frac{\partial \Phi^{(x)}}{\partial \rho_{klm}} + \frac{\partial \Phi^{(y)}}{\partial \rho_{klm}} + \frac{\partial \Phi^{(z)}}{\partial \rho_{klm}} \,. \tag{1.15}$$

Eventually we will need to minimize Φ along the line $(\rho, \mathbf{J}^{(x)}, \mathbf{J}^{(y)}, \mathbf{J}^{(z)}) + \alpha(\mathbf{u}, \mathbf{v}^{(x)}, \mathbf{v}^{(y)}, \mathbf{v}^{(z)})$ in function-space. α is a real number that parameterizes the line, $(\boldsymbol{\rho}, \mathbf{J}^{(x)}, \mathbf{J}^{(y)}, \mathbf{J}^{(z)})$ is a fixed starting point, and $(\mathbf{u}, \mathbf{v}^{(x)}, \mathbf{v}^{(y)}, \mathbf{v}^{(z)})$ is a direction-vector that will be determined by the conjugate-gradient algorithm. We will do this by differentiating Φ with respect to α, and then set the derivative equal to zero to determine the optimum α.

From (1.9), we have

$$\begin{aligned}
&\Phi(\boldsymbol{\rho} + \alpha\mathbf{u}, \mathbf{J}^{(x)} + \alpha\mathbf{v}^{(x)}, \mathbf{J}^{(y)} + \alpha\mathbf{v}^{(y)}, \mathbf{J}^{(z)} + \alpha\mathbf{v}^{(z)}) = \\
&\Phi^{(R)}(\mathbf{J}^{(x)} + \alpha\mathbf{v}^{(x)}, \mathbf{J}^{(y)} + \alpha\mathbf{v}^{(y)}, \mathbf{J}^{(z)} + \alpha\mathbf{v}^{(z)}) + \\
&\Phi^{(x)}(\boldsymbol{\rho} + \alpha\mathbf{u}, \mathbf{J}^{(x)} + \alpha\mathbf{v}^{(x)}, \mathbf{J}^{(y)} + \alpha\mathbf{v}^{(y)}, \mathbf{J}^{(z)} + \alpha\mathbf{v}^{(z)}) + \\
&\Phi^{(y)}(\boldsymbol{\rho} + \alpha\mathbf{u}, \mathbf{J}^{(x)} + \alpha\mathbf{v}^{(x)}, \mathbf{J}^{(y)} + \alpha\mathbf{v}^{(y)}, \mathbf{J}^{(z)} + \alpha\mathbf{v}^{(z)}) + \\
&\Phi^{(z)}(\boldsymbol{\rho} + \alpha\mathbf{u}, \mathbf{J}^{(x)} + \alpha\mathbf{v}^{(x)}, \mathbf{J}^{(y)} + \alpha\mathbf{v}^{(y)}, \mathbf{J}^{(z)} + \alpha\mathbf{v}^{(z)}) \,.
\end{aligned} \tag{1.16}$$

Clearly, $\Phi^{(R)}(\mathbf{J} + \alpha\mathbf{v})$ is quadratic in α, but the other three functionals are of the fourth-order in α, as can be seen by the bilinear function $\mathbf{E}(\boldsymbol{\rho}, \mathbf{J})$ that is defined in (1.3) and appears in (1.9).

The derivative of $\Phi^{(R)}$ is computed in a straight-forward (though slightly tedious) manner to be:

$$\frac{d\Phi^{(R)}(\mathbf{J}+\alpha\mathbf{v})}{d\alpha} = \Re\left\{\mathbf{v}^{(x)*}\cdot\frac{\partial\Phi^{(R)}(\mathbf{J})}{\partial\mathbf{J}^{(x)}} + \mathbf{v}^{(y)*}\cdot\frac{\partial\Phi^{(R)}(\mathbf{J})}{\partial\mathbf{J}^{(y)}} + \mathbf{v}^{(z)*}\cdot\frac{\partial\Phi^{(R)}(\mathbf{J})}{\partial\mathbf{J}^{(z)}}\right.$$
$$\left. +\alpha\left(\mathbf{v}^{(x)*}\cdot\frac{\partial\Phi^{(R)}(\mathbf{v})}{\partial\mathbf{J}^{(x)}} + \mathbf{v}^{(y)*}\cdot\frac{\partial\Phi^{(R)}(\mathbf{v})}{\partial\mathbf{J}^{(y)}} + \mathbf{v}^{(z)*}\cdot\frac{\partial\Phi^{(R)}(\mathbf{v})}{\partial\mathbf{J}^{(z)}}\right)\right\}. \tag{1.17}$$

The vector dot-product denotes a sum over the cell indices, klm, as before, and the components of the gradient vectors are given in (1.11).

Similarly,

$$\begin{aligned}
&\frac{d\Phi^{(x)}}{d\alpha}(\boldsymbol{\rho}+\alpha\mathbf{u},\mathbf{J}+\alpha\mathbf{v}) \\
&= \Re\left\{\mathbf{A}_{0x}^{*}\cdot\mathbf{B}_{0x} + \alpha\left(\mathbf{A}_{0x}^{*}\cdot\mathbf{B}_{1x} + \mathbf{A}_{1x}^{*}\cdot\mathbf{B}_{0x}\right)\right. \\
&\quad \left. +\alpha^{2}\left(\mathbf{A}_{0x}^{*}\cdot\mathbf{B}_{2x} + \mathbf{A}_{1x}^{*}\cdot\mathbf{B}_{1x}\right) + \alpha^{3}\mathbf{A}_{1x}^{*}\cdot\mathbf{B}_{2x}\right\} \\
&\frac{d\Phi^{(y)}}{d\alpha}(\boldsymbol{\rho}+\alpha\mathbf{u},\mathbf{J}+\alpha\mathbf{v}) \\
&= \Re\left\{\mathbf{A}_{0y}^{*}\cdot\mathbf{B}_{0y} + \alpha\left(\mathbf{A}_{0y}^{*}\cdot\mathbf{B}_{1y} + \mathbf{A}_{1y}^{*}\cdot\mathbf{B}_{0y}\right)\right. \\
&\quad \left. +\alpha^{2}\left(\mathbf{A}_{0y}^{*}\cdot\mathbf{B}_{2y} + \mathbf{A}_{1y}^{*}\cdot\mathbf{B}_{1y}\right) + \alpha^{3}\mathbf{A}_{1y}^{*}\cdot\mathbf{B}_{2y}\right\} \\
&\frac{d\Phi^{(z)}}{d\alpha}(\boldsymbol{\rho}+\alpha\mathbf{u},\mathbf{J}+\alpha\mathbf{v}) \\
&= \Re\left\{\mathbf{A}_{0z}^{*}\cdot\mathbf{B}_{0z} + \alpha\left(\mathbf{A}_{0z}^{*}\cdot\mathbf{B}_{1z} + \mathbf{A}_{1z}^{*}\cdot\mathbf{B}_{0z}\right)\right. \\
&\quad \left. +\alpha^{2}\left(\mathbf{A}_{0z}^{*}\cdot\mathbf{B}_{2z} + \mathbf{A}_{1z}^{*}\cdot\mathbf{B}_{1z}\right) + \alpha^{3}\mathbf{A}_{1z}^{*}\cdot\mathbf{B}_{2z}\right\}
\end{aligned} \tag{1.18}$$

where

$$\begin{aligned}
\mathbf{A}_{0x}^{*} &= \mathbf{J}^{(x)*}\cdot\mathbf{Q}^{(x)T}(\mathbf{u}) + \mathbf{v}^{(x)*}\cdot\mathbf{Q}^{(x)T}(\boldsymbol{\rho}) + \mathbf{v}^{(x)*}\cdot\mathbf{G}^{(xx)H} \\
&\quad +\mathbf{v}^{(y)*}\cdot\mathbf{G}^{(xy)H} + \mathbf{v}^{(z)*}\cdot\mathbf{G}^{(xz)H} \\
\mathbf{A}_{1x}^{*} &= 2\mathbf{v}^{(x)*}\cdot\mathbf{Q}^{(x)T}(\mathbf{u}) \\
\mathbf{B}_{0x} &= \mathbf{Q}^{(x)}(\boldsymbol{\rho})\cdot\mathbf{J}^{(x)} + \mathbf{G}^{(xx)}\cdot\mathbf{J}^{(x)} \\
&\quad +\mathbf{G}^{(xy)}\cdot\mathbf{J}^{(y)} + \mathbf{G}^{(xz)}\cdot\mathbf{J}^{(z)} - \mathbf{E}^{(0x)} \\
\mathbf{B}_{1x} &= \mathbf{Q}^{(x)}(\mathbf{u})\cdot\mathbf{J}^{(x)} + \mathbf{Q}^{(x)}(\boldsymbol{\rho})\cdot\mathbf{v}^{(x)} + \mathbf{G}^{(xx)}\cdot\mathbf{v}^{(x)} \\
&\quad +\mathbf{G}^{(xy)}\cdot\mathbf{v}^{(y)} + \mathbf{G}^{(xz)}\cdot\mathbf{v}^{(z)}
\end{aligned}$$

$$
\begin{aligned}
\mathbf{B}_{2x} &= \mathbf{Q}^{(x)}(\mathbf{u}) \cdot \mathbf{v}^{(x)} \\
\mathbf{A}_{0y}^{*} &= \mathbf{J}^{(y)*} \cdot \mathbf{Q}^{(y)T}(\mathbf{u}) + \mathbf{v}^{(y)*} \cdot \mathbf{Q}^{(y)T}(\boldsymbol{\rho}) + \mathbf{v}^{(x)*} \cdot \mathbf{G}^{(yx)H} \\
&\quad + \mathbf{v}^{(y)*} \cdot \mathbf{G}^{(yy)H} + \mathbf{v}^{(z)*} \cdot \mathbf{G}^{(yz)H} \\
\mathbf{A}_{1y}^{*} &= 2\mathbf{v}^{(y)*} \cdot \mathbf{Q}^{(y)T}(\mathbf{u}) \\
\mathbf{B}_{0y} &= \mathbf{Q}^{(y)}(\boldsymbol{\rho}) \cdot \mathbf{J}^{(y)} + \mathbf{G}^{(yx)} \cdot \mathbf{J}^{(x)} \\
&\quad + \mathbf{G}^{(yy)} \cdot \mathbf{J}^{(y)} + \mathbf{G}^{(yz)} \cdot \mathbf{J}^{(z)} - \mathbf{E}^{(0y)} \\
\mathbf{B}_{1y} &= \mathbf{Q}^{(y)}(\mathbf{u}) \cdot \mathbf{J}^{(y)} + \mathbf{Q}^{(y)}(\boldsymbol{\rho}) \cdot \mathbf{v}^{(y)} + \mathbf{G}^{(yx)} \cdot \mathbf{v}^{(x)} \\
&\quad + \mathbf{G}^{(yy)} \cdot \mathbf{v}^{(y)} + \mathbf{G}^{(yz)} \cdot \mathbf{v}^{(z)} \\
\mathbf{B}_{2y} &= \mathbf{Q}^{(y)}(\mathbf{u}) \cdot \mathbf{v}^{(y)} \\
\mathbf{A}_{0z}^{*} &= \mathbf{J}^{(z)*} \cdot \mathbf{Q}^{(z)T}(\mathbf{u}) + \mathbf{v}^{(z)*} \cdot \mathbf{Q}^{(z)T}(\boldsymbol{\rho}) + \mathbf{v}^{(x)*} \cdot \mathbf{G}^{(zx)H} \\
&\quad + \mathbf{v}^{(y)*} \cdot \mathbf{G}^{(zy)H} + \mathbf{v}^{(z)*} \cdot \mathbf{G}^{(zz)H} \\
\mathbf{A}_{1z}^{*} &= 2\mathbf{v}^{(z)*} \cdot \mathbf{Q}^{(z)T}(\mathbf{u}) \\
\mathbf{B}_{0z} &= \mathbf{Q}^{(z)}(\boldsymbol{\rho}) \cdot \mathbf{J}^{(z)} + \mathbf{G}^{(zx)} \cdot \mathbf{J}^{(x)} \\
&\quad + \mathbf{G}^{(zy)} \cdot \mathbf{J}^{(y)} + \mathbf{G}^{(zz)} \cdot \mathbf{J}^{(z)} - \mathbf{E}^{(0z)} \\
\mathbf{B}_{1z} &= \mathbf{Q}^{(z)}(\mathbf{u}) \cdot \mathbf{J}^{(z)} + \mathbf{Q}^{(z)}(\boldsymbol{\rho}) \cdot \mathbf{v}^{(z)} + \mathbf{G}^{(zx)} \cdot \mathbf{v}^{(x)} \\
&\quad + \mathbf{G}^{(zy)} \cdot \mathbf{v}^{(y)} + \mathbf{G}^{(zz)} \cdot \mathbf{v}^{(z)} \\
\mathbf{B}_{2z} &= \mathbf{Q}^{(z)}(\mathbf{u}) \cdot \mathbf{v}^{(z)} ,
\end{aligned}
\tag{1.19}
$$

and, once again, the vector dot-product denotes a sum over the cell indices, klm. Note that $\mathbf{B}_0$ is a residual-vector, which means that there may be some simplification in computing or using it, and finally, the order of some of the computations in (1.19) might be changed in order to take advantage of the convolutional or correlational nature of the matrices.

The total cubic polynomial for $\frac{d\Phi}{d\alpha}(\boldsymbol{\rho} + \alpha\mathbf{u}, \mathbf{J} + \alpha\mathbf{v}) = 0$ is given by setting the sum of (1.17) and (1.18) equal to zero. Note that, because the coefficients of the polynomial are real, any complex roots must occur in conjugate pairs. This guarantees at least one real root, and possibly three. We will take the smallest real root.

1.3 The Algorithm

Step 0: Initialization The user creates a starting point

$$\begin{pmatrix} \mathbf{J}_0^{(x)} \\ \mathbf{J}_0^{(y)} \\ \mathbf{J}_0^{(z)} \\ \boldsymbol{\rho}_0 \end{pmatrix}.$$

Step 1: Steepest Descent First calculate the gradient

$$\nabla\Phi(\boldsymbol{\rho}_0, \mathbf{J}_0) = \begin{pmatrix} \dfrac{\partial\Phi}{\partial\mathbf{J}^{(x)}} \\ \\ \dfrac{\partial\Phi}{\partial\mathbf{J}^{(y)}} \\ \\ \dfrac{\partial\Phi}{\partial\mathbf{J}^{(z)}} \\ \\ \dfrac{\partial\Phi}{\partial\boldsymbol{\rho}} \end{pmatrix} (\boldsymbol{\rho}_0, \mathbf{J}_0)$$

using (1.15). Then set the direction of movement at the first iteration to be

$$\begin{pmatrix} \mathbf{v}^{(x)} \\ \mathbf{v}^{(y)} \\ \mathbf{v}^{(z)} \\ \mathbf{u} \end{pmatrix} = -\nabla\Phi(\boldsymbol{\rho}_0, \mathbf{J}_0)$$

where $(\mathbf{v}^{(x)}, \mathbf{v}^{(y)}, \mathbf{v}^{(z)}) = \mathbf{v} \in \mathbf{C}^{N_c} \times \mathbf{C}^{N_c} \times \mathbf{C}^{N_c}$, and $\mathbf{u} \in \mathbf{R}^{N_c}$. Note that we will later use $\mathbf{f}_1$ to denote this direction. The problem now is to minimize Φ in the direction $\begin{pmatrix} \mathbf{v} \\ \mathbf{u} \end{pmatrix}$ from the point $(\boldsymbol{\rho}_0, \mathbf{J}_0)$. We should normalize the direction vector $\begin{pmatrix} \mathbf{v}^{(x)} \\ \mathbf{v}^{(y)} \\ \mathbf{v}^{(z)} \\ \mathbf{u} \end{pmatrix}$ before calculating the coefficients of the cubic equation that gives the minimum value. Therefore, define the new direction vector to be the unit vector

$$\frac{1}{\sqrt{(\|\mathbf{v}\|^2 + \|\mathbf{u}\|^2)}} \begin{pmatrix} \mathbf{v}^{(x)} \\ \mathbf{v}^{(y)} \\ \mathbf{v}^{(z)} \\ \mathbf{u} \end{pmatrix}$$

and note that we will use the same notation, $\begin{pmatrix} \mathbf{v}^{(x)} \\ \mathbf{v}^{(y)} \\ \mathbf{v}^{(z)} \\ \mathbf{u} \end{pmatrix}$, for the normalized (unit) direction vector. Now find the smallest positive α_0 that satisfies

$$a_0 + a_1\alpha + a_2\alpha^2 + a_3\alpha^3 = 0$$

where the coefficients are given by the appropriate sums of (1.17) and (1.18), with $\mathbf{J} = \mathbf{J}_0$ and $\boldsymbol{\rho} = \boldsymbol{\rho}_0$. Then define the new approximation to be

$$\begin{pmatrix} \mathbf{J}_1 \\ \boldsymbol{\rho}_1 \end{pmatrix} = \begin{pmatrix} \mathbf{J}_0 \\ \boldsymbol{\rho}_0 \end{pmatrix} + \alpha_0 \begin{pmatrix} \mathbf{v} \\ \mathbf{u} \end{pmatrix} .$$

Step $k+1$ Assume that the kth ($k \geq 1$) iteration, $\begin{pmatrix} \mathbf{J}_k \\ \boldsymbol{\rho}_k \end{pmatrix}$, has been determined. Let $\{\mathbf{f}_k\}$ be the sequence of direction vectors that is generated by defining $\mathbf{f}_1 = -\nabla\Phi(\mathbf{J}_0, \boldsymbol{\rho}_0)$ and for $k \geq 1$,

$$\mathbf{f}_{k+1} = -\nabla\Phi(\mathbf{J}_k, \boldsymbol{\rho}_k) + \beta_{k+1}\mathbf{f}_k ,$$

where $\nabla\Phi(\mathbf{J}_k, \boldsymbol{\rho}_k)$ and β_{k+1} are determined as follows:

$$\nabla\Phi(\boldsymbol{\rho}_k, \mathbf{J}_k) = \begin{pmatrix} \dfrac{\partial\Phi}{\partial\mathbf{J}^{(x)}} \\ \dfrac{\partial\Phi}{\partial\mathbf{J}^{(y)}} \\ \dfrac{\partial\Phi}{\partial\mathbf{J}^{(z)}} \\ \dfrac{\partial\Phi}{\partial\boldsymbol{\rho}} \end{pmatrix} (\boldsymbol{\rho}_k, \mathbf{J}_k) ,$$

and the parameter, β_{k+1}, depends upon one of the following iteration methods:

Steepest Descent This technique simply continues as **Step 1**, and defines the direction vector to always be the steepest descent direction. Hence, β_{k+1} will always be set equal to 0.

Fletcher-Reeves Here, β_{k+1} is defined as

$$\beta_{k+1} = \frac{\|\nabla\Phi(\mathbf{J}_k, \boldsymbol{\rho}_k)\|^2}{\|\nabla\Phi(\mathbf{J}_{k-1}, \boldsymbol{\rho}_{k-1})\|^2} .$$

Polak-Ribière In this algorithm, β_{k+1} is defined as

$$\beta_{k+1} = \frac{\left(\nabla\Phi(\mathbf{J}_k, \boldsymbol{\rho}_k) - \nabla\Phi(\mathbf{J}_{k-1}, \boldsymbol{\rho}_{k-1})\right)^* \cdot \nabla\Phi(\mathbf{J}_k, \boldsymbol{\rho}_k)}{\|\nabla\Phi(\mathbf{J}_{k-1}, \boldsymbol{\rho}_{k-1})\|^2} .$$

Fletcher-Reeves with Restart This method is similar to the Fletcher-Reeves algorithm except that after every r iterations, we restart the algorithm with the steepest descent step. Therefore, we can define β_{k+1} as

$$\beta_{k+1} = \begin{cases} 0 & \text{if } k \text{ is a multiple of } r \\ \dfrac{\|\nabla\Phi(\mathbf{J}_k, \boldsymbol{\rho}_k)\|^2}{\|\nabla\Phi(\mathbf{J}_{k-1}, \boldsymbol{\rho}_{k-1})\|^2} & \text{otherwise} \end{cases}$$

Hybrid 1 This hybrid algorithm tries to take into account the best aspects of some of the algorithms presented above.

$$\beta_{k+1} = \begin{cases} 0 \quad \text{if } \|\nabla\Phi(\mathbf{J}_k, \boldsymbol{\rho}_k)\|^2 > (0.2)^k \times 10^8 \\ \beta_{\text{FR}} \text{ if } \beta_{\text{PR}} < 0 \\ \beta_{\text{PR}} \text{ if } \beta_{\text{PR}} \leq 5\dfrac{\|\nabla\Phi(\mathbf{J}_k, \boldsymbol{\rho}_k)\|^2}{\|\nabla\phi(\mathbf{J}_{k-1}, \boldsymbol{\rho}_{k-1})\|^2} = 5\beta_{\text{FR}} \\ \beta_{\text{FR}} \text{ otherwise} \end{cases}$$

Hybrid 2 In this algorithm, we try to control the orthogonality of the gradients and define β_{k+1} as

$$\beta_{k+1} = \begin{cases} 0 & \text{if } |\nabla\Phi(\mathbf{J}_{k-1}, \boldsymbol{\rho}_{k-1})^* \cdot \nabla\Phi(\mathbf{J}_k, \boldsymbol{\rho}_k)| > 0.2 \\ \dfrac{\|\nabla\Phi(\mathbf{J}_k, \boldsymbol{\rho}_k)\|^2}{\|\nabla\Phi(\mathbf{J}_{k-1}, \boldsymbol{\rho}_{k-1})\|^2} & \text{otherwise} \end{cases}$$

After determining β_{k+1}, we set the new direction vector

$$\begin{pmatrix} \mathbf{v}^{(x)} \\ \mathbf{v}^{(y)} \\ \mathbf{v}^{(z)} \\ \mathbf{u} \end{pmatrix} = \mathbf{f}_{k+1} = -\nabla\Phi(\boldsymbol{\rho}_k, \mathbf{J}_k) + \beta_{k+1}\mathbf{f}_k$$

where, as before $(\mathbf{v}^{(x)}, \mathbf{v}^{(y)}, \mathbf{v}^{(z)}) \in \mathbf{C}^{N_c} \times \mathbf{C}^{N_c} \times \mathbf{C}^{N_c}$, and $\mathbf{u} \in \mathbf{R}^{N_c}$. As in **Step 1**, we should normalize $\begin{pmatrix} \mathbf{v} \\ \mathbf{u} \end{pmatrix}$ to make it a unit vector, and then find the smallest positive α_k that satisfies

$$a_0 + a_1\alpha + a_2\alpha^2 + a_3\alpha^3 = 0$$

with coefficients gotten from (1.17)–(1.19) and the various functions evaluated at the current point, $(\mathbf{J}_k, \boldsymbol{\rho}_k)$.

Since we are able to use exact line searches in this algorithm, most of the methods which we use will always yield a direction in which Φ is decreasing. In fact, since we normalize the direction vector, the constant in the cubic equation, a_0, is equal to the directional derivative of Φ at the point $(\mathbf{J}_k, \rho_k)$ in the direction $\begin{pmatrix} \mathbf{v} \\ \mathbf{u} \end{pmatrix}$. Therefore, a_0 should always be negative. We have found, however, that even in the methods which theoretically always yield descent directions, we sometimes will obtain a positive a_0. Usually this phenomenon occurs when we are 'close' to the actual solution. We suggest that if $a_0 \geq 0$, then the algorithm should be restarted by going to the initialization step, and setting $(\mathbf{J}_0, \rho_0) = (\mathbf{J}_k, \rho_k)$. This is equivalent to using the steepest descent direction (i.e., letting $\beta_{k+1} = 0$) at **Step** $k + 1$.

Now define the new iteration to be

$$\begin{pmatrix} \mathbf{J}_{k+1}^{(x)} \\ \mathbf{J}_{k+1}^{(y)} \\ \mathbf{J}_{k+1}^{(z)} \\ \rho_{k+1} \end{pmatrix} = \begin{pmatrix} \mathbf{J}_{k}^{(x)} \\ \mathbf{J}_{k}^{(y)} \\ \mathbf{J}_{k}^{(z)} \\ \rho_{k} \end{pmatrix} + \alpha_k \begin{pmatrix} \mathbf{v}^{(x)} \\ \mathbf{v}^{(y)} \\ \mathbf{v}^{(z)} \\ \mathbf{u} \end{pmatrix},$$

check the residuals for convergence, and either quit or continue with the iteration. This requires a stopping rule, which could be a variation of what we do now in **VIC-3D®**, or perhaps stopping when the relative error, $\frac{\Phi(\mathbf{J}, \rho)}{\|(\mathbf{J}, \rho)\|}$, is less than some prescribed value.

1.4 Example: Raster Scan at Three Frequencies[2]

Consider Fig. 1.1, which shows a $4 \times 4 \times 4$ mm^3 through-wall anomaly and a T/R-scan system in which the transmitter occupies two positions, $(-14, -14)$mm and $(14, -14)$ mm, and the receiver undergoes a two-dimensional raster scan of 11 points, with equal intervals of 0.75 mm in each direction. The anomaly grid consists of $16 \times 16 \times 4 = 1024$ cells, with the four z-layers numbered 0,1,2,3. The flaw that is to be reconstructed is shown in Fig. 1.2.

The model coil parameters are given in Table 1.1. It is clear that the transmit coil is much larger than the grid cell-size, and is quite remote from the anomalous region. The receive coil scans over the region with dimensions that are comparable to the cell-size, though this is not a requirement.

The transmitter is excited at three frequencies of 10^2, 10^4 and 10^5 Hz, which, with the two-point transmitter scan, produces six 'experiments,' in the language of set-theoretic estimation. Thus, the total impedance data set that is submitted to the bilinear conjugate-gradient inversion algorithm comprises $6 \times 11 \times 11 = 726$ values.

[2]See [102–104] for additional examples.

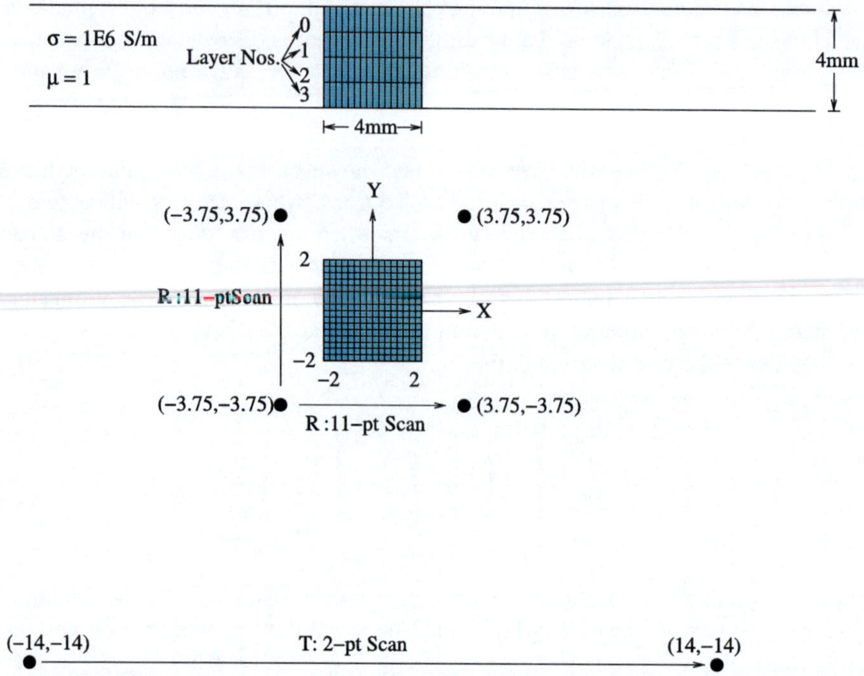

Fig. 1.1 Illustrating the first example. All unlabeled coordinates are in mms. The 11 × 11-point scan of the receiver probe is shown, as well as the 2-point scan of the transmitter probe. Top: Side view. Bottom: Top view

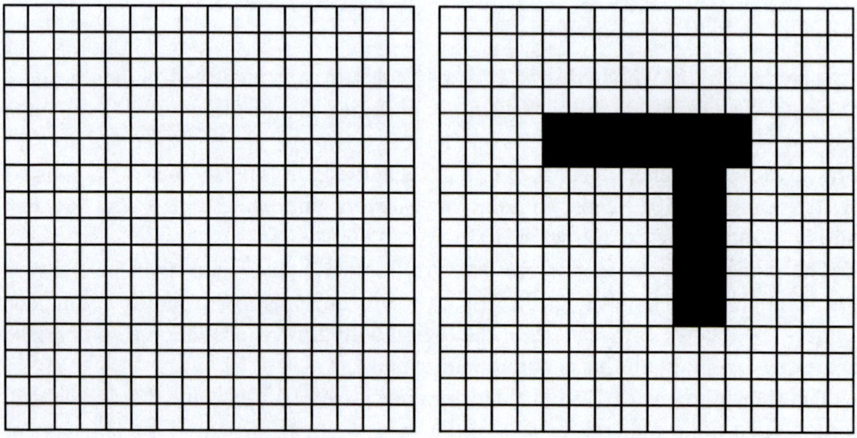

Fig. 1.2 Illustrating the flaw that is to be reconstructed. Left: Layers 0 and 3. Right: Layers 1 and 2. A solid cell has a volume-fraction of 1 (empty), and a blank cell a volume-fraction of 0 (host material)

Table 1.1 Coil parameters for Example 1

Parameter	Transmit	Receive
Inner radius (mm)	15.0	0.25
Outer radius (mm)	25.0	0.50
Height (mm)	3.0	1.0
Turns	100	100

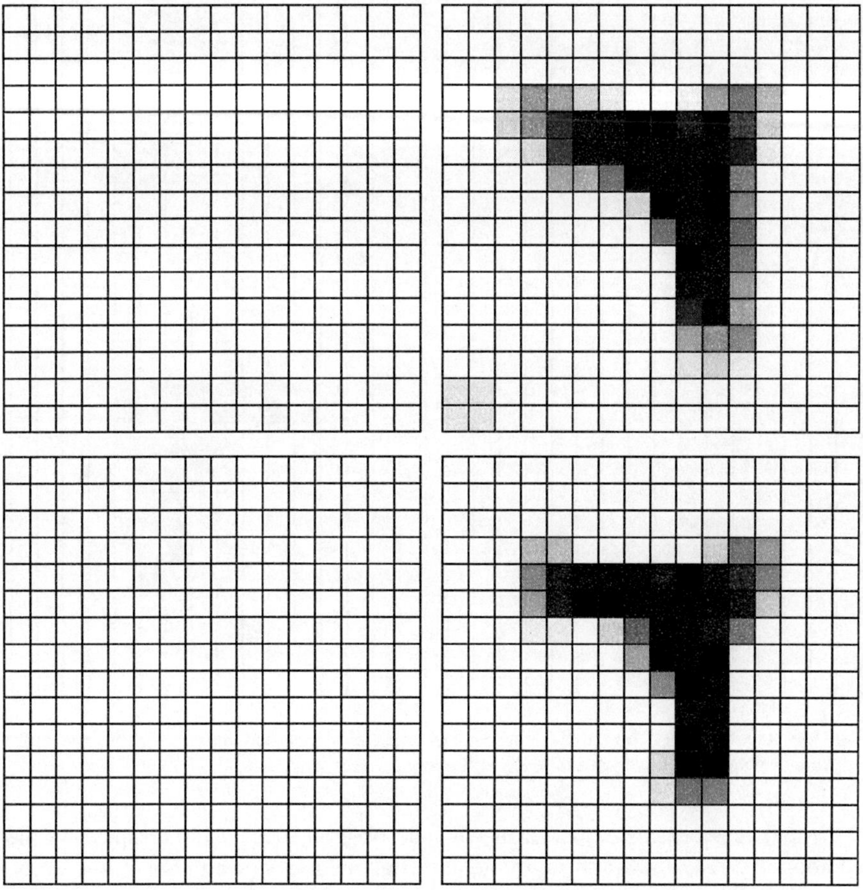

Fig. 1.3 Reconstruction of the flaw in Fig. 1.2. Left: Layers 0 and 3. Right: Layers 1 and 2

The unknowns are 1024 values of conductivity and 3072 values of anomalous current for each experiment, giving a total of 19,456 unknowns to be determined by the inversion algorithm.

In starting the inversion process, we assume that the anomaly is nonexistent, which means that the initial conductivity is that of the host region, and the initial anomalous currents are zero for each experiment. The Polak-Ribière algorithm is used to compute β_{k+1}. The results are shown in Fig. 1.3.

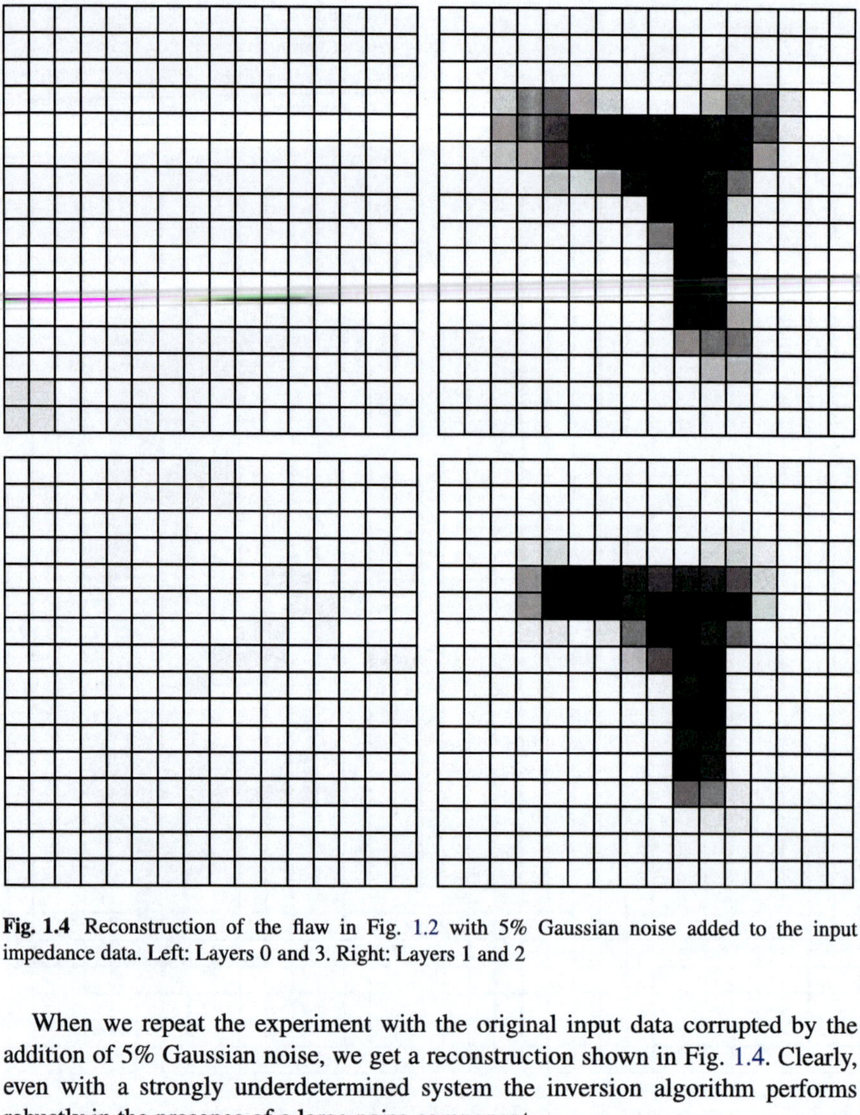

Fig. 1.4 Reconstruction of the flaw in Fig. 1.2 with 5% Gaussian noise added to the input impedance data. Left: Layers 0 and 3. Right: Layers 1 and 2

When we repeat the experiment with the original input data corrupted by the addition of 5% Gaussian noise, we get a reconstruction shown in Fig. 1.4. Clearly, even with a strongly underdetermined system the inversion algorithm performs robustly in the presence of a large noise component.

Though we have not done it here, it is possible to introduce adaptive preconditioning to reduce the effects of noise during the inversion process by introducing a scaling operator, $\mathscr{B}$, that selectively eliminates those components of the reconstructed vector that are less than a certain value. The threshold value is determined early in the iterative process, and requires a certain amount of intuition. It can be made more precise by using statistical decision theory. This is in contrast to post-reconstruction image processing, where weighted averaging filters are used to remove unwanted artifacts after the image has been reconstructed. An example of this approach is given in [14] in the context of a linearized Born approximation to the conjugate-gradient algorithm.

Chapter 2
Voxel-Based Inversion Via Set-Theoretic Estimation

2.1 The Electromagnetic Model Equations

The starting point for any electromagnetic inverse problem is Maxwell's equations:

$$\begin{aligned}
\nabla \times \mathbf{E}(\mathbf{r}) &= -j\omega\mu_0\mathbf{H}(\mathbf{r}) \\
\nabla \times \mathbf{H}(\mathbf{r}) &= j\omega\epsilon_0\mathbf{E} + \sigma(\mathbf{r})\mathbf{E}(\mathbf{r}) \\
&= j\omega\epsilon_0\mathbf{E}(\mathbf{r}) + (\sigma_f(\mathbf{r}) - \sigma_h)\mathbf{E}(\mathbf{r}) + \sigma_h\mathbf{E}(\mathbf{r}) \\
&= j\omega\epsilon_0\mathbf{E}(\mathbf{r}) + \sigma^{(a)}(\mathbf{r})\mathbf{E}(\mathbf{r}) + \sigma_h\mathbf{E}(\mathbf{r}) \,,
\end{aligned} \tag{2.1}$$

where $\sigma_f(\mathbf{r})$ is the flaw conductivity, σ_h is the uniform host conductivity, and $\sigma^{(a)}(\mathbf{r})$ is the anomalous conductivity. The product, $\sigma^{(a)}(\mathbf{r})\mathbf{E}(\mathbf{r})$ defines the anomalous electric current density, $\mathbf{J}(\mathbf{r})$.

The formal solution of Maxwell's equations can be obtained by equating the total electric field, $\mathbf{E}(\mathbf{r}) = \mathbf{J}(\mathbf{r})/\sigma_a(\mathbf{r})$, to the sum of the incident field, that is produced by the current in the exciter coil, and the scattered field, that is due to the anomalous electric current:

$$\begin{aligned}
E_x^{(i)}(\mathbf{r}) &= \frac{J_x(\mathbf{r})}{\sigma_a(\mathbf{r})} - E_x^{(s)}(\mathbf{r})\,[\mathbf{J}] \\
E_y^{(i)}(\mathbf{r}) &= \frac{J_y(\mathbf{r})}{\sigma_a(\mathbf{r})} - E_y^{(s)}(\mathbf{r})\,[\mathbf{J}] \\
E_z^{(i)}(\mathbf{r}) &= \frac{J_z(\mathbf{r})}{\sigma_a(\mathbf{r})} - E_z^{(s)}(\mathbf{r})\,[\mathbf{J}] \,.
\end{aligned} \tag{2.2}$$

H. A. Sabbagh et al., *Advanced Electromagnetic Models for Materials Characterization and Nondestructive Evaluation*, Scientific Computation,
https://doi.org/10.1007/978-3-030-67956-9_2

The second term on the right-hand side of each of the equations in (2.2) stands for a linear functional, whose kernel is a Green's function. In order to discretize the integral relations implied in (2.2) by means of the method of moments, we define a regular three-dimensional grid of cells, each of dimension $\delta x \times \delta y \times \delta z$, and expand the current vector on this grid as

$$
\begin{aligned}
J_x(\mathbf{r}) &= \sum_{KLM} J^{(x)}_{KLM} T^{(x)}_{KLM}(\mathbf{r}) \\
J_y(\mathbf{r}) &= \sum_{KLM} J^{(y)}_{KLM} T^{(y)}_{KLM}(\mathbf{r}) \\
J_z(\mathbf{r}) &= \sum_{KLM} J^{(z)}_{KLM} T^{(z)}_{KLM}(\mathbf{r})\,. \qquad (2.3)
\end{aligned}
$$

The expressions for the $T^{(q)}_{klm}(\mathbf{r})$ are:

$$
\begin{aligned}
T^{(x)}_{klm}(\mathbf{r}) &= \pi_{2k}(x/\delta x)\pi_{1l}(y/\delta y)\pi_{1m}(z/\delta z) \\
T^{(y)}_{klm}(\mathbf{r}) &= \pi_{1k}(x/\delta x)\pi_{2l}(y/\delta y)\pi_{1m}(z/\delta z) \\
T^{(z)}_{klm}(\mathbf{r}) &= \pi_{1k}(x/\delta x)\pi_{1l}(y/\delta y)\pi_{2m}(z/\delta z)\,, \qquad (2.4)
\end{aligned}
$$

where $\pi_{1m}(y/\delta y)$ is the mth unit pulse function, and $\pi_{2k}(x/\delta x)$ is the kth tent function, which is the convolution of $\pi_{1k}(x/\delta x)$ with itself (see Fig. 2.1).

The present version of **VIC-3D®** uses the Galerkin variant of the method of moments, in which testing is done with the same basis set that is used to expand the unknown currents. This differs from the earlier version, which used point-matching to complete the discretization. The implication for the present inverse problem is that we can no longer assume that the electric fields are known at the center of each cell; rather, we are given the moments of the electric field throughout each cell, as well as the expansion coefficients for the currents. We will now show how this knowledge can be used to determine the conductivity of each cell, if the anomalous currents are given.

The field moments for the x and y components of the (total) electric field are

$$E^x_{klm} = \int\int\int E^x(x,y,z)T^x_{klm}(x,y,z)dxdydz \qquad (2.5)$$

$$E^y_{klm} = \int\int\int E^y(x,y,z)T^y_{klm}(x,y,z)dxdydz\,. \qquad (2.6)$$

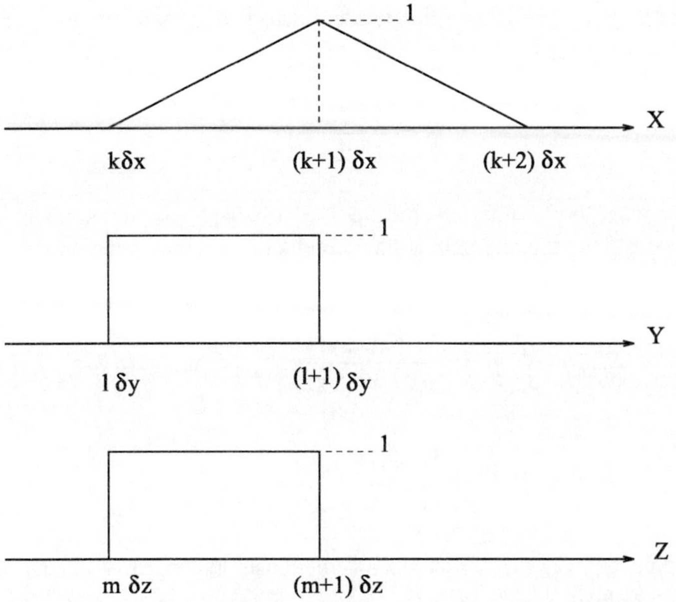

Fig. 2.1 Showing the location of the tent and pulse functions for the element $T^{(x)}_{klm}(x, y, z) = \pi_{2k}(x/\delta x)\pi_{1l}(y/\delta y)\pi_{1m}(z/\delta z)$

We can compute these from the discretized form of the volume-integral equations

$$\begin{aligned} E^x_{klm} &= E^{(i)(x)}_{klm} + \sum_{KLM} G^{(xx)}_{klm,KLM} J^x_{KLM} + \sum_{KLM} G^{(xy)}_{klm,KLM} J^y_{KLM} \\ &\quad + \sum_{KLM} G^{(xz)}_{klm,KLM} J^z_{KLM} \\ E^y_{klm} &= E^{(i)(y)}_{klm} + \sum_{KLM} G^{(yx)}_{klm,KLM} J^x_{KLM} + \sum_{KLM} G^{(yy)}_{klm,KLM} J^y_{KLM} \\ &\quad + \sum_{KLM} G^{(yz)}_{klm,KLM} J^z_{KLM} , \end{aligned} \tag{2.7}$$

where $E^{(i)}_{klm}$ are the field moments of the incident field, and the J_{KLM}'s are the expansion coefficients of the anomalous electric currents. The $G_{klm,KLM}$'s are the matrices that result from the discretization of the functionals with the Green's function kernels, and, thus, do not depend upon the cell conductivities.

We relate the cell conductivities to the electric field moments in the following way:

$$\mathbf{E} = \frac{\mathbf{J}^{(a)}(x, y, z)}{\sigma^{(a)}(x, y, z)}, \tag{2.8}$$

where $\mathbf{E}(x, y, z)$ is the total electric field, $\mathbf{J}^{(a)}(x, y, z)$ is the anomalous electric current, and $\sigma^{(a)}$ is the anomalous flaw conductivity. Upon substituting (2.8) into (2.5) and (2.6), we get

$$E^x_{klm} = \int\int\int \frac{J^{(a)(x)}(x, y, z)}{\sigma^{(a)}(x, y, z)} T^x_{klm}(x, y, z) dx dy dz \tag{2.9}$$

$$E^y_{klm} = \int\int\int \frac{J^{(a)(y)}(x, y, z)}{\sigma^{(a)}(x, y, z)} T^y_{klm}(x, y, z) dx dy dz\,. \tag{2.10}$$

Expanding $\mathbf{J}^{(a)}(x, y, z)$ in terms of the basis functions $T^x_{KLM}(x, y, z)$ and $T^y_{KLM}(x, y, z)$ gives

$$E^x_{klm} = \int\int\int \frac{\sum_{KLM} J^x_{KLM} T^x_{KLM}(x, y, z)}{\sigma^{(a)}(x, y, z)} T^x_{klm}(x, y, z) dx dy dz \tag{2.11}$$

$$E^y_{klm} = \int\int\int \frac{\sum_{KLM} J^y_{KLM} T^y_{KLM}(x, y, z)}{\sigma^{(a)}(x, y, z)} T^y_{klm}(x, y, z) dx dy dz\,. \tag{2.12}$$

We need to rewrite T^x_{KLM}, T^y_{KLM}, T^x_{klm}, and T^y_{klm} in terms of functions whose support is a single flaw cell. Thus, we write

$$T^x_{klm}(x, y, z) = R^x_{klm}(x, y, z) + S^x_{k+1lm}(x, y, z) \tag{2.13}$$

$$T^y_{klm}(x, y, z) = R^y_{klm}(x, y, z) + S^y_{kl+1m}(x, y, z)\,, \tag{2.14}$$

with the definitions

$$R^x_{klm}(x, y, z) = \begin{cases} \dfrac{x - k\delta x}{\delta x} & \text{for } k\delta x \le x \le (k+1)\delta x, \\ & l\delta y \le y \le (l+1)\delta y,\ m\delta z \le z \le (m+1)\delta z \\ 0 & \text{otherwise} \end{cases} \tag{2.15}$$

$$S^x_{klm}(x, y, z) = \begin{cases} \dfrac{(k+1)\delta x - x}{\delta x} & \text{for } k\delta x \le x \le (k+1)\delta x, \\ & l\delta y \le y \le (l+1)\delta y,\ m\delta z \le z \le (m+1)\delta z \\ 0 & \text{otherwise} \end{cases} \tag{2.16}$$

$$R^y_{klm}(x,y,z) = \begin{cases} \dfrac{y - l\delta y}{\delta y} & \text{for } k\delta x \le x \le (k+1)\delta x, \\ & l\delta y \le y \le (l+1)\delta y,\ m\delta z \le z \le (m+1)\delta z \\ 0 & \text{otherwise} \end{cases} \tag{2.17}$$

$$S^y_{klm}(x,y,z) = \begin{cases} \dfrac{(l+1)\delta y - y}{\delta y} & \text{for } k\delta x \le x \le (k+1)\delta x, \\ & l\delta y \le y \le (l+1)\delta y,\ m\delta z \le z \le (m+1)\delta z \\ 0 & \text{otherwise}\,. \end{cases} \tag{2.18}$$

Rewriting (2.11) and (2.12) in terms of these new functions gives

$$\begin{aligned} E^x_{klm} &= \iiint \sum_{KLM} J^x_{KLM} \left[\frac{R^x_{KLM}(x,y,z)}{\sigma_{KLM}} + \frac{S^x_{K+1LM}(x,y,z)}{\sigma_{K+1LM}} \right] \\ &\quad \left[R^x_{klm}(x,y,z) + S^x_{k+1lm}(x,y,z) \right] dxdydz \\ E^y_{klm} &= \iiint \sum_{KLM} J^y_{KLM} \left[\frac{R^y_{KLM}(x,y,z)}{\sigma_{KLM}} + \frac{S^y_{KL+1M}(x,y,z)}{\sigma_{KL+1M}} \right] \\ &\quad \left[R^y_{klm}(x,y,z) + S^y_{kl+1m}(x,y,z) \right] dxdydz\,, \end{aligned} \tag{2.19}$$

where σ_{KLM} is the (uniform) conductivity of flaw cell KLM.

Because of the compact support of the basis functions, we have

$$\iiint R^x_{KLM}(x,y,z) R^x_{klm}(x,y,z) dxdydz = \frac{1}{3}\delta x \delta y \delta z \delta_{kK}\delta_{lL}\delta_{mM} \tag{2.20}$$

$$\iiint R^x_{KLM}(x,y,z) S^x_{klm}(x,y,z) dxdydz = \frac{1}{6}\delta x \delta y \delta z \delta_{kK}\delta_{lL}\delta_{mM} \tag{2.21}$$

$$\iiint R^y_{KLM}(x,y,z) R^y_{klm}(x,y,z) dxdydz = \frac{1}{3}\delta x \delta y \delta z \delta_{kK}\delta_{lL}\delta_{mM} \tag{2.22}$$

$$\iiint R^y_{KLM}(x,y,z) S^y_{klm}(x,y,z) dxdydz = \frac{1}{6}\delta x \delta y \delta z \delta_{kK}\delta_{lL}\delta_{mM}\,. \tag{2.23}$$

Putting these results into (2.19) gives

$$E^x_{klm} = \left(\frac{1}{3}\frac{J^x_{klm}}{\sigma_{klm}} + \frac{1}{6}\frac{J^x_{k+1lm}}{\sigma_{k+1lm}} + \frac{1}{6}\frac{J^x_{k-1lm}}{\sigma_{klm}} + \frac{1}{3}\frac{J^x_{klm}}{\sigma_{k+1lm}} \right) \delta x \delta y \delta z \tag{2.24}$$

$$E^y_{klm} = \left(\frac{1}{3}\frac{J^y_{klm}}{\sigma_{klm}} + \frac{1}{6}\frac{J^y_{kl+1m}}{\sigma_{kl+1m}} + \frac{1}{6}\frac{J^y_{kl-1m}}{\sigma_{klm}} + \frac{1}{3}\frac{J^y_{klm}}{\sigma_{kl+1m}} \right) \delta x \delta y \delta z \,. \tag{2.25}$$

By adding the following equations for E^x_{kl+1m} and E^y_{k+1lm},

$$E^x_{kl+1m} = \left(\frac{1}{3}\frac{J^x_{kl+1m}}{\sigma_{kl+1m}} + \frac{1}{6}\frac{J^x_{k+1l+1m}}{\sigma_{k+1l+1m}} + \frac{1}{6}\frac{J^x_{k-1l+1m}}{\sigma_{kl+1m}} + \frac{1}{3}\frac{J^x_{kl+1m}}{\sigma_{k+1l+1m}} \right) \delta x \delta y \delta z \tag{2.26}$$

$$E^y_{k+1lm} = \left(\frac{1}{3}\frac{J^y_{k+1lm}}{\sigma_{k+1lm}} + \frac{1}{6}\frac{J^y_{k+1l+1m}}{\sigma_{k+1l+1m}} + \frac{1}{6}\frac{J^y_{k+1l-1m}}{\sigma_{k+1lm}} + \frac{1}{3}\frac{J^y_{k+1lm}}{\sigma_{k+1l+1m}} \right) \delta x \delta y \delta z \tag{2.27}$$

we obtain four equations in the four unknowns, σ_{klm}, σ_{k+1lm}, σ_{kl+1m}, $\sigma_{k+1l+1m}$. These equations can be written in matrix form as:

$$\begin{bmatrix} x_{00} & x_{10} & 0 & 0 \\ y_{00} & 0 & y_{01} & 0 \\ 0 & 0 & x_{01} & x_{11} \\ 0 & y_{10} & 0 & y_{11} \end{bmatrix} \begin{bmatrix} \rho_{klm} \\ \rho_{k+1lm} \\ \rho_{kl+1m} \\ \rho_{k+1l+1m} \end{bmatrix} \delta x \delta y \delta z = \begin{bmatrix} E^x_{klm} \\ E^y_{klm} \\ E^x_{kl+1m} \\ E^y_{k+1lm} \end{bmatrix} , \tag{2.28}$$

where $\rho_{klm} = 1/\sigma_{klm}$ and

$$x_{00} = \frac{J^x_{klm}}{3} + \frac{J^x_{k-1lm}}{6} \,, \quad y_{00} = \frac{J^y_{klm}}{3} + \frac{J^y_{kl-1m}}{6} \tag{2.29}$$

$$x_{10} = \frac{J^x_{klm}}{3} + \frac{J^x_{k+1lm}}{6} \,, \quad y_{01} = \frac{J^y_{klm}}{3} + \frac{J^y_{kl+1m}}{6} \tag{2.30}$$

$$x_{01} = \frac{J^x_{kl+1m}}{3} + \frac{J^x_{k-1l+1m}}{6} \,, \quad y_{10} = \frac{J^y_{k+1lm}}{3} + \frac{J^y_{k+1l-1m}}{6} \tag{2.31}$$

$$x_{11} = \frac{J^x_{kl+1m}}{3} + \frac{J^x_{k+1l+1m}}{6} \,, \quad y_{11} = \frac{J^y_{k+1lm}}{3} + \frac{J^y_{k+1l+1m}}{6} \,. \tag{2.32}$$

Note that on the borders of the flaw, some of the expansion coefficients for the anomalous currents vanish. In particular

$$J^x_{klm} = \begin{cases} \text{nonzero } 0 \le k \le N_x - 2, \; 0 \le l \le N_y - 1, \; 0 \le m \le N_z - 1 \\ \text{zero} \quad \text{otherwise} \end{cases} \tag{2.33}$$

$$J^y_{klm} = \begin{cases} \text{nonzero } 0 \le k \le N_x - 1, \; 0 \le l \le N_y - 2, \; 0 \le m \le N_z - 1 \\ \text{zero} \quad \text{otherwise} \end{cases} \tag{2.34}$$

Inverting matrix equation (2.28) gives

$$\delta x\delta y\delta z\begin{bmatrix}\rho_{klm}\\ \rho_{k+1lm}\\ \rho_{kl+1m}\\ \rho_{k+1l+1m}\end{bmatrix}=\begin{bmatrix}\frac{x_{11}y_{01}y_{10}}{D} & \frac{-y_{11}x_{01}x_{10}}{D} & \frac{y_{01}y_{11}x_{10}}{D} & \frac{-y_{01}x_{11}x_{10}}{D}\\ \frac{-y_{00}y_{11}x_{01}}{D} & \frac{y_{11}x_{01}x_{00}}{D} & \frac{-y_{01}y_{11}x_{00}}{D} & \frac{x_{00}x_{11}y_{01}}{D}\\ \frac{-x_{11}y_{00}y_{10}}{D} & \frac{x_{00}x_{11}y_{10}}{D} & \frac{-y_{00}y_{11}x_{10}}{D} & \frac{y_{00}x_{11}x_{10}}{D}\\ \frac{x_{01}y_{00}y_{10}}{D} & \frac{-x_{00}x_{01}y_{10}}{D} & \frac{x_{00}y_{01}y_{10}}{D} & \frac{-y_{00}x_{01}x_{10}}{D}\end{bmatrix}\begin{bmatrix}E^x_{klm}\\ E^y_{klm}\\ E^x_{kl+1m}\\ E^y_{k+1lm}\end{bmatrix}, \tag{2.35}$$

where $D = x_{00}x_{11}y_{01}y_{10} - y_{00}y_{11}x_{01}x_{10}$. The solution for the σ's is

$$\begin{aligned}
\sigma_{klm} &= \frac{(x_{00}x_{11}y_{01}y_{10} - y_{00}y_{11}x_{01}x_{10})\,\delta x\delta y\delta z}{x_{11}y_{01}y_{10}E^x_{klm} - y_{11}x_{01}x_{10}E^y_{klm} + y_{01}y_{11}x_{10}E^x_{kl+1m} - y_{01}x_{11}x_{10}E^y_{k+1lm}}\\
\sigma_{k+1lm} &= \frac{(x_{00}x_{11}y_{01}y_{10} - y_{00}y_{11}x_{01}x_{10})\,\delta x\delta y\delta z}{-y_{00}x_{01}y_{11}E^x_{klm} + y_{11}x_{01}x_{00}E^y_{klm} - y_{01}y_{11}x_{00}E^x_{kl+1m} + y_{01}x_{11}x_{00}E^y_{k+1lm}}\\
\sigma_{kl+1m} &= \frac{(x_{00}x_{11}y_{01}y_{10} - y_{00}y_{11}x_{01}x_{10})\,\delta x\delta y\delta z}{-x_{11}y_{00}y_{10}E^x_{klm} + x_{11}x_{00}y_{10}E^y_{klm} - y_{00}y_{11}x_{10}E^x_{kl+1m} + y_{00}x_{11}x_{10}E^y_{k+1lm}}\\
\sigma_{k+1l+1m} &= \frac{(x_{00}x_{11}y_{01}y_{10} - y_{00}y_{11}x_{01}x_{10})\,\delta x\delta y\delta z}{x_{01}y_{00}y_{10}E^x_{klm} - y_{10}x_{01}x_{00}E^y_{klm} + y_{01}y_{10}x_{00}E^x_{kl+1m} - y_{00}x_{01}x_{10}E^y_{k+1lm}}\,.
\end{aligned} \tag{2.36}$$

This solution for the σ's will not exist unless the system, (2.28), is independent. There are many situations that are likely to arise which will result in dependence among these equations, and we, therefore, turn to a different method of 'solving' (2.28), which we call the 'two-cell' hypothesis. Consider a 2×2 array of adjacent cells in the x- and y-directions. We call this array a 'window pane,' and ask for the solution of (2.28) under the hypothesis that two adjacent cells in the window pane have equal conductivities. The answer is:

$$\begin{aligned}
\sigma_{klm} &= \sigma_{k+1lm} = \frac{(x_{00}+x_{10})}{E^x_{klm}}\\
\sigma_{klm} &= \sigma_{kl+1m} = \frac{(y_{00}+y_{01})}{E^y_{klm}}\\
\sigma_{kl+1m} &= \sigma_{k+1l+1m} = \frac{(x_{01}+x_{11})}{E^x_{kl+1m}}\\
\sigma_{k+1lm} &= \sigma_{k+1l+1m} = \frac{(y_{10}+y_{11})}{E^y_{k+1lm}}
\end{aligned} \tag{2.37}$$

Of course, this results in non-unique solutions; each cell in the window pane has two solutions. In fact, because the σ's are known to be real, we can separate the x's and y's and E's into their real and imaginary parts, and get four solutions for each cell of the window pane. Each solution, then, is an outcome of an experiment, in the language of set-theoretic estimation theory, and is a candidate for processing by the robust statistical estimator that is described in the next chapter.

The integral relation that allows us to determine the anomalous current is derived from the measurement process. Typically, we measure the perturbation of the probe impedance, ΔZ, due to the flaw. Adopting the probe current $\mathbf{J}_p$ as the phase reference, this impedance is given by:

$$I^2 \Delta Z = -\int_{coil} \mathbf{E}^{(s)}(\mathbf{r}) \cdot \mathbf{J}_p(\mathbf{r}) d\mathbf{r}$$
$$= -\int_{flaw} \mathbf{E}^{(i)}(\mathbf{r}) \cdot \mathbf{J}(\mathbf{r}) d\mathbf{r}, \tag{2.38}$$

where I is the total driving-current in the probe coil. In arriving at the final expression, we used a reciprocity theorem (see [111, Chapter 5]) that relates the scattered field, $\mathbf{E}^{(s)}$, at the primary source (the eddy-current probe coil) to the incident field at the secondary source (the anomalous current source due to the flaw).

The integral expression, (2.38), is discretized by substituting the expansions, (2.3), for the currents, and making use of the definitions of the field moments, (2.5), (2.6). The result is quickly obtained:

$$\Delta Z = -\sum_{KLM} \left(J^{(x)}_{KLM} E^{(i)(x)}_{KLM} + J^{(y)}_{KLM} E^{(i)(y)}_{KLM} \right), \tag{2.39}$$

where we assume that $I = 1$, and that the exciting coils produce an incident field that is oriented in the (x, y)-plane; i.e., $E^{(i)(z)}_{KLM} = 0$.

2.2 Set-Theoretic Estimation

We quote Combettes [27, p. 202]

> The basic philosophical motivation for the set theoretic approach is that more reliable solutions can be obtained by exploiting known information rather than imposing an often subjective notion of optimality. Thus, in the set theoretic framework, the emphasis is placed on the feasibility of a solution rather than its optimality, as is done in the conventional approach. The goal is not to produce a "best" solution but one that is consistent with all available information. In set theoretic estimation, all the members of the feasibility set are acceptable solutions. They can be regarded as the objects that, in light of all available information, may have given rise to the observed data. The only way to restrict objectively the feasibility set is to incorporate more information in the formulation. If some of the feasible solutions are not acceptable, then it must be the case that the formulation fails to

include some constraint that has not been identified. Once the set based on this constraint is incorporated, any point in the feasibility set should be acceptable; if not the cycle is repeated. Usually, there is more than one solution, which may be counterintuitive from the standpoint of conventional point estimation theory where, to extract a single solution, an objective function with a unique extremum is employed. On the other hand, because of the arbitrariness in the selection of such an objective function, the result is, at best, nothing but a qualitative selection of a feasible solution.

In set theoretic estimation, we do not seek a unique solution, as in optimization, but we seek the set of feasible solutions; all such solutions are acceptable [27]. A feasible solution is one that is consistent with all available information, such as the field equation, (2.41), and the data equation, (2.40), both shown below.

The algorithm that we are developing is based on statistical decision theory applied to the outcomes of a number of "experiments." Each experiment is labeled by the view-index v that is associated with the frequency of excitation and position of the exciting probe in the eddy-current NDE process. The outcomes of the experiments are elements of the "feasibility set," in the language of set theoretic estimation; that is, they satisfy all known information about the problem.

We write the data equation and the field equation as

$$Z(v) = -\sum_{LMJ} \mathbf{E}^{(i)}_{LMJ}(v) \cdot \mathbf{J}_{LMJ}(v) \tag{2.40}$$

$$\mathbf{E}^{(i)}_{lmj}(v) = \mathbf{E}_{lmj}(v) - \sum_{LMJ} G^{(ee)}_{jJ}(l-L, m-M; \omega) \cdot \mathbf{J}_{LMJ}(v) \ , \tag{2.41}$$

respectively, where $G^{(ee)}$ is an 'electric-electric' Green's dyadic, that transforms an electric current into an electric field moment.[1] Note that (2.40) and (2.41) are linear in all the variables.

We define the vth "experiment" to be the pair $(Z(v), \mathbf{E}^{(i)}_{lmj}(v))$, and the "outcomes" of this experiment to be the pair $(\mathbf{J}_{lmj}(v), \mathbf{E}_{lmj}(v))$, which satisfy (2.40) and (2.41). Thus, the outcomes are feasible because they satisfy all known information about the problem. Clearly, we cannot talk about a unique solution, because the feasible set contains many points.

In the problems that are described in this chapter, we model a single transmitting coil to excite the system, and a single receiver coil that is scanned over the region of interest. In place of a single, movable, receiver coil, it is possible to use a fixed array of receivers. In either case, this is an example of a transmit-receive (T/R) configuration, which is becoming more widely used in the NDE industry. The use of a T/R configuration allows us to gain more information from each experiment (i.e., from each viewing).

For example, if we have a single excitation source (the transmitter), and a single receiver sensor that is scanned over N_s points, then each view, v, produces N_s results, $\{Z_1(v), \ldots, Z_{N_s}(v)\}$. The actual current $\mathbf{J}_{LMJ}(v)$, however, is associated

[1] We consider only problems in which the host and anomalies are nonmagnetic.

only with the transmitter, because the transmitter is the sole exciter of the system (the receiver is assumed to carry no current). Hence, (2.40) is replaced by

$$
\begin{aligned}
Z_1(v) &= -\sum_{LMJ} \mathbf{E}^{(1)}_{LMJ}(v) \cdot \mathbf{J}_{LMJ}(v) \\
Z_2(v) &= -\sum_{LMJ} \mathbf{E}^{(2)}_{LMJ}(v) \cdot \mathbf{J}_{LMJ}(v) \\
&\vdots \\
Z_{N_s}(v) &= -\sum_{LMJ} \mathbf{E}^{(N_s)}_{LMJ}(v) \cdot \mathbf{J}_{LMJ}(v) \,, \qquad (2.42)
\end{aligned}
$$

where $\mathbf{E}^{(n)}_{LMJ}(v)$ is the "incident" field produced by the sensor when it is in its nth scan position, during the vth view. This field is known a priori, because we know the location and geometry of the receiving sensor during the vth view.

The minimum-norm solution of (2.42) is given by

$$
\tilde{\mathbf{J}}_{LMJ}(v) = \mathscr{M}^{\dagger}(v) \begin{bmatrix} Z_1(v) \\ \vdots \\ Z_{N_s}(v) \end{bmatrix} , \qquad (2.43)
$$

where $\mathscr{M}^{\dagger}(v)$ is the pseudoinverse of the matrix in (2.42). In the examples of this chapter, we used the QR-decomposition [62] to compute the minimum-norm solution. The algorithm that is presented in [62] to solve for the minimum-norm solution allows the user to define a tolerance level, from which an effective pseudo-rank is obtained for the system matrix. The tolerance is chosen to produce a stable solution in the presence of noisy data.

The electric field produced by this current is gotten by substituting (2.43) into (2.41):

$$
\tilde{\mathbf{E}}_{lmj}(v) = \mathbf{E}^{(i)}_{lmj}(v) + \sum_{LMJ} G^{(ee)}_{jJ}(l-L, m-M; \omega) \cdot \tilde{\mathbf{J}}_{LMJ}(v) \,. \qquad (2.44)
$$

Note that this is the correct electric field-moment corresponding to the current $\tilde{\mathbf{J}}$, and **VIC-3D®** computes this field very quickly and accurately. This gives **VIC-3D®** the advantage over other field-solving methods, such as finite-elements or finite-differences. More important, however, is that (2.44) is the unique field-moment associated with the current, and is calculated to the same precision as the current, which will ensure that the final step in the algorithm, namely the statistical decision step that is described below, will be meaningful.

Given the electric field moments, we can then calculate an expansion for the electric field within the flaw. The resulting electric field expansion coefficients will be called $\widetilde{\mathbf{e}}_{lmj}$. Hence, for this value of the view-index, v, we associate the triple, $(\widetilde{J}_x, \widetilde{e}_x)_{lmj}(v)$, $(\widetilde{J}_y, \widetilde{e}_y)_{lmj}(v)$, $(\widetilde{J}_z, \widetilde{e}_z)_{lmj}(v)$, with the lmjth cell, and when we take the ratio of the electric current to the electric field at the middle of the lmjth cell, we arrive at the conductivity, σ_{lmj} of the lmjth cell, which is our final goal. We then perform another experiment by choosing another value of v, thereby generating another triplet. This ensemble of triplets constitutes the feasible set for each cell.

2.3 Statistical Analysis of the Feasible Set

One of the principal efforts of research in Set-Theoretic Estimation is to determine the number and nature of the experiments, $(Z(v), \mathbf{E}^{(i)}_{lmj}(v))$, and receiver scans, $\mathbf{E}^{(n)}_{LMJ}$, that produce a good feasible set for statistical analysis. As might be expected, this depends upon the complexity of the flaw that is to be reconstructed, and upon the resolution desired in the reconstruction.

After deducing the feasibility set, we are then faced with the task of assigning a single number for the conductivity of each cell; this is a problem of data analysis, which uses robust regression, as described in [111, Chapter 13]. The aim of the statistical analysis of the feasible set is to fit a constant through the data, which are produced by the algorithm described in the preceding section, for all views (or experiments), v.

A beneficial feature of this algorithm is that the analysis of the data set for each cell is done independently of every other cell; i.e., a decision is made on a cell-by-cell basis. Since the decision to be made for each cell involves a nonlinear (robust) estimator, the computational burden is greatly reduced when compared to using a nonlinear estimator to solve for many cells jointly. (The bilinear conjugate-gradient algorithm is an exception.) Furthermore, this leads us to a constrained iterative ('layer-stripping') algorithm, that uses the known and accepted results for some cells to determine the results for others at a later stage of the iteration.

2.4 A Layer-Stripping Algorithm

A classical layer-stripping algorithm consists of solving an inverse problem layer-by-layer, when the physical system permits such a reconstruction. If the excitation and detection methods permit only one layer to be detected, say due to timing arrangements in a pulsed system, then only that layer will be reconstructed. This reconstruction, then, constitutes known data for the reconstruction of the next layer.

In our situation, we have discovered that we can reconstruct certain cells accurately, similar to the reconstruction of a layer, and we then want to use this result in the reconstruction of the remaining cells. We do not assume that an entire layer has been reconstructed, nor that we will reconstruct on a layer-by-layer basis.

The proposed algorithm proceeds as follows: suppose that we have successfully reconstructed the conductivity of each cell of a 2×2 window pane. That means that we know σ_{klm}, σ_{k+1lm}, σ_{kl+1m}, and$\sigma_{k+1l+1m}$; hence, all of the coefficients in (2.24) and (2.25) are well-defined. Then, we substitute these equations into the left-hand sides of the field equations, (2.7), and derive the constraint equations:

$$\left(\frac{1}{3}\frac{J^x_{klm}}{\sigma_{klm}} + \frac{1}{6}\frac{J^x_{k+1lm}}{\sigma_{k+1lm}} + \frac{1}{6}\frac{J^x_{k-1lm}}{\sigma_{klm}} + \frac{1}{3}\frac{J^x_{klm}}{\sigma_{k+1lm}}\right)\delta x \delta y \delta z =$$
$$E^{(i)(x)}_{klm} + \sum_{KLM} G^{(xx)}_{klm,KLM} J^x_{KLM} + \sum_{KLM} G^{(xy)}_{klm,KLM} J^y_{KLM} + \sum_{KLM} G^{(xz)}_{klm,KLM} J^z_{KLM}$$

$$\left(\frac{1}{3}\frac{J^y_{klm}}{\sigma_{klm}} + \frac{1}{6}\frac{J^y_{kl+1m}}{\sigma_{kl+1m}} + \frac{1}{6}\frac{J^y_{kl-1m}}{\sigma_{klm}} + \frac{1}{3}\frac{J^y_{klm}}{\sigma_{kl+1m}}\right)\delta x \delta y \delta z =$$
$$E^{(i)(y)}_{klm} + \sum_{KLM} G^{(yx)}_{klm,KLM} J^x_{KLM} + \sum_{KLM} G^{(yy)}_{klm,KLM} J^y_{KLM}$$
$$+ \sum_{KLM} G^{(yz)}_{klm,KLM} J^z_{KLM}, \qquad (2.45)$$

The unknowns in this equation are the currents in all the cells (including klm); everything else is known. This equation, which is of the form that **VIC-3D®** solves, constitutes a constraint on the system of data equations (2.42). Of course, we get an equation that is identical to (2.45) for each successfully reconstructed window-pane. The system of such equations is consistent, and generally well conditioned. Note, in particular, that if $\sigma_{klm} = 0$ then the constraint equations become $J^x_{klm} = J^x_{k-1lm} = J^y_{klm} = J^y_{kl-1m} = 0$.

Thus, our problem now is to determine a minimum-norm (least-squares) solution of (2.42), subject to the linear constraint(s) of (2.45). We appeal to well-known algorithms in linear least-squares problems. Lawson and Hanson [62, pp.134–157] give three algorithms, together with supporting code, to solve this problem. Each of the three methods consists of three stages:

1. Derive a lower-dimensional unconstrained least squares problem from the given problem.
2. Solve the derived problem.
3. Transform the solution of the derived problem to obtain the solution of the original constrained problem.

In Lawson and Hanson's first method, one makes use of an orthogonal basis for the null space of the matrix of the constraint equations. If the problem does not have a unique solution, this method will produce the unique minimum norm solution of the original constrained problem. Furthermore, this method has the very attractive feature of being amenable to numerically stable updating techniques, which will be necessary as we add more rows to the constraint equation; this will happen as we satisfactorily reconstruct more and more cells.

The second method uses direct elimination by premultiplication using both orthogonal and nonorthogonal transformation matrices, and the third solves the problem by using weighted least-squares. We have tested the second and third methods, and have found them to produce good results. We used the second method in the examples reported later in this chapter.

Once the minimum-norm solution of the constrained least-squares problem has been computed, the algorithm proceeds as before. The currents are substituted into (2.41), thereby determining the total field in each cell (we can exclude cell klm, unless we believe that it could be improved). Then the feasibility set is fitted with a straight line by means of robust estimation; the slope of this line is accepted as the normalized conductivity of the cell.

If we are still not satisfied with the results of one or more cells, we will perform a new experiment. The experiment could be simply a new frequency of excitation, which would make the process a multifrequency algorithm.

2.5 Some Examples of the Inversion Algorithm

Introduction The model examples in this section include surface-breaking and buried flaws in a half-space. We simulate a transmit-receive (T/R) probe configuration, which includes a single transmitting and receiving probe. Each probe is scanned independently of the other, but the receiving probe is assumed to be connected to an infinite-impedance amplifier. This means that it carries no current; hence, the excitation of the flaw is accomplished solely by means of the transmitter probe. In all cases the transmitting and receiving coils are identical.

In all the problems considered in this chapter, the host region is a half-space, whose conductivity is 10^5 S/m. The anomalous conductivity is normalized to have a value $\sigma_a = \sigma_f/\sigma_h - 1$, where σ_f is the conductivity of the flaw, and σ_h is the conductivity of the host. Hence, for a void, in which the conductivity of the flaw, $\sigma_f = 0$, we have $\sigma_a = -1$, and this is the smallest value that the anomalous conductivity can take, because $\sigma_f \geq 0$. For an unflawed cell, $\sigma_f = \sigma_h$, which implies that $\sigma_a = 0$.

In principle, there is no upper limit to the value that σ_a can assume, but in practice the conductivity of the flaw is not likely to exceed that of copper ($= 5.8 \times 10^7$), which means that for typical host materials, whose conductivities are in the range 10^6 to 10^7, σ_a is likely to be bounded by 9 or so. In the numerical experiments presented here, we used this as our upper bound in searching for the best fit for the conductivity of each cell.

A Surface-Breaking Slot at 50 kHz Consider a cubic region intersecting the surface of the half-space and extending 3 mm vertically (the z-direction) into the half-space. The (x, y) dimensions are also 3 mm, centered at the origin in (x, y)-

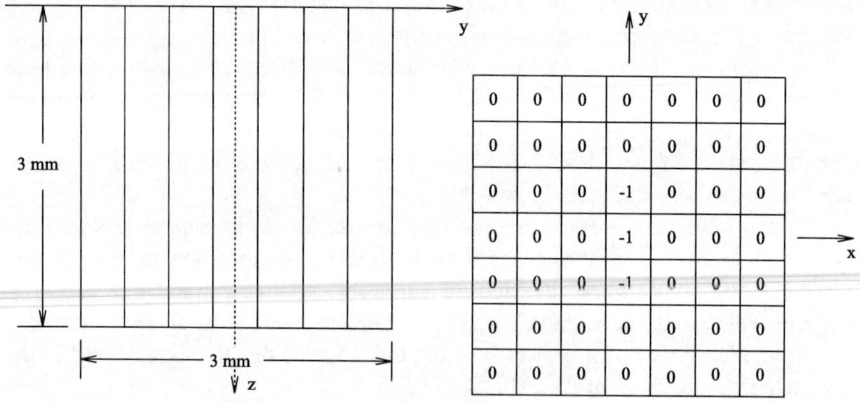

Fig. 2.2 A surface-breaking slot. Those cells labeled '0' are unflawed, and those labeled '-1' are empty (zero conductivity)

space. This cube is partitioned into forty-nine cells in the (x, y)-plane, and one cell in the z-direction. The middle three cells are filled with air ($\sigma_a = -1$), and the remaining forty-six are filled with host material ($\sigma_a = 0$), as shown in Fig. 2.2. This arrangement constitutes a surface-breaking slot, whose precise location in the (x, y)-plane is uncertain. The frequency of excitation is 50 kHz, which produces a skin-depth of

$$\begin{aligned}\delta_s &= \sqrt{\frac{2}{2\pi f \mu \sigma_h}} \\ &= \sqrt{\frac{2}{2\pi \times 5 \times 10^4 \times 4\pi \times 10^{-7} \times 10^5}} \\ &= 7.12 \text{ mm} .\end{aligned} \tag{2.46}$$

The transmitting coil is scanned over the (x, y)-plane, using 16 equi-spaced points in each direction, starting at coordinates $(-3.0, -3.0)$ and ending at $(3.0, 3.0)$. This scan constitutes 256 'experiments.' The outcome of each of these experiments is obtained by scanning the receiver coil over the same (x, y)-raster as for the transmitter, when the transmitter is fixed at each of its points. This gives us 256 complex equations, with 98 complex unknowns (the x and y-components of the anomalous current in each cell), which are to be massaged by the QR-decomposition, producing a single least-squares estimate of the complex current in each cell. The algorithm then produces the x and y-components of the corresponding complex electric field within each cell; these are the feasibility sets that are defined above.

We display the feasibility sets for cells numbered 18, 25, and 32 in Fig. 2.3. Clearly, the data favor a conductivity value of -1, and that is what the LMS-

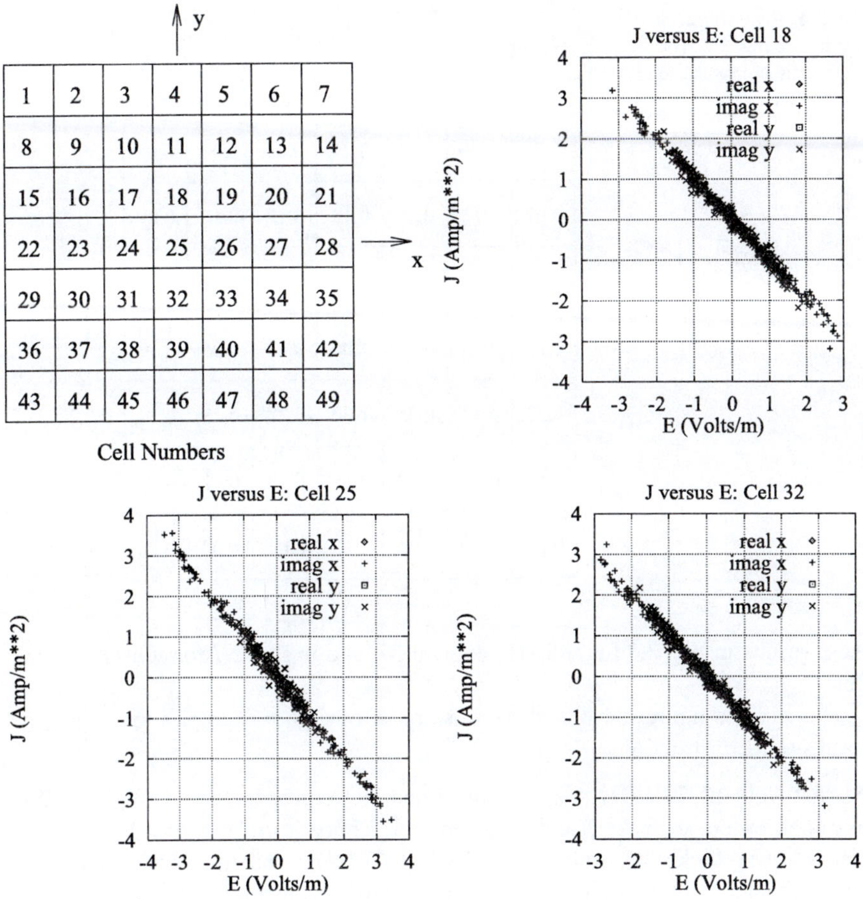

Fig. 2.3 Feasibility sets for cells numbered 18, 25, and 32

estimator produces in the reconstruction shown in Fig. 2.4. This, of course, is the exact answer for these three cells.

In Fig. 2.5 we show feasibility sets for cells numbered 1, 17, and 26. These cells are unflawed, but only cell number 1 is reconstructed exactly. Despite this fact, this reconstruction can be useful in practical NDE, because it indicates clearly the presence of a flaw, and gives a reasonable estimate of its size. An application of the S-estimator as well as the classical least-squares estimator produced essentially the same results.

We can get an idea of the quality of the reconstruction of each cell (or, to put it roughly, the confidence we can place on the results), by plotting the logarithm of the median of the squares of the residuals (if we are using the LMS-estimator) versus theta. The algorithm requires us to choose the minimum of this function, so we wish to determine the global picture to determine the 'quality' of this minimum. We show

Fig. 2.4 Reconstruction of the flaw of Fig. 2.2. The LMS-estimator was used for this reconstruction

0.00	0.00	-0.01	-0.21	-0.13	-0.01	0.00
-0.07	-0.13	-0.13	-0.18	-0.12	-0.16	-0.09
-0.16	-0.11	-0.25	-1.00	-0.22	-0.12	-0.14
-0.10	-0.25	-0.30	-1.00	-0.32	-0.25	-0.10
-0.13	-0.12	-0.25	-1.00	-0.22	-0.23	-0.14
-0.08	-0.12	-0.12	-0.16	-0.12	-0.12	-0.08
-0.01	0.00	-0.04	-0.20	-0.01	0.00	0.00

these results in Fig. 2.6 for cells 18, 25, and 32, and in Fig. 2.7 for cells 1, 17, and 26.

We will use figures such as these in some of our other examples to explain the results of reconstructions.

Buried Void at 50 MHz We take the same slot configuration of the preceding example, except to make it only 1.5 mm deep, bury it under a host layer that is also 1.5 mm thick, and excite this system at 50 MHz. Our use of 50 MHz stems from our interest in improving the resolution of the reconstructions of the preceding example, and is guided by the following argument. The (x, y) cell dimensions are 0.429 mm, and the skin depth at 50 kHz is 7.12 mm. At 50 MHz, the skin depth is 0.225 mm, which is about one-half the cell dimensions. The skin effect is isotropic, which means that a localized source radiating in the host material will have its signal reduced to $1/e = 0.368$ in 0.225 mm, in any direction, at 50 MHz. Thus, at this frequency, we expect two nearby cells to be well distinguished as compared to the situation at 50 kHz. The use of the over-layer of host material is merely to make the problem more challenging and realistic. We used the same raster scan for both transmitter and receiver as in Section (a), but the inversion process now involves 98 cells, since we are attempting to reconstruct the top layer, as well as the buried void. That is, we are assuming that we know nothing about the depth of the void. The system is shown in Fig. 2.8; the upper grid corresponds to the 49 cells of the host layer, and the lower grid shows the flaw.

The results of the inversion support our conjecture about the smaller skin depth aiding the resolution of the reconstruction. Each cell of the host layer is reconstructed exactly (to machine precision) as 0. This exact result is probably due

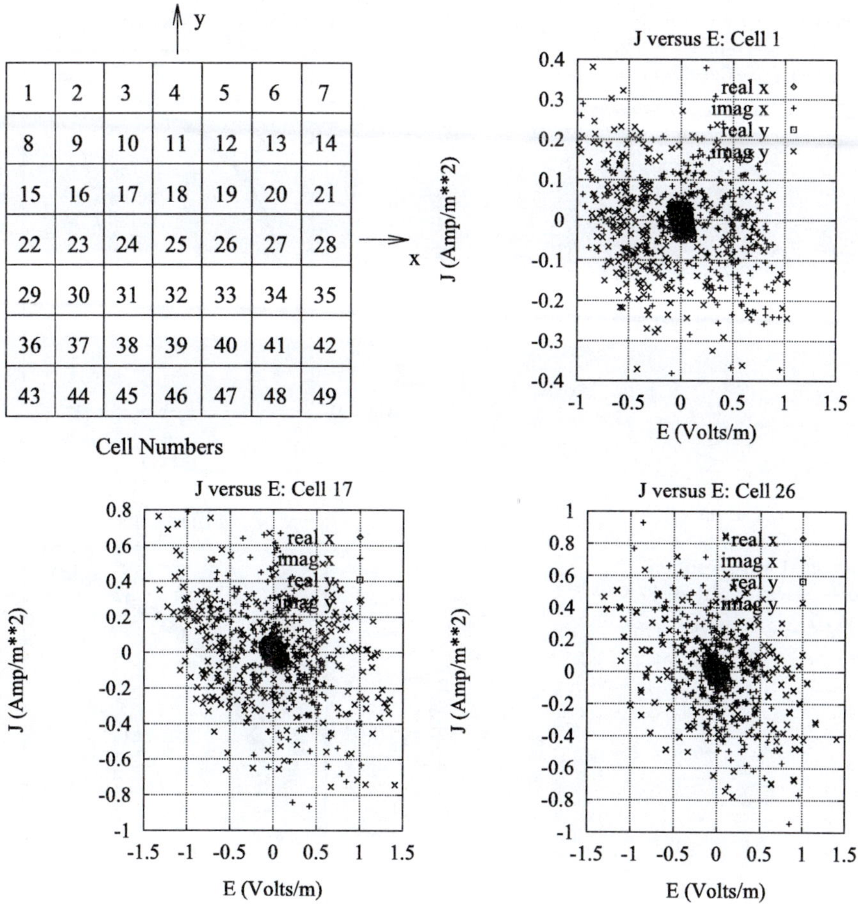

Fig. 2.5 Feasibility sets for the unflawed cells numbered 1, 17, and 26

to the fact that this part of the 'flaw' is uniform, which means that resolution is no problem. Furthermore, those parts of the flaw that are nearer the exciting source are generally reconstructed more accurately than those that are buried.

As for the buried part, it, too, is reconstructed well. We show in Fig. 2.9 several of the cells surrounding the slot that were not well reconstructed in Fig. 2.4; clearly, there is significant improvement in the reconstructions of these cells.

In performing this reconstruction we used the entire data set of 1024 points. If we use only the partial data sets associated with the real and imaginary parts of the x and y-components of the current density and electric field within each cell, we continued to get a good reconstruction, using either the S-estimator or the LMS-estimator. If we consider a performance criterion (or, perhaps we should say a parameter of confidence), to be the ratio of the largest value of the median (or estimate of scale in the case of the S-estimator) to the smallest, as θ (the conductivity

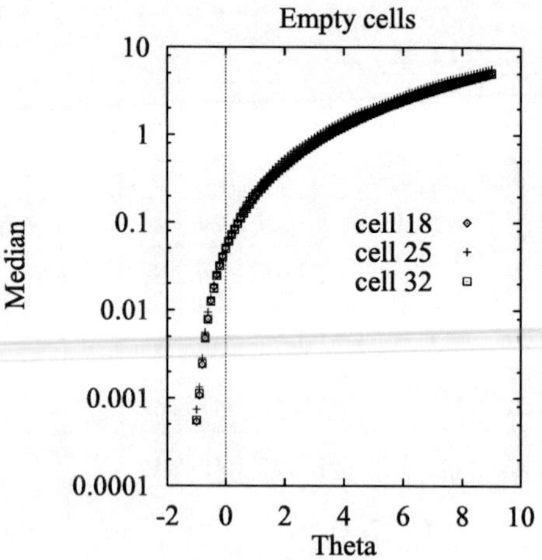

Fig. 2.6 Logarithm of the median-of-the-squares of the residual (LMS-algorithm) versus theta for cells 18, 25, 32

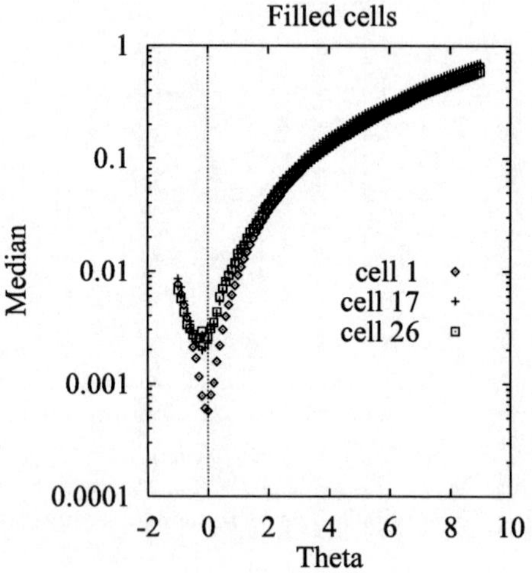

Fig. 2.7 Logarithm of the median-of-the-squares of the residual (LMS-algorithm) versus theta for cells 1, 17, and 26

slope) is varied from -1 to 0, then we found the use of the real and imaginary parts of the y-components to give, respectively, ratios of 100 and 200 for cell 73 (conductivity = -0.02 in Fig. 2.9). The use of the real and imaginary parts of the x-components gave, respectively, 40 and 180. When we used the entire data set of 1024 in cell 73, we computed a ratio of 105. In this sense, we can say that the imaginary part of both components provides superior data.

A Surface-Breaking Checkerboard at 50 MHz In this experiment we replace the simple slot of the first example by a complex checkerboard flaw that is shown in Fig. 2.10, and, again, perform a 16×16 raster scan with the transmitter and receiver probes at 50 MHz.

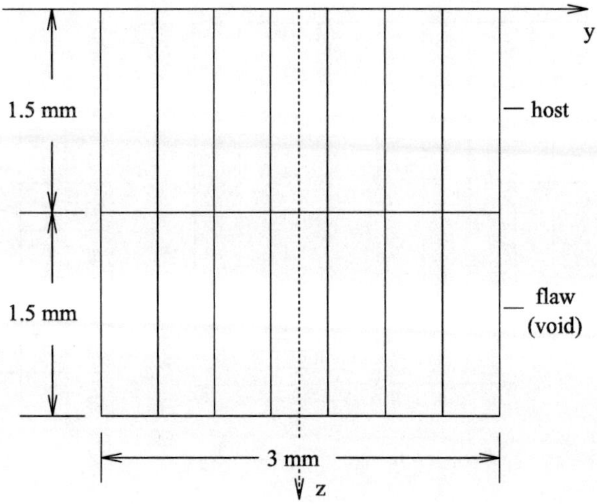

Fig. 2.8 A buried void. The upper grid corresponds to the 49 cells of the host layer, and the lower grid shows the flaw

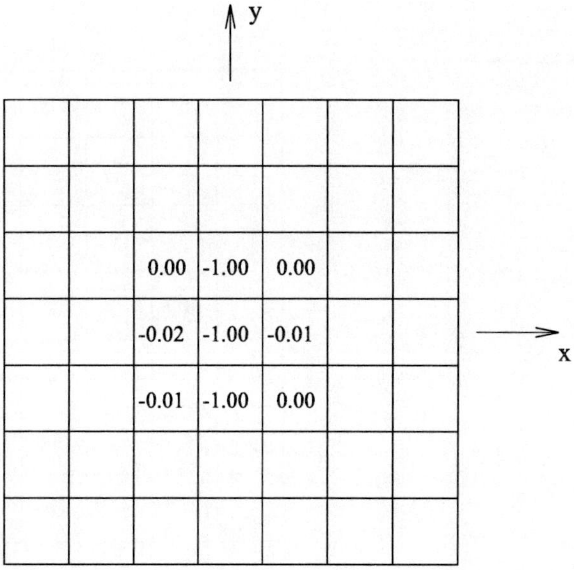

Fig. 2.9 Showing the reconstruction of several cells surrounding the slot, using the S-estimator. These are well reconstructed when compared to Fig. 2.4

The checkerboard is a difficult flaw to reconstruct, because the 'scene' changes so rapidly; i.e., it contains high spatial frequencies and the receiver scan must be fine enough to reconstruct these frequencies. The reconstruction using the LMS-estimator is quite good (corresponding results were obtained using the S-estimator and classical estimator). We show the reconstruction of the middle row and column in Fig. 2.11.

A Buried Checkerboard When we bury the checkerboard below a layer of host material we get the added complication of increasing the number of unknowns that

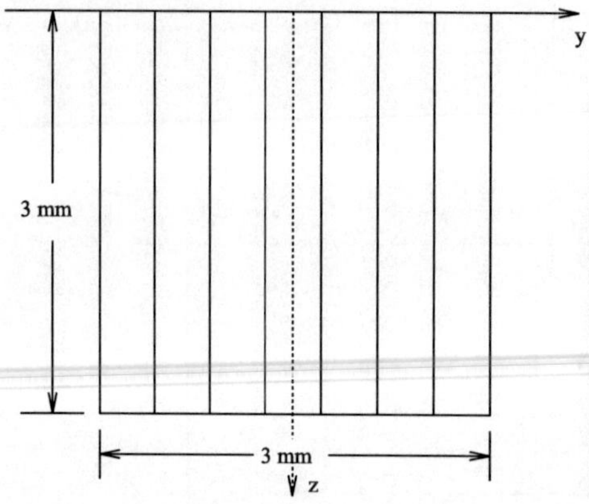

x

-1	0	-1	0	-1	0	-1
0	-1	0	-1	0	-1	0
-1	0	-1	0	-1	0	-1
0	-1	0	-1	0	-1	0
-1	0	-1	0	-1	0	1
0	-1	0	-1	0	-1	0
-1	0	-1	0	-1	0	-1

y

Fig. 2.10 A surface-breaking checkerboard

Fig. 2.11 Reconstruction of the middle row and column of the surface-breaking checkerboard

x

			0.00			
			-1.00			
			-0.01			
-0.01	-1.00	0.00	-1.00	0.00	-1.00	-0.01
			-0.01			
			-1.00			
			0.00			

y

are to be determined during the reconstruction. This, together with the fact that the checkerboard scene contains high spatial-frequencies, suggests that we use a 31×31 raster scan for the receiver probe, rather than the 16×16 scan that was used in the preceding examples, but covering the same area. The resolution of the receiver scan was thereby improved to 0.2 over 0.4 mm.

When we do this we get an excellent reconstruction; the top layer is reconstructed exactly (zero anomalous conductivity), and the reconstruction of the middle row and column of the bottom layer are shown in Fig. 2.12.

Fig. 2.12 Reconstruction of the middle row and column of the buried checkerboard

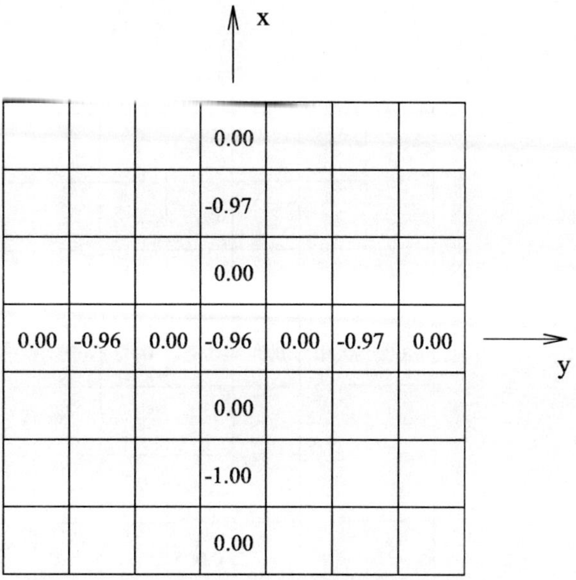

When we attempted to reconstruct the buried checkerboard using the previous 16×16 raster scan with the receiver, we obtained a reasonable reconstruction– one that was good enough to indicate the presence of a checkerboard flaw beneath a layer of host material, but not as good as Fig. 2.12. See Fig. 2.13 for these results.

A Double Checkerboard When we stack two checkerboards with the opposite polarity on top of each other, and reconstruct at 50 MHz using a 31×31 receiver raster scan, we have difficulty obtaining an accurate reconstruction, except for the zeros of the top layer.

We have developed an empirical rule for determining the quality of a reconstructed cell, when using either the LMS or S-estimators. We plot the median of the squares of the residuals, in the case of the LMS-estimator, or the scale factor, in the case of the S-estimator, versus the sought-for parameter, θ, for $-1 \le \theta \le 9$ (say, or some other upper bound, perhaps 0). If the minimum value of the minimum median or scale factor is two or more orders of magnitude smaller than the maximum value of median or scale, then the resulting answer is reliable; otherwise, it is suspect. For example, in Fig. 2.14 we plot the curve of the logarithm of the median versus θ, for $-1 \le \theta \le 0$, for cell 1 (corner cell, top layer, conductivity = 0) and cell 2 (conductivity = -1). The curve for cell 1 clearly satisfies our empirical rule for a reliable reconstruction, whereas that for cell 2 does not. The reconstructed value for cell 1 is zero, whereas it is not well defined for cell 2.

The source of this problem in our inversion algorithm is the quantity and quality of data that are presented to the QR-decomposer in the first stage of the algorithm. Typically, the QR-method in least-squares analysis produces reliable results if the system of equations is quite over-determined. Hence, there are three possible ways

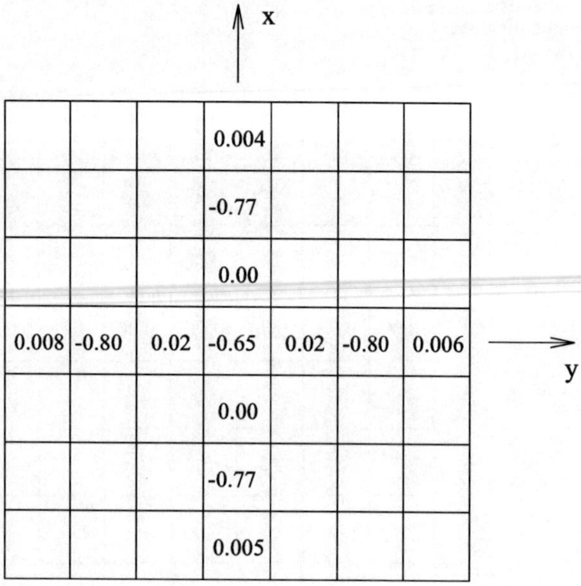

Fig. 2.13 Reconstruction of a buried checkerboard, when a 16×16 receiver scan is used; the top layer was reconstructed perfectly. The classical (least-squares) estimator produced these results

Fig. 2.14 Log-median versus θ for cells 1 and 2

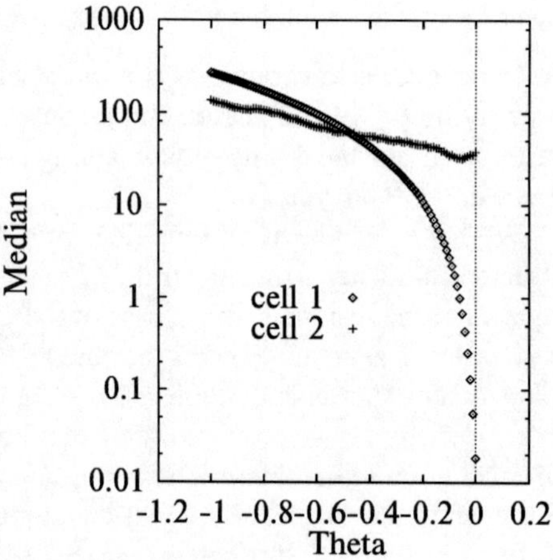

of improving the reconstruction of complex flaws of the double-checkerboard type: (1) increase the amount of input data by scanning the receiver probe over a larger range, with increased resolution, (2) using a different least-squares algorithm, such as Kaczmarz' [24, 27, 51] algorithm, that works well for even strongly underdetermined systems, or (3) modify our inversion algorithm to account for known constraints on certain correctly reconstructed cells.

2.6 Application to Aircraft Structures

The Canonical Problems The problems that will be solved in this section are:

1. Use **VIC-3D®** and our inversion algorithm to model the detection and characterization of metal corrosion in hidden or inaccessible airframe locations, such as double- or triple-layer airframes.
2. Use **VIC-3D®** and our inversion algorithm to model the detection and characterization of cracking or multisite damage in metallic airframe structures.
3. Use **VIC-3D®** and our inversion algorithm to model the detection, imaging, and characterization of surface and bulk anomalies in metallic airframe structures or engine components.

We can model corrosion as a weakly conducting region within a host material, and a crack as a nonconducting region within the same host. In fact, any region within the host material that is electrically distinct from the host will be called an anomalous region, and includes corrosion, cracks, fasteners, etc. We will take the host material to be aluminum, with a conductivity of 3.06×10^7 S/m, and a relative magnetic permeability of unity. In the problems that we will solve, we will replace 'corrosion' by a 'fastener,' whose conductivity is 16% that of aluminum. This is typical of brass, say.

The canonical structure for the inversion problems is shown in Fig. 2.15, which we will call a fastener with multisite damage. The structure consists of two layers, with the damage (or crack) emanating from opposite sides of the fastener in each layer.

Recall that the anomalous conductivity is given by $\sigma_f(\mathbf{r}) - \sigma_h$, where σ_f is the conductivity of the flaw, and σ_h the host conductivity. The normalized anomalous conductivity is simply the anomalous conductivity divided by the host conductivity. The conductivity of the crack will be zero, which means that the normalized anomalous conductivity of the crack will be -1. The normalized anomalous conductivity of the fastener is -0.84. Of course, the normalized anomalous conductivity of the host region is zero. Hence, the normalized anomalous conductivity-map of the structure is as shown in Fig. 2.16. We number the cells of the grid starting with the bottom layer, and working to the top. Cell number 1 is in the upper-left corner of the bottom layer, and cell number 64 is in the lower-right corner of the bottom layer. Cell number 65 is in the upper-left corner of the top layer (for a two-layer structure),

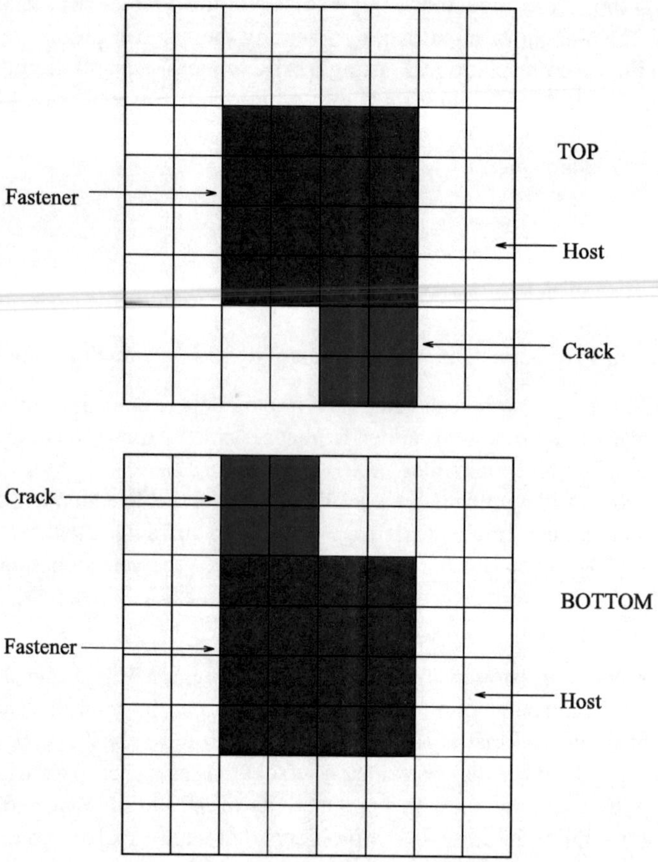

Fig. 2.15 Showing a fastener with multisite damage (the 'crack') embedded in a host region

and cell number 128 is in the lower-right corner. The length and width of the flaw will be 10 mm, and the depth 2.5 mm.

The data for the inversion experiments are gotten by assuming a transmit-receive system, in which the transmitter and receiver coils move independently of each other. The transmitter coil is taken to have an inner radius of 5 mm, an outer radius of 10 mm, a height of 2 mm, and to contain 200 turns. It undergoes a 4×4 raster scan in the (x, y) plane, which means that there are sixteen 'experiments,' in the language of set-theoretic estimation theory. The scan increment in the x- and y-directions is 3.3333 mm, which means that the raster scan covers the entire range of 10×10 square-millimeters of the flaw. The lift-off of the transmitter coil is 1 mm.

The receiver coil undergoes a 32×32 raster scan in the (x, y)-plane, again covering the 10×10 square-millimeters of the flaw. The inner radius of the receive coil is assumed to be 1 mm, the outer radius 2 mm, and the height 0.5 mm. The receive coil contains 200 turns, and its scan lift-off is zero.

0	0	0	0	0	0	0	0
0	0	0	0	0	0	0	0
0	0	-.84	-.84	-.84	-.84	0	0
0	0	-.84	-.84	-.84	-.84	0	0
0	0	-.84	-.84	-.84	-.84	0	0
0	0	-.84	-.84	-.84	-.84	0	0
0	0	0	0	-1.0	-1.0	0	0
0	0	0	0	-1.0	-1.0	0	0

TOP

0	0	-1.0	-1.0	0	0	0	0
0	0	-1.0	-1.0	0	0	0	0
0	0	-.84	-.84	-.84	-.84	0	0
0	0	-.84	-.84	-.84	-.84	0	0
0	0	-.84	-.84	-.84	-.84	0	0
0	0	-.84	-.84	-.84	-.84	0	0
0	0	0	0	0	0	0	0
0	0	0	0	0	0	0	0

BOTTOM

Fig. 2.16 Normalized anomalous conductivity-map of the fastener-with-multisite-damage structure

We have added noise to all the model impedance data used in our inversions. We assume that the noise has a zero-mean Gaussian distribution for both the magnitude and the phase of the impedance. We use a 1% variance in the magnitude and a 1° variance in the phase angle. These figures are realistic for contemporary instruments and data gathering techniques. As for the robust estimators, we have experimented with both, the Least Median of Squares (LMS) and the Scale (S) estimators, and

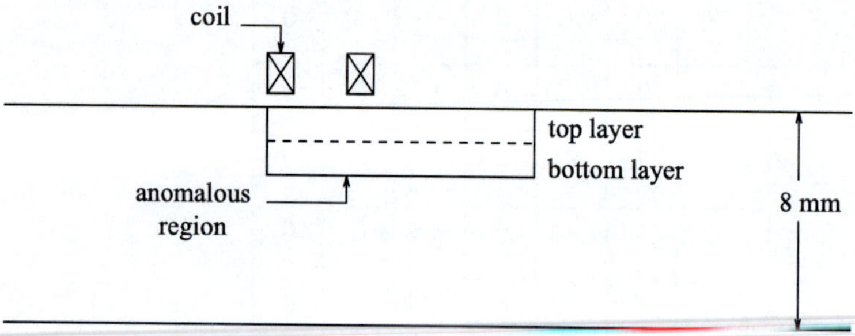

Fig. 2.17 Fastener with multisite damage as a surface-breaking anomaly

will report only the results of the S-estimator, which we call 'stest' in the captions of several figures.

Problem No. 1: Fastener with Multisite Damage in a Slab In the first problem, the anomalous region breaks the surface of a slab, as shown in Fig. 2.17. The data are gathered at 200 Hz.

The results for this problem are shown in Fig. 2.18.

Problem No. 2: Layer-Stripping Using Multifrequencies In an attempt to improve the results of Problem No. 1, we redo that problem by adopting a strategy, that we call layer-stripping using multifrequencies. The idea is to isolate the top layer of the anomalous region of Fig. 2.17 from the bottom layer. We do this by gathering data at a very high frequency, such that the incident field produced by the sensors does not penetrate to the bottom layer. This means that the anomalous currents in the cells of the bottom layer will not contribute to the measured impedances. Hence, we simply constrain these currents to be zero when we do the reconstruction starting with the QR-decomposition. We used **VIC-3D®** to determine suitable frequencies, trying 85, 170, and 340 kHz; we will report the results for 340 kHz.

The results of the inversion are shown in Fig. 2.19. The important point to note here is that the 'zeroes' in the upper layer are very sensitive to the systematic and random errors, and are therefore not well reconstructed, whereas the flaw and fastener cells are well reconstructed. This numerical experiment (and others which we will not report here) is the basis for determining a *heuristic rule*, which will be discussed now.

The stest-estimator generates a curve of scale, S, versus the normalized anomalous conductivity. The conductivity-estimate for each cell is chosen to be that which produces the (unique) minimum of this curve. In Fig. 2.20 we plot S versus conductivity for cells 86 and 87 of Fig. 2.19. Cell no. 86 contains flawed material, whereas no. 87 contains host material.

The minimum for cell no. 87 would, ideally, be sharply peaked at zero, but is quite broad, and lies too far to the left. Furthermore, it is a shallow null; i.e., the

-.06	-.16	-.26	-.24	-.12	-.08	+.02	-.02	
-.10	-.14	-.18	-.18	-.16	-.08	-.02	-.04	
-.22	-.22	-.75	-.88	-.80	-.72	-.12	-.08	TOP
-.30	-.40	-.72	-.74	-.72	-.76	-.36	-.26	
.20	-.42	-.72	-.76	-.74	-.74	-.52	-.48	
-.10	-.12	-.70	-.78	-.72	-.72	-.40	-.34	
-.08	-.02	-.08	-.18	-.68	-.72	-.38	-.26	
-.02	0.0	-.06	-.16	-.82	-.88	-.04	+.02	

-.06	-.08	-.16	-.26	-.18	-.16	-.08	-.02	
-.10	-.12	-.22	-.30	-.22	-.18	-.10	-.06	
-.14	-.16	-.36	-.42	-.42	-.32	-.16	-.10	BOTTOM
-.16	-.16	-.40	-.56	-.60	-.34	-.18	-.12	
-.14	-.16	-.38	-.60	-.60	-.48	-.18	-.14	
-.10	-.14	-.32	-.36	-.50	-.40	-.18	-.12	
-.04	-.06	-.18	-.26	-.36	-.30	-.18	-.18	
+.04	0.0	-.10	-.18	-.32	-.24	-.08	0.0	

Fig. 2.18 Fastener with multisite damage. Reconstructed with 4 × 4 transmitter scan and 32 × 32 receiver scan, using the stest estimator. Noise: 1% relative magnitude error; 1° phase error

ratio of the largest to the smallest value of S is not large, typically five or less. On the other hand, the minimum for cell no. 86 *is* quite sharply peaked at a value of −0.86, which is quite close to its 'correct' value of −0.84. Furthermore, the minimum is much deeper than that for cell no. 87. This, then, is the basis of our **heuristic rule: if host cells and anomalous cells occupy the same layer, the host cells are quite poorly resolved compared to the anomalous cells. Hence, when we see curves such as those in Fig. 2.20, we will assume that cells, such as no. 87, for which**

?	?	?	?	?	?	?	?
?	?	?	?	?	?	?	?
?	?	-0.84	-0.84	-0.84	-0.86	?	?
?	?	-0.86	-0.84	-0.84	-0.86	?	?
?	?	-0.84	-0.84	-0.84	-0.84	?	?
?	?	-0.84	-0.82	-0.86	-0.88	?	?
?	?	?	?	-1.0	-1.0	?	?
?	?	?	?	-1.0	-1.0	?	?

0	0	0	0	0	0	0	0
0	0	0	0	0	0	0	0
0	0	0	0	0	0	0	0
0	0	0	0	0	0	0	0
0	0	0	0	0	0	0	0
0	0	0	0	0	0	0	0
0	0	0	0	0	0	0	0
0	0	0	0	0	0	0	0

Fig. 2.19 Fastener with multisite damage. Reconstructed with 4×4 transmitter scan and 32×32 receiver scan. This is an example of layer-stripping with multifrequencies, using the stest estimator. Noise: 1% relative magnitude error; 1° phase error. Cells of the top layer, that are marked with a '?', have poorly resolved zeroes. Cells of the bottom layer, marked with a zero and shown in black, are forced to have a zero conductivity

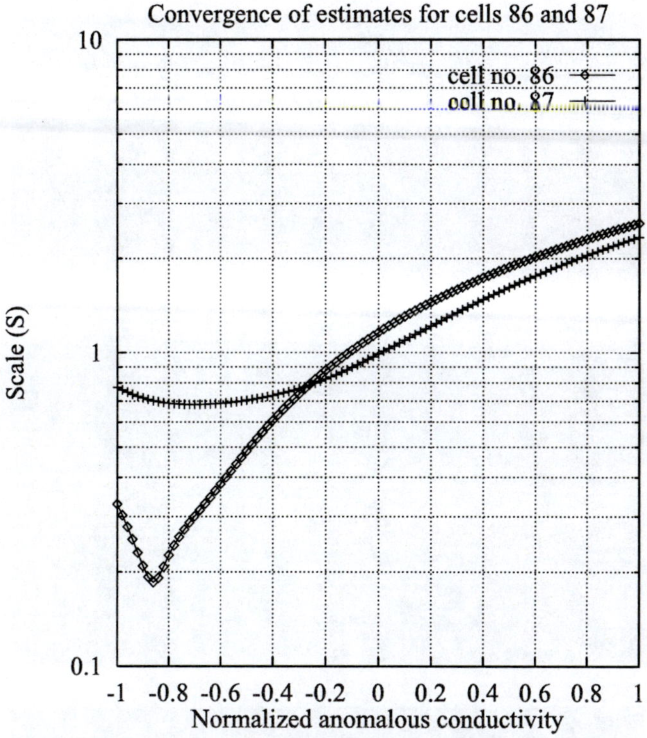

Fig. 2.20 Scale, S, versus normalized anomalous conductivity for cells 86 and 87 of Fig. 2.19. Cell no. 86 contains flawed material, whereas no. 87 contains host material

the ratio of the largest to the smallest value of S is five (5) or less, should have a zero value for the normalized anomalous conductivity.

When we apply the heuristic rule, and force the appropriate cells in the upper layer to have zero anomalous conductivity, we obtain the improved results of Fig. 2.21. Note that by a rather straightforward post-processing algorithm, we have virtually eliminated the effects of noise.

Furthermore, we have even improved the solution for cell no. 86, as we see in Fig. 2.22. Here we plot S vs. conductivity for cell no. 86, when the conditions of Figs. 2.19 (unforced zeros) and 2.21 (forced zeros) hold. Note that the curve is even more sharply peaked downward when the appropriate cells of the top layer are forced to be zero than when they are not. Furthermore, the new estimate of the value of the normalized anomalous conductivity *is* now −0.84, which agrees with the original model.

At this point, the logic of the layer-stripping algorithm would have us use the results for the top layer as a constraint when reconstructing the bottom layer at a lower frequency, say 200 Hz. This is why the algorithm is called a multifrequency layer-stripping algorithm. We cannot proceed to do this, however, because the

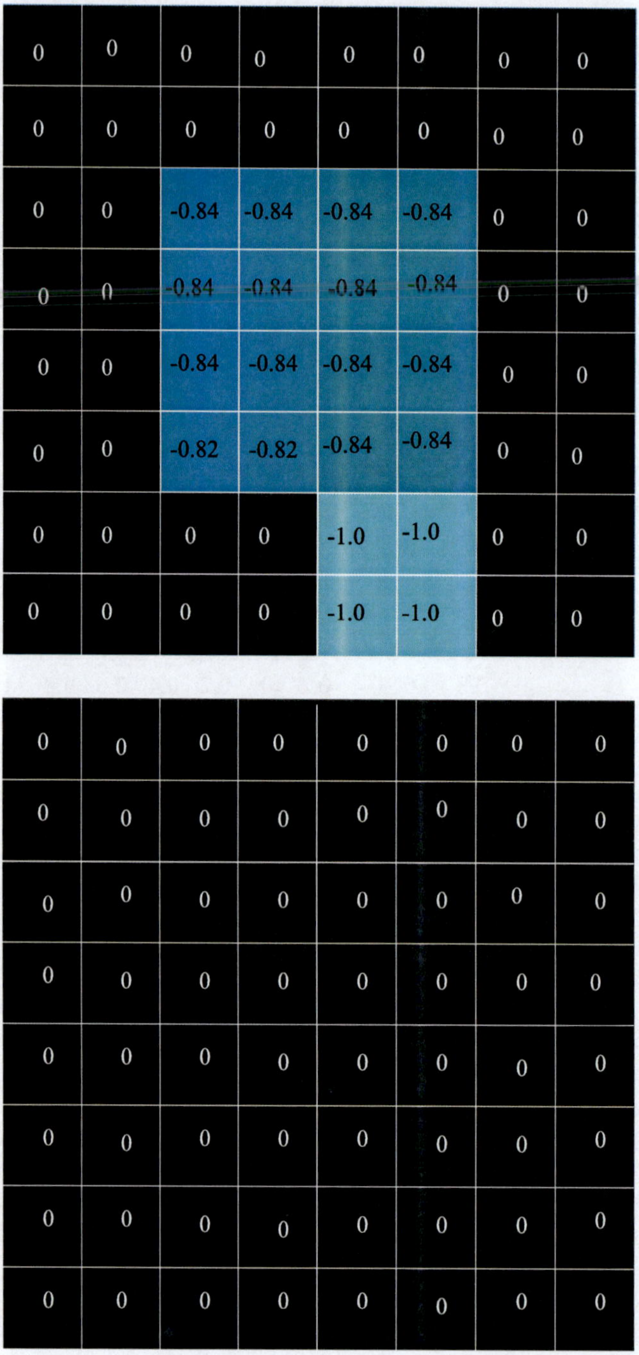

Fig. 2.21 Results for the same system as Fig. 2.19, except that the appropriate cells in the upper layer are forced to have zero anomalous conductivity. Hence, all cells labeled '0', which are shown in black, have been forced to have zero anomalous conductivity

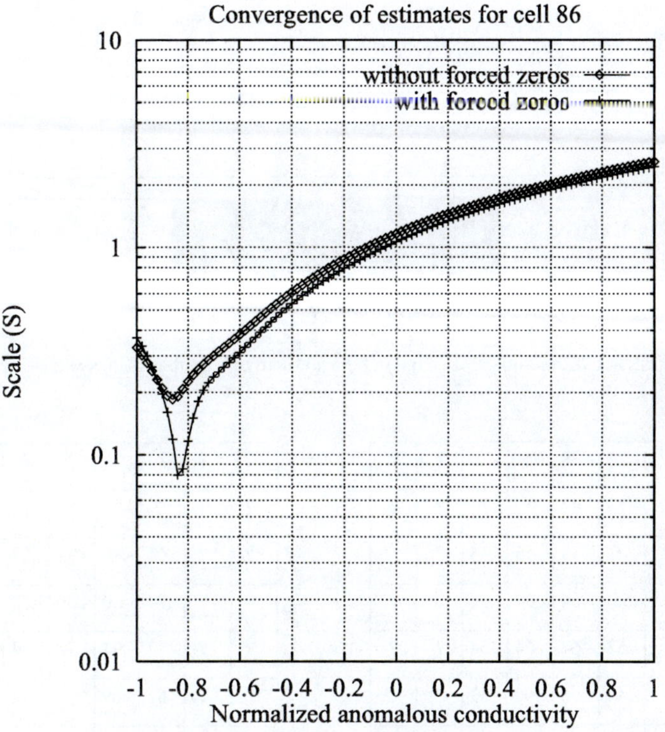

Fig. 2.22 Scale, S, versus normalized anomalous conductivity for cell no. 86 before (diamond) and after (+) forcing zeros for the appropriate cells in the upper layer

current version of **VIC-3D®** allows us only to apply constraints when a cell has a zero anomalous conductivity.

Problem No. 3: Fastener with Multisite Damage in Bottom Plate of a Double-Plate System In this problem, we take the same anomaly as before (the fastener with multisite damage) and put in the bottom plate of a double-plate system, as shown in Fig. 2.23. Each plate is 2.5 mm thick, and the distance between them is 1 mm. The anomaly, therefore, penetrates the bottom layer (a 'throughwall' anomaly).

In order to make the problem more interesting, we consider the anomaly to lie in four layers, rather than two. The bottom two layers repeat the bottom layer of Fig. 2.15, and the top two layers repeat the top layer of that figure. The numerical experiment is run at 200 Hz, and the results shown in Fig. 2.24. As before, the '?' denotes a poorly resolved cell, which, according to our heuristic rule, will be interpreted as being a cell containing host material, so that its conductivity will be constrained to be zero.

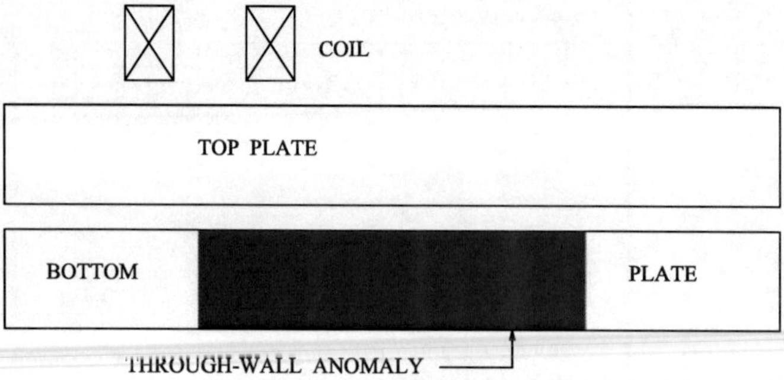

Fig. 2.23 Fastener with multisite damage in bottom plate of a double-plate system

2

?	?	-.24	-.38	?	?	?	?
?	?	-.33	-.42	?	?	?	?
?	?	-.46	-.58	-.55	-.44	?	?
?	?	-.55	-.64	-.64	-.54	?	?
?	?	-.53	-.64	-.64	-.54	?	?
?	?	-.44	-.56	-.60	-.52	?	?
?	?	?	?	?	?	?	?
?	?	?	?	?	?	?	?

4

?	?	?	?	?	?	?	?
?	?	?	?	?	?	?	?
?	?	-.64	-.80	-.81	-.57	?	?
?	?	-.84	-1.0	-1.0	-.78	?	?
?	?	-.77	-.91	-.92	-.74	?	?
?	?	-.51	-.69	-.85	-.72	?	?
?	?	?	?	-.87	-.78	?	?
?	?	?	?	-.56	-.52	?	?

1

?	?	-.26	-.40	?	?	?	?
?	?	-.32	-.41	?	?	?	?
?	?	-.41	-.49	-.48	-.38	?	?
?	?	-.48	-.54	-.56	-.46	?	?
?	?	-.46	-.56	-.58	-.48	?	?
?	?	-.38	-.48	-.52	-.45	?	?
?	?	?	?	?	?	?	?
?	?	?	?	?	?	?	?

3

?	?	?	?	?	?	?	?
?	?	?	?	?	?	?	?
?	?	-.54	-.65	-.65	-.49	?	?
?	?	-.65	-.82	-.80	-.64	?	?
?	?	-.63	-.76	-.76	-.62	?	?
?	?	-.48	-.62	-.72	-.62	?	?
?	?	?	?	-.61	-.58	?	?
?	?	?	?	-.41	-.38	?	?

Fig. 2.24 Fastener with multisite damage in second plate of double-plate system. Reconstructed with 4×4 transmitter scan and 32×32 receiver scan, using the stest estimator. Noise: 1% relative magnitude error; 1° phase error. The layers are numbered from bottom to top

4

0	0	0	0	0	0	0	0
0	0	0	0	0	0	0	0
0	0	-.15	-.85	-.82	-.28	0	0
0	0	-.86	-1.0	-1.0	-.74	0	0
0	0	-.80	-.84	-.87	-.72	0	0
0	0	-.24	-.57	-.84	-.78	0	0
0	0	0	0	-.94	-.78	0	0
0	0	0	0	-.57	-.52	0	0

These zeros are forced

Fig. 2.25 Results for the fourth (top) layer of Fig. 2.24, when cells in only the fourth layer are constrained

When we force the appropriate cells of only the fourth (top) layer to be zero, we get the results shown in Fig. 2.25 for the fourth layer. There is a significant improvement compared to Fig. 2.24.

If, however, we constrain the appropriate cells in all of the layers, then we get the dramatic improvement shown in Fig. 2.26.

Even though the results are excellent, there remains an interesting anomaly in them. Note the twelve circled cells; their conductivities are too low. This is a manifestation of the failure of the 'two-cell hypothesis,' because two of the four nearest neighbor cells have perfect zero anomalous conductivities. Hence, the hypothesis must fail in 50% of the cases for these cells. Nevertheless, this anomaly is easily recognized, and easily cured. We simply redo the two-cell hypothesis using the two nearest neighbors that have nonzero conductivities, and recalculate the conductivities using the robust estimator. When we do this we get excellent agreement with the true value for these cells.

Problem No. 4: Another Layer-Stripping Example Using Multifrequencies
This problem concerns a buried flaw in an aluminum slab. The anomalous region is again modeled as a two-layer system, as in Fig. 2.17, but the top layer is simply host material (anomalous conductivity of zero for each cell). The bottom layer is identical to the bottom layer of Fig. 2.15.

Because we do not know a priori that the top layer is host material, we run the problem first at 85 kHz, in order to reconstruct the top layer independently of the bottom, as in Problem No. 2. This is done by forcing the bottom cells to have zero anomalous conductivity, as in Fig. 2.27.

The reconstruction of the top layer produces excellent zeros for all of the cells, as we would hope. This is not a violation of the heuristic rule that was defined in

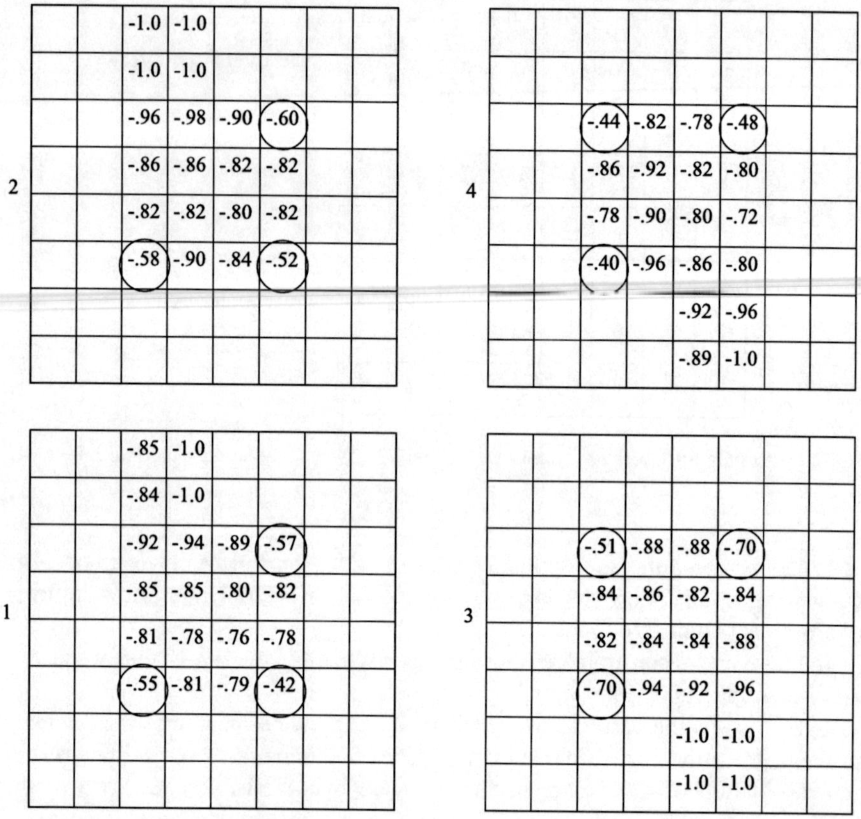

Fig. 2.26 Results for all four layers of Fig. 2.24 when the appropriate cells (which are unmarked in this figure) in all four layers are constrained to have zero relative anomalous conductivity. The twelve circled cells have values that are too low because of the failure of the 'two-cell hypothesis.'

Problem No. 2, because there are only host cells in the top layer, not a mixture of host and anomalous cells.

Next, we rerun the problem at 200 Hz, constraining the top cells to have zero conductivity, because that is what the reconstruction produced. This is the 'layer-stripping' step, and we can rigorously apply it in this example because the layer stripped-off has only zero-conductivity cells. The result is shown in Fig. 2.28.

When we apply the heuristic rule to the bottom layer of Fig. 2.28, and force the appropriate cells to take the value of zero, we get the excellent reconstruction shown in Fig. 2.29. The cells shown in black are forced to have zero conductivity. The anomalously low values of three cells are due to the 'two-cell' hypothesis, and can be corrected as described before.

0	0	0	0	0	0	0	0
0	0	0	0	0	0	0	0
0	0	0	0	0	0	0	0
0	0	0	0	0	0	0	0
0	0	0	0	0	0	0	0
0	0	0	0	0	0	0	0
0	0	0	0	0	0	0	0
0	0	0	0	0	0	0	0

0	0	0	0	0	0	0	0
0	0	0	0	0	0	0	0
0	0	0	0	0	0	0	0
0	0	0	0	0	0	0	0
0	0	0	0	0	0	0	0
0	0	0	0	0	0	0	0
0	0	0	0	0	0	0	0
0	0	0	0	0	0	0	0

Fig. 2.27 Reconstruction at 85 kHz, with the bottom layer constrained to have zero anomalous conductivity in each cell (shown in black). The cells of the top layer are freely reconstructed in accordance with the inversion algorithm, and produce very well-resolved zeros

0	0	0	0	0	0	0	0
0	0	0	0	0	0	0	0
0	0	0	0	0	0	0	0
0	0	0	0	0	0	0	0
0	0	0	0	0	0	0	0
0	0	0	0	0	0	0	0
0	0	0	0	0	0	0	0
0	0	0	0	0	0	0	0

0.0	-0.08	-0.76	-0.82	-0.18	-0.06	-0.08	-0.02
-0.06	-0.22	-0.68	-0.76	-0.26	-0.12	-0.06	-0.04
-0.12	-0.26	-0.63	-0.68	-0.60	-0.50	-0.14	-0.06
-0.20	-0.26	-0.54	-0.56	-0.59	-0.59	-0.14	-0.04
-0.28	-0.30	-0.59	-0.66	-0.66	-0.66	-0.18	-0.03
-0.10	-0.18	-0.48	-0.61	-0.62	-0.54	-0.20	-0.06
-0.04	-0.04	-0.14	-0.07	-0.02	-0.03	-0.04	-0.02
-0.02	-0.06	-0.04	-0.02	-0.02	-0.03	-0.08	-0.06

Fig. 2.28 Reconstruction of buried flaw, using layer-stripping. Only the cells of the top layer (shown in black) have been forced to have zero conductivity. The frequency is 200 Hz

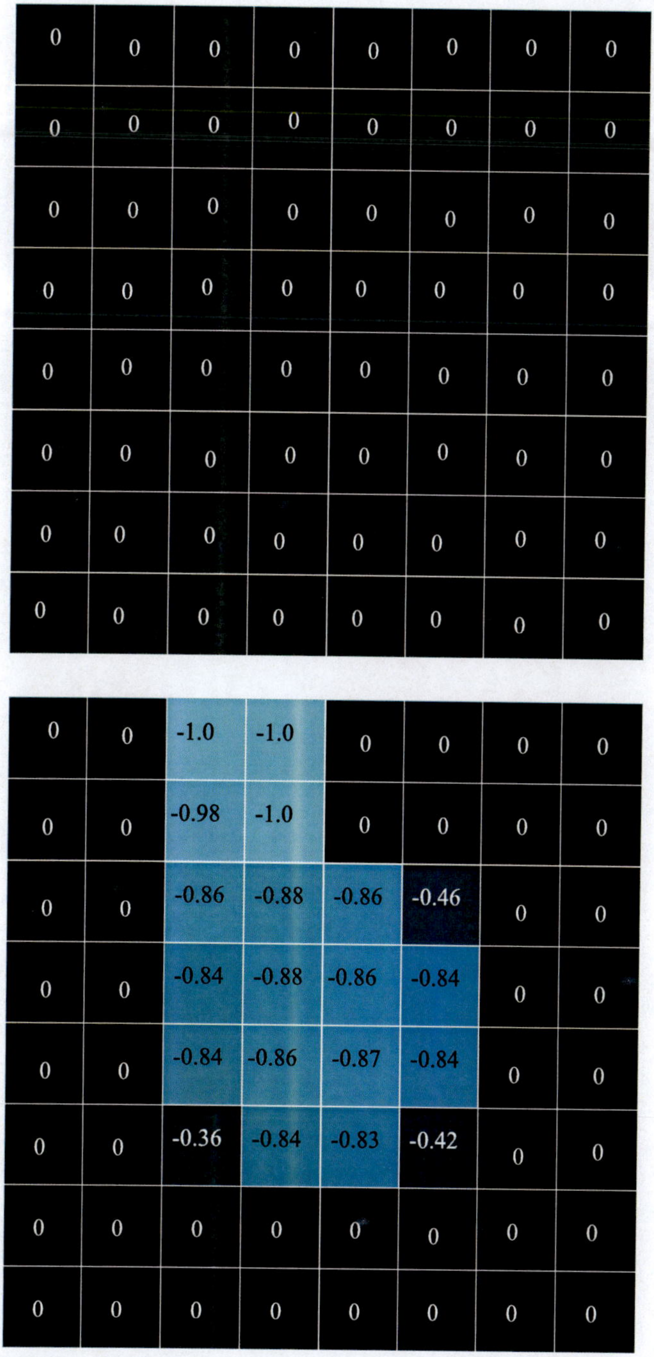

Fig. 2.29 Reconstruction of buried flaw, using layer-stripping. All of the cells labeled zero, and shown in black, have been forced to have zero anomalous conductivity. The color map in the bottom figure has a resolution of only 5%

Part II
Materials Characterization

Chapter 3
Modeling Composite Structures

3.1 Background

Composite materials in the form of fiber-reinforced matrix materials as, for example, graphite-epoxy, are being increasingly used in critical structures and structural components because of their high strength-to-weight ratio. In order to assess the integrity of these structures, it is necessary to employ suitable methods for quantitative NDE. One method uses eddy-currents; composite materials, however, are inherently anisotropic, which means that many of the classical eddy-current technology and design procedures are not applicable. In addition, composite materials vary widely in their permittivities and conductivities, which means that new analyses must be carried out to develop effective strategies for using eddy-currents in quantitative NDE. A final problem is that there is a variety of potential failure modes in composites, such as delaminations, fiber-breakage due to impact damage, flaws, etc., some of which may not be readily detectable by eddy-currents [85]. In order to complement empirical studies it is necessary to embark upon a rigorous quantitative NDE program for composites to assess the role that eddy-currents play in it, and especially to determine suitable inversion algorithms.

Eddy-current methods for the examination of carbon-fiber reinforced epoxy resins and other composite materials have been discussed and analyzed by Owston and Prakash [82, 83, 86, 87]. These analyses have been based on an *ad hoc* equivalent circuit in which the composite test piece is regarded as being inductively coupled to the probe, much as in the classical treatment of eddy-current evaluation of metals. Though the technique gives a useful indication of the form of the results, a more satisfactory approach, as Owston [83] points out, is to use a field-theoretic analysis which is capable of giving exact results for a given model. A field-theoretic analysis is also desirable when computing electromagnetic interactions for shielding effectiveness of advanced composites in aircraft [132].

H. A. Sabbagh et al., *Advanced Electromagnetic Models for Materials Characterization and Nondestructive Evaluation*, Scientific Computation,
https://doi.org/10.1007/978-3-030-67956-9_3

The heart of the problem, as we have emphasized throughout this book, is to determine a Green's function for the composite material. Much work has been done in recent years on the subject of electromagnetic interactions with composite materials, mostly in the context of electromagnetic shielding of avionics equipment from electromagnetic pulses [4, 5, 45, 120, 132]. Some of this work is directly applicable to the problem of computing eddy-current flow within composites, but the Green's function problem must be attacked by applying rigorous electromagnetic theory to anisotropic media.

3.2 Constitutive Relations for Advanced Composites

Advanced composite materials are laminates made up of a number of individual layers bonded together. Each layer consists of a unidirectional array of long fibers embedded in, and firmly bonded to, a matrix. The basic building blocks of any specific composite are defined by the types of fibers and matrix involved. Some fiber-matrix systems are: boron-epoxy, graphite-epoxy, Kevlar-epoxy, graphite-polymide and graphite-thermoplastic [5]. The matrix for each of these materials is normally a good dielectric, whereas the fibers vary in electrical conductivity from modest (graphite) to a poor dielectric (boron) to a good dielectric (Kevlar). These materials are nonmagnetic, so that the magnetic permeability is μ_0.

Composites have anisotropic conductivities because of the unidirectional arrays of fibers within. For example, for graphite-epoxy the average macroscopic conductivity along the fiber direction is 20,000 S/m, whereas in the direction transverse to the fibers, the conductivity is 100 S/m. It may be surprising to find a nonzero transverse conductivity in graphite-epoxy, in view of the earlier statement that the matrix is a good dielectric. The fact is that there is enough local fiber-to-fiber contact that the average macroscopic conductivity is not zero, as illustrated in Fig. 3.1 [83]. Other materials, of course, have different longitudinal and transverse conductivities, as shown in Table 3.1 [5]. The reason that ϵ_R for graphite-epoxy is indeterminate is because the fiber-to-fiber contact effectively shunts the capacitors between fibers with a fairly low resistance path, making it impossible to measure dielectric permittivities at frequencies less than 100 MHz, or so. Thus, in Fig. 3.1b, which shows a possible AC equivalent circuit for eddy-current flow, the capacitors are effectively short-circuited by the fiber-to-fiber resistors at the lower frequencies.

The anisotropy of the composite manifests itself in a complex-permittivity tensor, the tensor being diagonal in a coordinate system (ξ_1, ξ_2, ξ_3), where ξ_1 is parallel to the average fiber direction, ξ_2 is perpendicular to the average fiber direction, but lies in the plane of the composite layer, and ξ_3 is perpendicular to both fibers and the plane of the layer:

$$\epsilon = \begin{bmatrix} \hat{\epsilon}_{11} & 0 & 0 \\ 0 & \hat{\epsilon}_{22} & 0 \\ 0 & 0 & \hat{\epsilon}_{33} \end{bmatrix}, \tag{3.1}$$

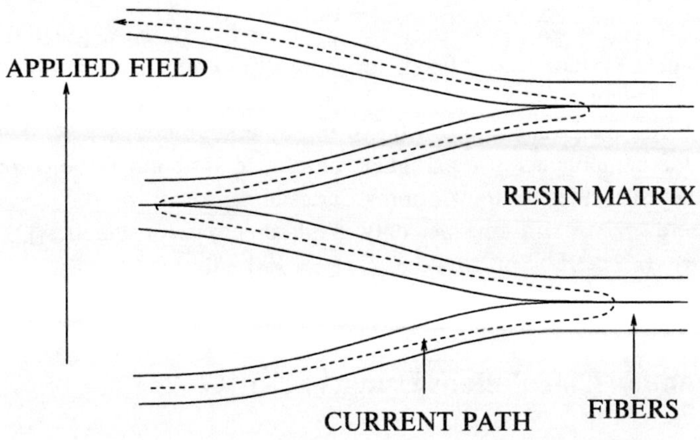

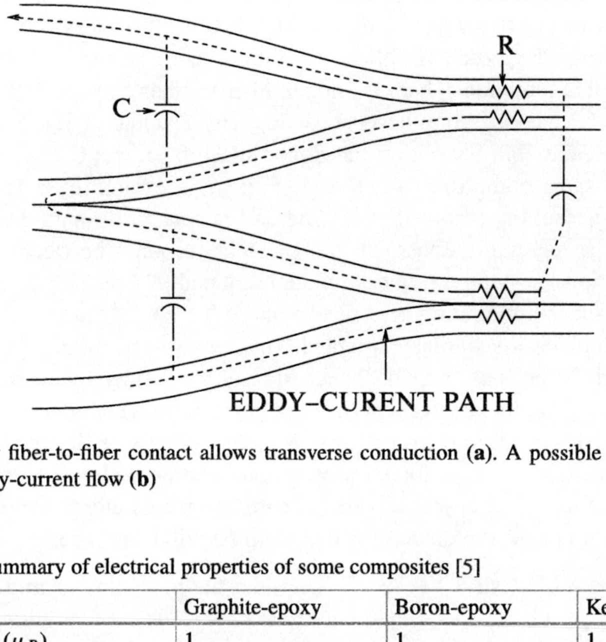

Fig. 3.1 How fiber-to-fiber contact allows transverse conduction (**a**). A possible AC equivalent circuit for eddy-current flow (**b**)

Table 3.1 Summary of electrical properties of some composites [5]

	Graphite-epoxy	Boron-epoxy	Kevlar
Permeability (μ_R)	1	1	1
Permittivity (ϵ_R)	Indeterminate	5.6	3.6
DC Conductivity (S/m)			
Longitudinal (σ_L)	2×10^4	30	6×10^{-9}
Transverse (σ_T)	100	2×10^{-8}	6×10^{-9}
Anisotropy ratio (σ_L/σ_T)	200	1.5×10^9	1

where $\hat{\epsilon}_{ii} = \epsilon_{ii} - j\sigma_{ii}/\omega$, with $j = \sqrt{-1}$, and ω is the angular frequency. The coordinate system just defined, for which the complex permittivity tensor is diagonal, is not necessarily the laboratory system, (x, y, z), in which the electromagnetic field vectors are defined. In any case, the tensor symbol will be used, and the components in a particular coordinate system may be computed by applying the usual rules for transforming Cartesian tensors.

From here on, we will consider only graphite-epoxy, for which $\epsilon_{11} = \epsilon_{22} = \epsilon_{33} = \epsilon_0$, $\sigma_{11} = 2 \times 10^4$ S/m, and $\sigma_{22} = \sigma_{33} = 100$ S/m.

3.3 Example Calculations Using VIC-3D®

Figure 3.2 illustrates the response of a two-layered composite structure with the bottom layer oriented at 0 and 90° with respect to the top layer. The principal-axis conductivities of the two layers are $[2 \times 10^4, 100, 100]$ S/m, which is typical of cfrp composites. The x-axis is aligned with the fibers, and the y- and z- axes are transverse to this direction. There is enough fiber-to-fiber contact in the transverse directions to yield a nonzero transverse conductivity. Excitation of this system is by means of a circular coil. Each layer has dimensions of $6 \times 6 \times 0.25$ in.

When the same composite structure of Fig. 3.2 is excited by a racetrack coil at 1 MHz, the resulting response when the coil is rotated through 360° is shown in Fig. 3.3. The relative 'flatness' of the reactance response between 60 and 120° for the situtation in which the bottom plate is rotated 90° relative to the top plate suggests that this configuration is somewhat more 'isotropic' than the configuration in which both plates are similarly oriented. This seems to be intuitively reasonable. Another metric is the peak-to-peak change in response divided by the mean value of the response. Again, this shows that the configuration in which the bottom layer is rotated 90° yields a 'more isotropic' response than otherwise. Indeed, the purpose of such a 'lay-up order' is to force the stress distribution within the structure to be more isotropic than with a single plate, or with two plates aligned parallel to each other. One result of the 0–90° order is that warping will be reduced.

A Second Example Using VIC-3D® Consider the preceding example with both plates aligned parallel to each other, but excited by a T/R configuration comprising a circular transmit and receive coil. The receive coil is raster-scanned in the (x, y)-plane symmetrically about the transmit coil, which is excited at 1 MHz. The host is isotropic with a conductivity of 100 S/m, and two types of anomalies will be presented. The first is an isotropic patch with conductivity 20,000 S/m, and the second a graphite-epoxy patch with the usual conductivity pattern of $[2 \times 10^4,\ 100,\ 100]$ S/m.

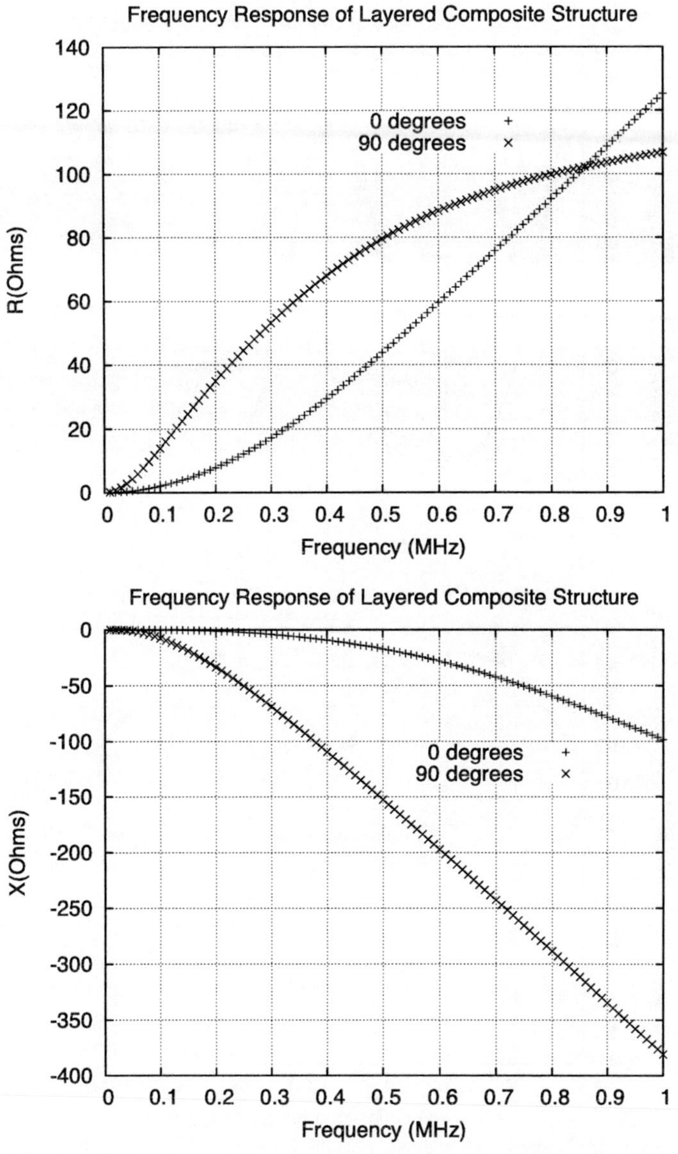

Fig. 3.2 Frequency response of a two-layered composite structure with the bottom layer oriented at 0 and 90° with respect to the top layer. The principal-axis conductivities of the two layers are $[2 \times 10^4, 100, 100]$ S/m, which is typical of cfrp composites. The x-axis is aligned with the fibers, and the y- and z-axes are transverse to this direction. There is enough fiber-to-fiber contact in the transverse directions to yield a nonzero transverse conductivity

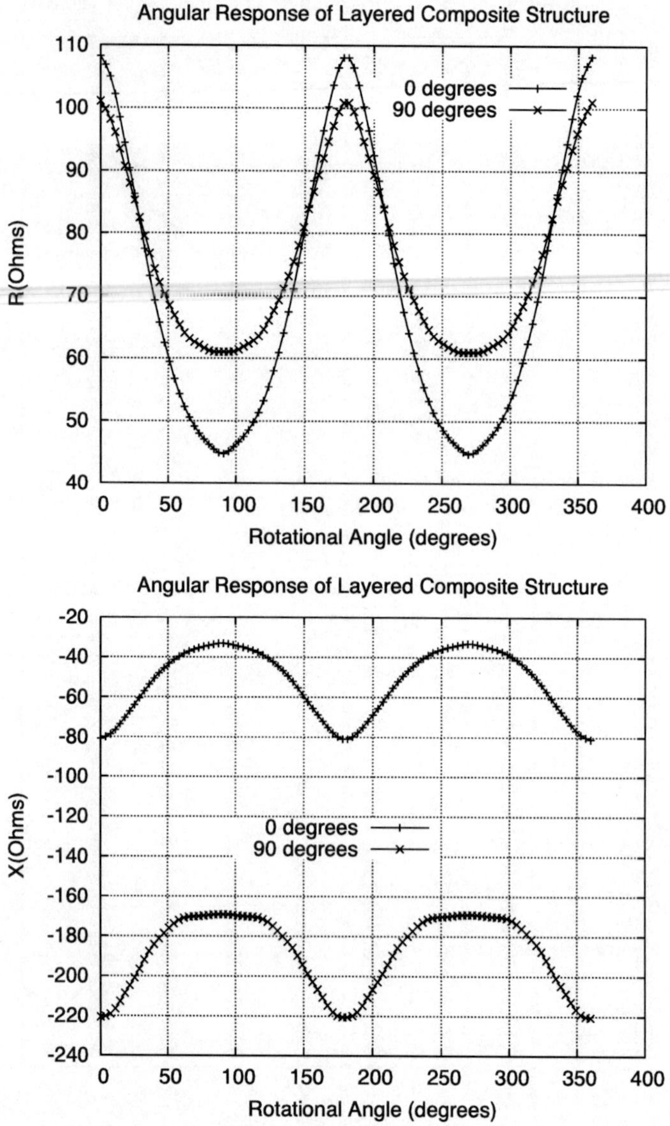

Fig. 3.3 Response of the two-layered structure of Fig. 3.2 when a racetrack coil excites the structure. The x-axis is the orientation of the coil, and the frequency of excitation is 1 MHz

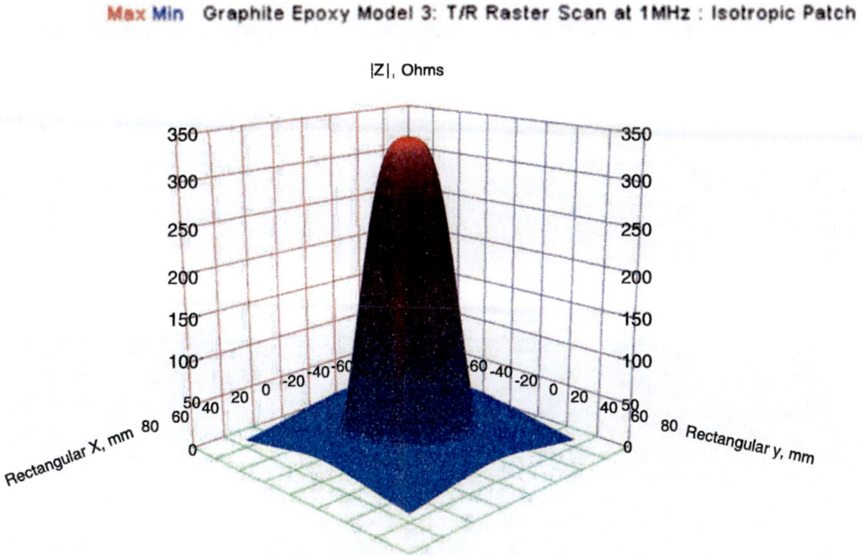

Fig. 3.4 Isotropic response of the T/R probe to the isotropic patch. The plotted data are the magnitude of the transfer impedance, $|Z|$, measured by the receive coil

The results for the isotropic patch are shown in Fig. 3.4, and for the graphite-epoxy patch in Fig. 3.5. These figures illustrate the the magnitude of the transfer impedance, $|Z|$, measured by the receive coil. The response to the isotropic patch is circular (isotropic), as expected, whereas the response to the graphite-epoxy patch is strongly oriented in the x-direction, but also includes significant 'side-lobes' extending in the y-direction.

The side-lobes are much less apparent in the model consisting of the lay-up order $[0000000090]_S$ shown in Fig. 3.6. The rest of the model is as shown in Fig. 3.5. By a 'lay-up order of $[0000000090]_S$' we mean a structure consisting of eighteen layers, the first eight of which are aligned in the 0° direction, the ninth rotated 90° relative to the first eight, and the remaining nine symmetrically placed relative to the first nine. This structure manifests itself as a sandwich in which the top and bottom layers are identical with a conductivity tensor of $[2 \times 10^4,\ 100,\ 100]$ S/m and the thin middle layer has a conductivity of $[100,\ 2 \times 10^4,\ 100]$ S/m. It is clear that the middle layer of the sandwich produces a more isotropic structure, thereby reducing the possibility of warpage. The electromagnetic response, therefore, shows a closer similarity to the isotropic response of Fig. 3.4 than to Fig. 3.5.

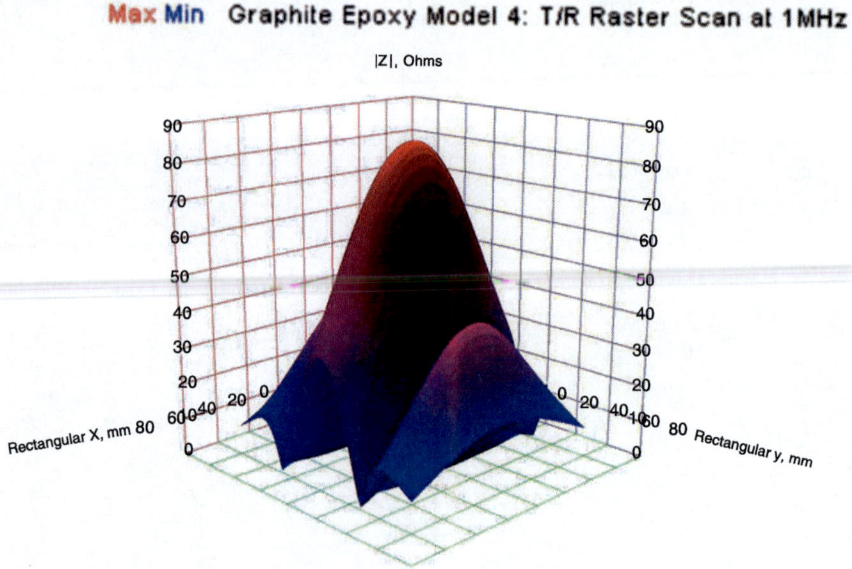

Fig. 3.5 Oriented response of the T/R probe to the graphite-epoxy patch with conductivity pattern, [2×10^4, 100, 100] S/m. The plotted data are the magnitude of the transfer impedance, $|Z|$, measured by the receive coil

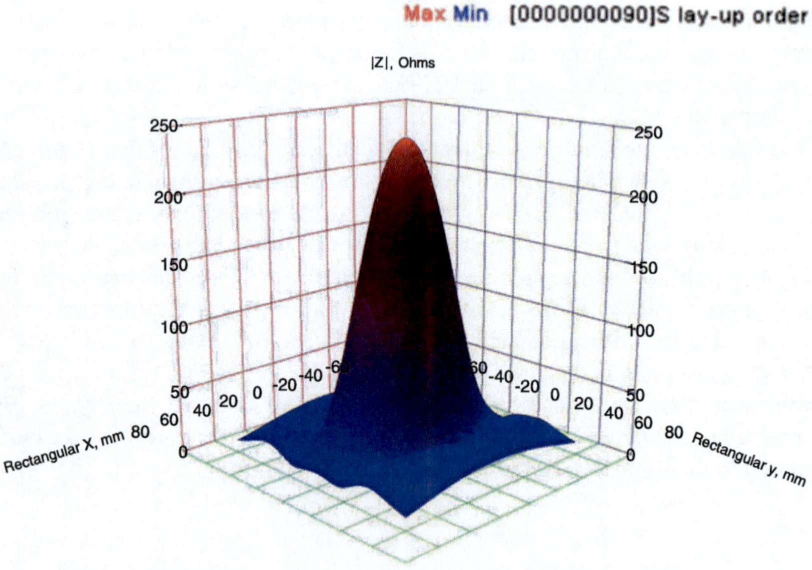

Fig. 3.6 Oriented response of the T/R probe to the graphite-epoxy patch with lay-up order $[0000000090]_S$. The rest of the model is as shown in Fig. 3.5. The plotted data are the magnitude of the transfer impedance, $|Z|$, measured by the receive coil

3.4 A Coupled-Circuit Model of Maxwell's Equations

Maxwell's dual equations are

$$\begin{aligned}\nabla \times E &= -j\omega\mu_0(\mathbf{H}+\mathbf{M}) \\ \nabla \times H &= j\omega\epsilon\mathbf{E}+\sigma\mathbf{E}\,,\end{aligned} \tag{3.2}$$

where the first equation is Faraday's law, and the second Ampere's circuital law. We call these 'dual equations,' in the sense that the sources of the first are magnetic currents, and those of the second are electric currents. In representing fields by means of electric circuits, we use duality in the same way. In a circuit, we would represent the sum of magnetic effects (voltages) by a series circuit, since voltages add in such a circuit, and the sum of electric currents by a parallel circuit, since currents add in such a circuit.

Thus, we could use Fig. 3.7 as a coupled-circuit model of a coil inducing electric currents within a composite plate. R_0 and L_0 are, respectively, the resistance and self-inductance of the coil in freespace, and L_μ is the increased inductance of the coil due to the permeability of the plate. L_1 is the 'virtual' secondary inductance that accounts for induction effects within the plate, M_0 is the mutual inductance between L_0 and L_1, and M_μ is the mutual inductance between L_μ and L_1. R is the effective 'secondary resistance' that is due to the transverse electrical conductivity of the plate and C is the effective 'secondary capacitance' that is due to the transverse electrical permittivity of the plate. That there may be magnetic effects in composite structures is made clear in [55, 56, 126, 139], at least in the case of composites made of carbon nanotubes.

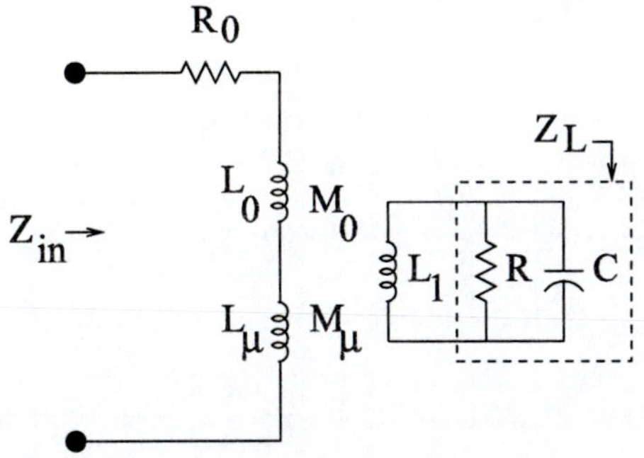

Fig. 3.7 A coupled-circuit model of the coil in the presence of the composite plate. R_0 and L_0 are, respectively, the resistance and self-inductance of the coil in freespace, and L_μ is the increased inductance of the coil due to the permeability of the plate. L_1 is the 'virtual' secondary inductance that accounts for induction effects within the plate, M_0 is the mutual inductance between L_0 and L_1, and M_μ is the mutual inductance between L_μ and L_1. R is the effective 'secondary resistance' that is due to the transverse electrical conductivity of the plate and C is the effective 'secondary capacitance' that is due to the electrical permittivity of the plate

We are going to apply this coupled-circuit model to the analysis of carbon-fiber refinforced polymers (cfrp), which do not exhibit ferromagnetic effects, so we will drop L_μ from here on. Furthermore, our interest is in the transverse effects of the electromagnetic interaction with the plate, i.e., effects that are orthogonal to the direction of the fibers, as in Fig. 3.1. Elementary coupled-circuit theory yields an expression for the driving-point impedance of the loaded coil:

$$Z_{\rm in} = R_0 + j\omega L_0 + \frac{\omega^2 M_0^2}{Z_L + j\omega L_1} . \tag{3.3}$$

From this we get the change in impedance due to the presence of the composite plate:

$$\begin{aligned}\delta Z_{\rm in} = Z_{\rm in} - R_0 - j\omega L_0 &= \frac{\omega^2 M_0^2}{Z_L + j\omega L_1} \\ &= \frac{\omega^2 M_0^2 R_L - j\omega^2 M_0^2(\omega L_1 + X_L)}{R_L^2 + (\omega L_1 + X_L)^2} ,\end{aligned} \tag{3.4}$$

where R_L and X_L are the real and imaginary parts of the load impedance, Z_L.

If $X_L \geq 0$ in (3.4), say due to a resistor or inductor, then $\delta X_{\rm in} < 0$, which is Lenz' law for inductively coupled circuits. If, on the other hand, $X_L \leq 0$, say due to a capacitor, then the sign of $\delta X_{\rm in}$ depends upon the relative value of ωL_1 and X_L. In the case illustrated in Fig. 3.7, which in turn is suggested by the physical picture of Fig. 3.1, we have a parallel RC circuit, for which

$$\begin{aligned} R_L &= \frac{R}{1 + \omega^2 R^2 C^2} \\ X_L &= -\frac{\omega C R^2}{1 + \omega^2 R^2 C^2} . \end{aligned} \tag{3.5}$$

If $\omega L_1 > \frac{\omega C R^2}{1 + \omega^2 R^2 C^2}$, then $\delta X_{\rm in}$ is negative in (3.4), as in Lenz' law, but if $\omega L_1 < \frac{\omega C R^2}{1 + \omega^2 R^2 C^2}$, then the change in reactance is positive. Finally, we have the interesting result that if $\omega L_1 = X_L = \frac{\omega C R^2}{1 + \omega^2 R^2 C^2}$, then the reactance change is zero, which is resonance.

A rigorous coupled-circuit model that includes the interactions between the driver coil, host and anomaly is shown in Fig. 3.8. The analysis of this circuit leads to the slightly more complicated expression for the driving-point impedance of (3.6), compared to (3.4).

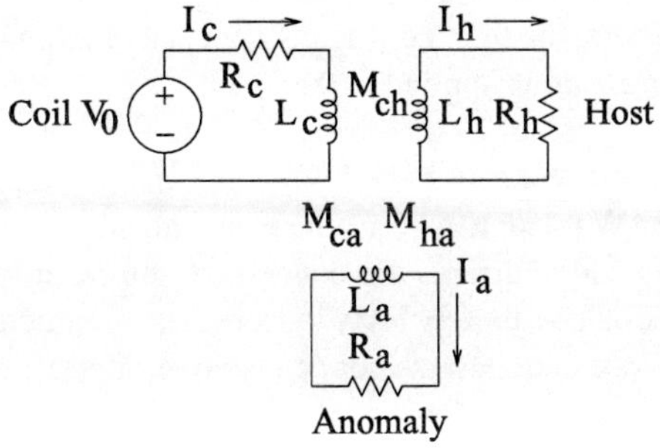

Fig. 3.8 A coupled-circuit model showing the three current systems, I_c, I_h, and I_a

$$
\begin{aligned}
Z_{in} &= \frac{V_0}{I_c} = R_c + j\omega L_c + \frac{\omega^2 M_{ch}^2 (R_a + j\omega L_a) - j\omega^3 M_{ch} M_{ha} M_{ca}}{(R_h + j\omega L_h)(R_a + j\omega L_a) + \omega^2 M_{ha}^2} \\
&+ \frac{\omega^2 M_{ca}^2 (R_h + j\omega L_h) - j\omega^3 M_{ch} M_{ha} M_{ca}}{(R_h + j\omega L_h)(R_a + j\omega L_a) + \omega^2 M_{ha}^2} \\
&= R_c + j\omega L_c + \frac{\omega^2 M_{ch}^2}{R_h + j\omega L_h} \\
&+ \frac{\omega^2 M_{ca}^2 (R_h + j\omega L_h)^2 - j2\omega^3 M_{ch} M_{ca} M_{ha} (R_h + j\omega L_h) - \omega^4 M_{ch}^2 M_{ha}^2}{(R_h + j\omega L_h)\left[(R_h + j\omega L_h)(R_a + j\omega L_a) + \omega^2 M_{ha}^2\right]} . \qquad (3.6)
\end{aligned}
$$

3.5 Eddy-Current Detection of Prepreg FAWT

Now we want to apply the previous model to the question of eddy-current detection of the fiber areal weight (FAWT) of graphite-epoxy prepreg.[1] The FAWT, measured in grams per square meter, is a way of expressing the fiber d e n s it y for a given material thickness. By 'fiber density' we mean a number between zero and one that indicates how densely distributed the fibers are in the material. If we know the specific gravity of the fiber material and the thickness of the material, we can convert FAWT to fiber density and vice-versa. The conversion between fiber density and FAWT may not be a simple task; we must take into account the resin content. Increasing the resin content is likely to increase the thickness of the material for a given FAWT, which will probably decrease the fiber density. The thickness of the material is not easily measured. Since the material is made up of many distributed

[1]Reference [130] should be consulted for more details on modeling and measurements for eddy-current detection of prepreg FAWT that were performed at Sabbagh Associates for the Hercules Company in 1990. That work did not use **VIC-3D®**, which was not fully developed at that time.

fibers, the thickness at any spot is a random variable, and also depends on how much pressure is applied to the layer.

It is not immediately clear exactly how changing FAWT changes the conductivity. We can make a few reasonable assumptions. First of all, we note that if the FAWT reaches zero, there are no fibers, and we must have simply resin. As far as eddy currents are concerned, the resin looks just like freespace. Therefore , we conclude that as FAWT decreases, conductivity decreases toward zero (freespace). If we take into account capacitive effects , we must look at the dielectric properties of the resin. Capacitive effects should not come into the picture for eddy-current inspection, as long as we keep the frequencies relatively low. As FAWT increases, the prepreg begins to look more and more like a slab of solid carbon. An increasing number of fibers contact, making the transverse conductivity increase. It seems reasonable to assume that as FAWT increases, both the transverse and longitudinal conductivity approach a limit determined by the conductivity of the graphite fibers. We can probably improve somewhat on these simple-minded observations by using models for graphite fiber interactions found in the literature.

We note that the transverse conductivity is likely to be more affected by a change in FAWT than the longitudinal conductivity, assuming a nominal value of 60% fiber. A simple argument is based on our previous comments. Assuming that the transverse conductivity, denoted σ_T, approaches the longitudinal conductivity, σ_L, as the fiber density approaches 100%, and given a typical anisotropy ratio of 200:1 (based on 60% fiber density) , one can get an idea of how σ_T and σ_L must change when the fiber density is increased toward 100%. We note that σ_T must change by at least a factor of 200, while σ_L is known to change by only a factor of about two. Such a refined estimate of conductivity change will be useful when we assess the accuracy of the proposed method of measurement. For now, it will be reasonable to assume that the conductivity scales by the same percentage change as the FAWT (the transverse and longitudinal conductivities). Actually, it is suggested that the change in transverse conductivity is a higher order than direct-proportion relationship; the simple model used in the reference predicts a squared relationship. If indeed the transverse conductivity depends on the FAWT squared, or a higher power of FAWT, then we are in a good position to measure the FAWT change. We know from our own experience and models that our laboratory measurements are sensitive to conductivity changes in the material. All that remains to be discovered is the relationship between conductivity and FAWT.

Resin content appears to affect the conductivity in a perhaps complicated way. It appears that resin content can dramatically affect the transverse conductivity, but may have little effect on the longitudinal conductivity. One way to explain the physical reason for the conductivity change with resin content is to think of the resin as an insulator partially shielding the adjacent fibers ("wires") from contacting each other, thereby reducing σ_T. Using this explanation also leads us to the conclusion that σ_L is not significantly changed by resin content, assuming that the resin content does not significantly change the overall volume of the material, since the "wires" conduct equally well when they are surrounded by insulation.

It is somewhat easier to predict the variation of a measured signal when the conductivity of the sample is changed than it is to predict the variation of the signal when the FAWT is changed. The problem is somewhat simplified if the sample thickness is known a priori, but knowledge of the sample thickness should not be a requirement; the information can presumably be extracted from multi-frequency data. Our model is useful in generating the prediction of signal variation when the conductivity changes. Another possible method, other than the computer model, for determining the relationship between conductivity and measured signal would be to measure the signal given known-conductivity samples. It is unlikely, however, that we would have enough samples to accurately determine the response to conductivity change using strictly experimental data. A useful approach might be to use the model in conjunction with experimental data to determine a better approximation of the relationship.

3.6 An Anisotropic Inverse Problem for Measuring FAWT

Following the previous discussion of eddy-current detection of prepreg FAWT, we consider the transverse conductivities, σ_y and σ_z, to be the FAWT metric. A simple model for inferring these conductivities using eddy-currents is shown in Fig. 3.9. Data are taken at 11 frequencies equally logarithmically spaced between 1 and 100 MHz over the FAWT test region, and these data are then presented to NLSE, **VIC-3D®**'s nonlinear least-squares estimator, for inversion. The unknowns are the transverse conductivities.

As is usual in doing model-based inversions, we must generate a surrogate interpolation table using nodes at prescribed values of the unknowns. The resulting responses are called 'blending functions', and the objective is to determine what combination of the functions best fits the measured data. For the proposed model shown in Fig. 3.9, we assume that $\sigma_y = \sigma_z = 100$ is the unknown transverse conductivity, and to infer this we use values of $\sigma_y = \sigma_z = 25, 75, 125$ to create the interpolation table. The blending functions that correspond to these nodes are

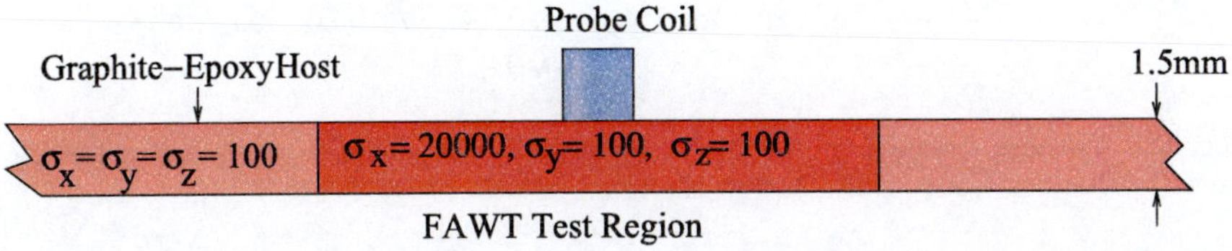

Fig. 3.9 Illustrating the model problem for analyzing FAWT. The host graphite-epoxy slab is isotropic, with the conductivities shown, whereas the FAWT region is anisotropic. The ratio of the longitudinal to the two transverse conductivities is typical for a FAWT=60% for this particular sample of graphite-epoxy. The probe coil is excited over a frequency range of 1–100 MHz in ten equal logarithmic steps

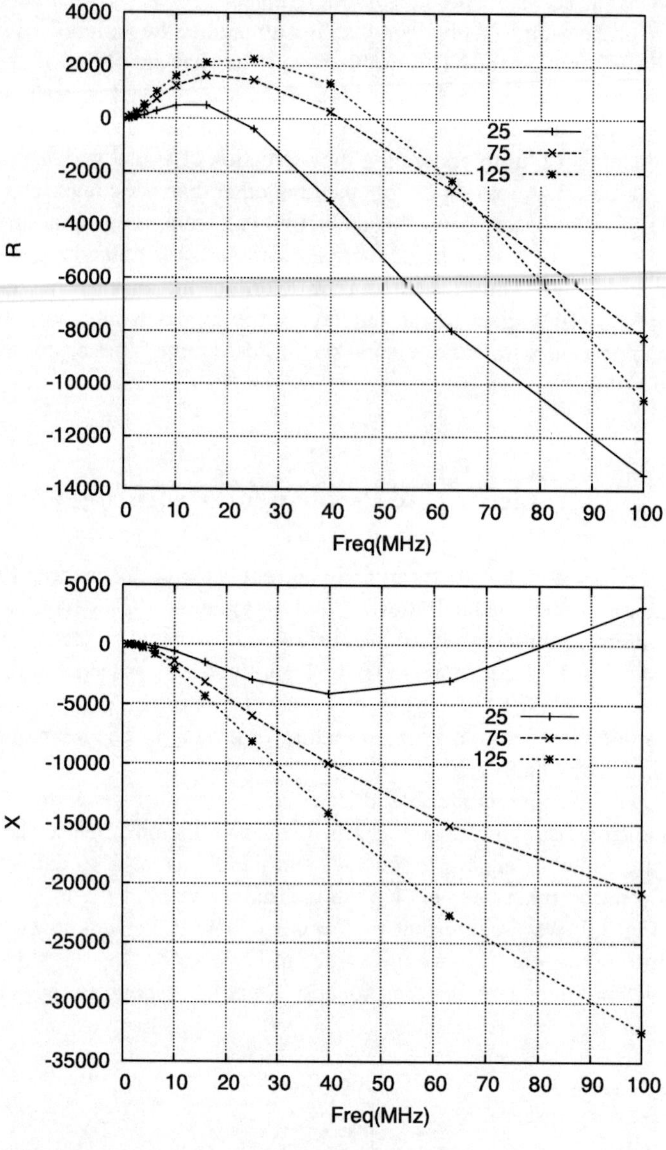

Fig. 3.10 Blending functions for the inverse problem of determining $\sigma_y = \sigma_z$ in Fig. 3.9. The nodes for the transverse conductivities are shown

shown in Fig. 3.10. These blending functions have a special structure that will be discussed in connection with the equivalent circuit of Fig. 3.7 shortly.

The result of the inversion is shown in Table 3.2. The result gives a reliable estimate that is off by 1.64%, with a reasonably low sensitivity coefficient as described in Chap. 6.

Table 3.2 Result of FAWT inversion for a 1.5 mm-thick slab: $\sigma_y = \sigma_z = [25, 75, 125]$

Φ	σ_y/sensit	σ_z/sensit	No. pts.
0.212(3)	101.64/0.798	101.64/0.798	500

Table 3.3 Result of FAWT inversion for a 1.5 mm-thick slab: $\sigma_y = [25, 75, 125]$, $\sigma_z = [25, 75, 125]$

Φ	σ_y/sensit	σ_z/sensit	No. pts.
0.206(3)	101.53/0.781	125.0/164.9	14

Table 3.4 Result of FAWT inversion for a 12.7 mm-thick slab: $\sigma_y = \sigma_z = [25, 75, 125]$

Φ	σ_y/sensit	σ_z/sensit	No. pts.
0.239(3)	97.18/2.126	97.18/2.126	500

Table 3.5 Result of FAWT inversion for a 12.7 mm-thick slab: $\sigma_y = [25, 75, 125]$, $\sigma_z = [25, 75, 125]$

Φ	σ_y/sensit	σ_z/sensit	No. pts.
0.183(3)	99.09/1.834	75.92/13.05	500

It seems reasonable to treat the transverse conductivities as a package in which $\sigma_y = \sigma_z$. If, however, we wish to consider problems in which they may not be equal, perhaps due to the presence of fiber tows, then the corresponding inverse problem treats σ_y and σ_z as independent unknowns, so the inverse problem is two-dimensional. If we use the same one-dimensional nodes as before, the interpolation table becomes the Cartesian product $[25, 75, 125] \otimes [25, 75, 125]$. Using the same data, over the same frequency range as before, the result of the inversion is shown in Table 3.3. While σ_y is reliably reconstructed, σ_z is not. It seems that the 1.5mm thick sample is too thin to generate a significant E_z field that would be necessary to interrogate the structure to produce σ_z. The challenge in measuring σ_z has been discussed in [130].

To test this idea, we reran the inverse problem for a 12.7 mm-thick sample, using the same data as before. The result for the restricted case of $\sigma_y = \sigma_z$, i.e., a one-dimensional problem, is shown in Table 3.4. The results are still quite good, with perhaps a little less certainty because of the larger sensitivity coefficient.

Finally, we consider the two-dimensional case with $\sigma_y = [25, 75, 125]$, $\sigma_z = [25, 75, 125]$, independently. The result of this inversion is shown in Table 3.5. The improvement here, over Table 3.3, is in the estimation of σ_z and its reliability, as indicated by the much smaller sensitivity coefficient. This lends credence to the hypothesis that there must be sufficient thickness in the sample to allow a significant value of E_z to evolve. The manner in which such a component is derived can be argued by the following simple model. In an infinite, homogeneous, isotropic plate, the induced electric field has a vanishing z-component, where the z-direction is normal to the surface, so the resulting eddy-currents flow in planes parallel to the surface.

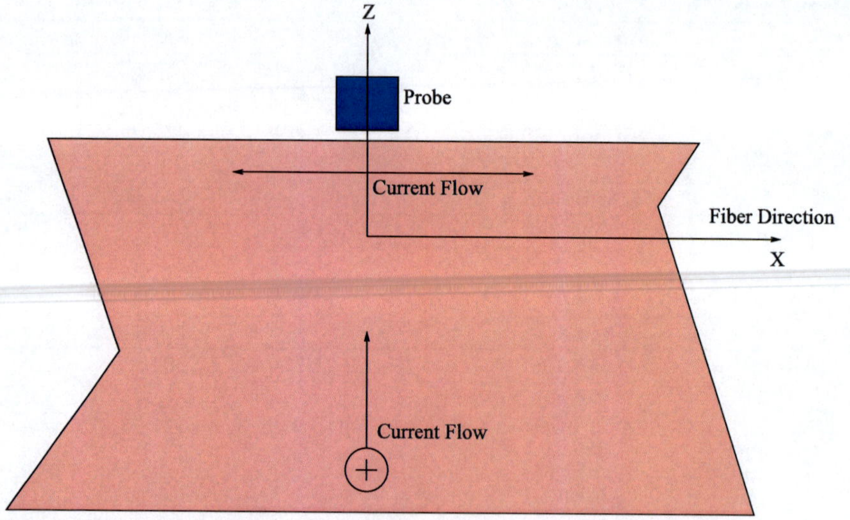

Fig. 3.11 Currents flowing orthogonally to the fiber direction penetrate more deeply into the structure because of the extended skin depth

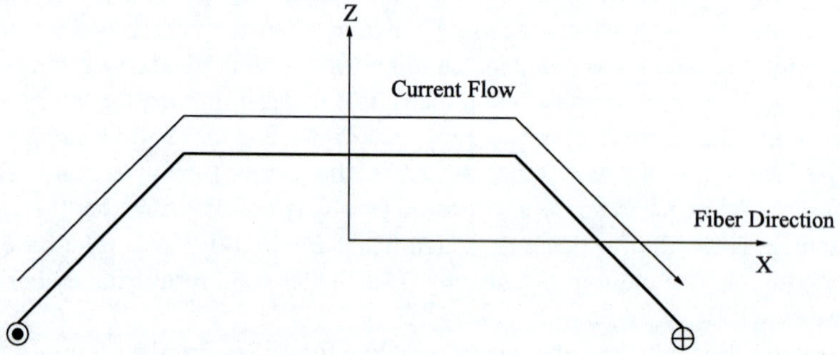

Fig. 3.12 The 'bent-plate' phenomenon: currents flowing orthogonally to the fiber direction penetrate more deeply into the structure because of the extended skin depth. This gives rise to a z-directed current component. If the structure is thick enough in the z-direction, this component should be large enough to produce a reliable estimate of the zz-component of the conductivity tensor

In the anisotropic plate with which we are concerned, however, the x-component of the electric field experiences a skin depth much smaller than that experienced by the y-component (see Fig. 3.11), which means that the resulting eddy-currents flow in a 'bent-plate' mode, as suggested by Fig. 3.12. This gives rise to a z-directed field and current. It is this field, and the resulting current that allows us to infer the zz-component of the conductivity tensor. We expect that the z-field becomes larger with a thicker structure, allowing a more reliable estimate of σ_z.

3.6.1 Return to an Analysis of Fig. 3.10

The blending functions of this figure exhibit a feature that supports the circuit model of Fig. 3.7 and the analysis surrounding it, namely Eqs. (3.3)–(3.5).

The reciprocal of the conductivities shown in Fig. 3.10 corresponds to the resistor, R, that shunts the capacitor, C, in Fig. 3.1. These two 'circuit elements' are repeated in the load impedance, Z_L, of Fig. 3.7. Hence, the smaller the conductance value in Fig. 3.10, the larger is R, relative to C, which means that C will dominate in Z_L at a lower frequency. According to our analysis in (3.4) and (3.5) the driving-point reactance, δX_{in}, may go positive earlier than for larger conductivities, which we clearly see in Fig. 3.10. In fact, we see what appears to be a resonance, following the discussion of (3.5) at 80 MHz.

There is a caveat, however, and that is that δR_{in} in (3.4) is always positive, as we would expect for a passive system, whereas the resistance in Fig. 3.10 is, for the most of the frequency range, negative. This is due to the fact that R and X in the figure correspond to the 'FAWT Test Region' in Fig. 3.9, and **VIC-3D®** treats this region as an 'anomaly' relative to the Graphite-Epoxy Host. Hence, the 'true' δZ_{in} of (3.4) is the sum of the host response plus the responses shown in Fig. 3.10. This response can be inferred in Fig. 3.13, where we plot the host response together with the anomalous responses (the blending functions) of Fig. 3.10.

It is clear that when we add the host response to each of the blending functions, we will get a positive R for all frequencies, but that X will not go positive in this frequency range, though it is tending that way. When using this model in the lab, we would subtract the host-only response from the total response to get the anomalous response which contains the important information for determining the parameter values in an inverse problem.

3.7 Further Results for Permittivity

To further test our ability to sense capacitive effects in the composite, we model the problem shown in Fig. 3.9, except that we excite the coil from 0.1–10 GHz, and we fix $\sigma_y = \sigma_z = 25$ S/m. We assume the dielectric constant of the composite to be isotropic, since we have no justification for choosing an anisotropic model. We choose principal values of the dielectric constant tensor to be $[1, 3, 5]$, and compute the response, which is shown in Fig. 3.14.

We clearly see a distinct difference in both R and X at the higher frequencies, as we would expect for the small value of σ_y and σ_z. Had we chosen larger transverse values, it is unlikely that we would see dielectric effects at these frequencies. The oscillations in the responses over this large frequency range are due to the complex frequency response of (3.6).

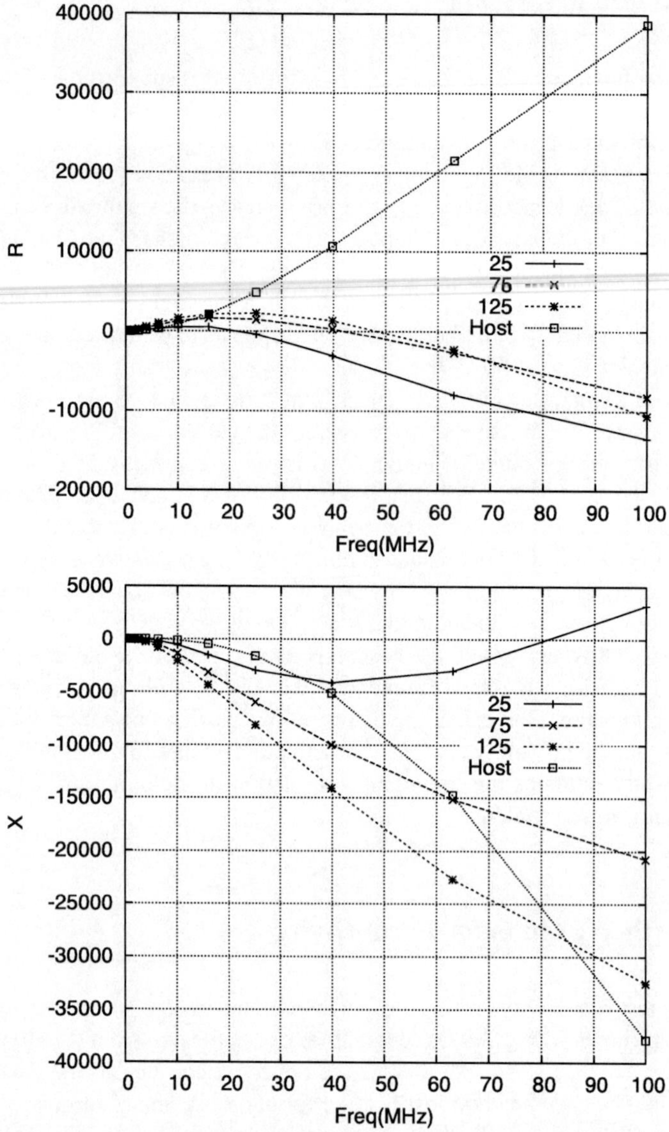

Fig. 3.13 Showing the isotropic host response, together with the blending functions. The sum of the host response with each of the blending functions produces the response of (3.4)

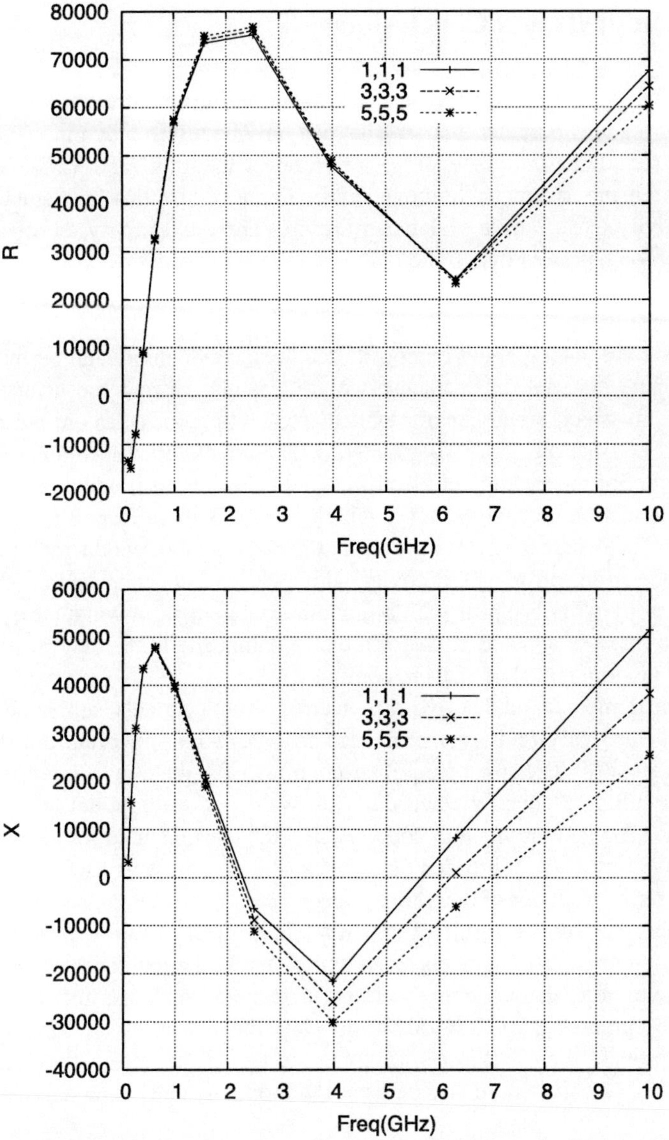

Fig. 3.14 Anomalous response of the FAWT Test Region of Fig. 3.9 over the frequency range 0.1–10 GHz. The parameters are $\sigma_x = 20000$ S/m and $\sigma_y = \sigma_z = 25$ S/m. The test parameter is the relative isotropic dielectric constant, which has values [1, 1, 1], [3, 3, 3], and [5, 5, 5]

3.8 Comments and Conclusions

We should make clear at this point that the 'permittivity' that we are computing is not that of the host polymer matrix. That can only be inferred by measurements on a pristine sample before embedding carbon fibers within it. The **VIC-3D®** model that we are using assumes a homogeneous sample of cfrp that is characterized by a conductivity tensor and a scalar permittivity. The conductivity tensor allows an anisotropy to be present due to the fibers, but we are assuming a scalar permittivity because we have no justification not to. In any case, it is a trivial matter to add a tensor permittivity.

Thus, we are talking about composite features, conductivity and permittivity, of the composite material. In this sense we are treating the material as an 'artificial dielectric', in which a number of identical conducting obstacles are arranged in a regular pattern in a dielectric filler material. The net effect of such an arrangement under the action of an applied electromagnetic field is to produce a net average dipole polarization per unit volume, which increases the effective permittivity of the system [26, Chapter 12]. With a regular arrangement of conductors, the system will produce an anisotropic permittivity, which we have not considered in this paper. The anisotropy will disappear with a random arrangement of conductors, which is probably a more reasonable model of the cfrp composite in the first place. We will attack the random model of cfrps in coming work.

We have demonstrated that rigorous electromagnetic models, supplemented with equivalent electrical circuits, can be used to analyze carbon-fiber reinforced polymer (cfrp) composites. The use of equivalent circuits facilitates the interpretation of the field solution, especially when the 'observable' of each model is an electrical impedance. The volume-integral code, **VIC-3D®**, is well suited for these analyses, and we have further demonstrated that it is capable of providing useful results in the gigahertz range, well out of the normal range for eddy-current models.

An important further result of this modeling effort is the demonstration that simple eddy-current NDE models, even with a standard circular probe, can produce useful results for characterizing certain properties, such as fiber-areal weight (FAWT) of prepreg cfrp. **VIC-3D®** also has the capability of modeling more complex probes, such as those that are designed to operate in the gigaherts or terahertz range, which could also be useful in modeling and characterizing cfrp.

3.9 Eigenmodes of Anisotropic Media

The previous calculations in this chapter were based on a model in which the anisotropy was a bounded anomaly within an isotropic host. This model allowed us to use the usual isotropic Green's function, which has been the basis of our work to this point. There may be occasions in which one might wish to generalize the model, so that the host, itself, is anisotropic, and any departures from this host would be anomalies that would be gridded by themselves, without the need to grid the host anisotropy. Furthermore, one may be interested in other problems of electromagnetic scattering from composite materials, without reference to nondestructive evaluation. For these reasons, we will outline the development of a Green's function for aniostropic planar layered media, that follows closely the spirit of [111, Chapter 2].

We will consider plane-parallel bodies of infinite extent in the (x, y) plane, which are made up of layers of homogeneous, anisotropic material. To be specific, we consider host materials that are characterized by the following biaxial generalized electrical permittivity matrix:

$$\epsilon_h = \begin{bmatrix} \epsilon_x & \epsilon_{xy} & 0 \\ \epsilon_{yx} & \epsilon_y & 0 \\ 0 & 0 & \epsilon_z \end{bmatrix}, \tag{3.7}$$

where the entries are generalized permittivities $\epsilon + \sigma/j\omega$.

Maxwell's equations for an electrically anisotropic body are

$$\begin{aligned} \nabla \times \mathbf{E} &= -j\omega\mu_h\mathbf{H} - j\omega(\mu(\mathbf{r}) - \mu_h)\mathbf{H} \\ &= -j\omega\mu_h\mathbf{H} + \mathbf{J}_m \\ \nabla \times \mathbf{H} &= j\omega\epsilon_h \cdot \mathbf{E} + j\omega(\epsilon(\mathbf{r}) - \epsilon_h) \cdot \mathbf{E} \\ &= j\omega\epsilon_h \cdot \mathbf{E} + \mathbf{J}_e, \end{aligned} \tag{3.8}$$

where $\mathbf{J}_m$ and $\mathbf{J}_e$ are anomalous magnetic and electric currents that account for the presence of flaws, or anomalies, in the otherwise-uniform host material. From here on we drop the subscript h on the generalized host permittivity and permeability.

Because of the material anisotropy, it is convenient to work with a matrix formulation of these equations that has been useful in crystal optics, plasmas and microwave devices [7–9, 17, 59–61, 115, 125]. If the body is homogeneous with respect to (x, y), then Maxwell's equations can be Fourier transformed with respect to (x, y), and written as the following four-vector matrix differential equation in the spectral domain:

$$\frac{d\tilde{\mathbf{e}}}{dz} = \mathbf{S} \cdot \tilde{\mathbf{e}} + \mathbf{U} \cdot \tilde{\mathbf{J}} \tag{3.9}$$

$$\tilde{E}_z = \frac{k_y}{\epsilon_z\omega}\tilde{H}_x - \frac{k_x}{\epsilon_z\omega}\tilde{H}_y + \frac{j}{\epsilon_z\omega}\tilde{J}_{ez} \tag{3.10}$$

$$\tilde{H}_z = \frac{-k_y}{\mu\omega}\tilde{E}_x + \frac{k_x}{\mu\omega}\tilde{E}_y - \frac{j}{\mu\omega}\tilde{J}_{mz}, \tag{3.11}$$

where the tilde denotes a function defined in the transform domain (k_x, k_y), and

$$\tilde{\mathbf{e}} = \begin{bmatrix} \tilde{E}_x \\ \tilde{E}_y \\ \tilde{H}_x \\ \tilde{H}_y \end{bmatrix} ; \tilde{\mathbf{J}} = \begin{bmatrix} \tilde{J}_{ex} \\ \tilde{J}_{ey} \\ \tilde{J}_{ez} \\ \tilde{J}_{mx} \\ \tilde{J}_{my} \\ \tilde{J}_{mz} \end{bmatrix} . \tag{3.12}$$

The subscript e denotes an electric current, and m denotes a magnetic current. The matrices in (3.9) are given by

$$\mathbf{S} = -\begin{bmatrix} 0 & 0 & a & b \\ 0 & 0 & c & d \\ \alpha & \beta & 0 & 0 \\ \gamma & \delta & 0 & 0 \end{bmatrix} ; \mathbf{U} = \begin{bmatrix} 0 & 0 & k_x/\omega\epsilon_z & 0 & 1 & 0 \\ 0 & 0 & k_y/\omega\epsilon_z & -1 & 0 & 0 \\ 0 & 1 & 0 & 0 & 0 & -k_x/\omega\mu \\ -1 & 0 & 0 & 0 & 0 & -k_y/\omega\mu \end{bmatrix} . \tag{3.13}$$

The entries of $\mathbf{S}$ are given in terms of the entries of (3.7) by

$$\begin{aligned} a &= \frac{j}{\omega\epsilon_z} k_x k_y \; ; \; \alpha = \frac{j}{\omega\mu}(-\mu\epsilon_{yx}\omega^2 - k_x k_y) \\ b &= \frac{j}{\omega\epsilon_z}(\mu\epsilon_z\omega^2 - k_x^2) \; ; \; \beta = \frac{j}{\omega\mu}(-\mu\epsilon_y\omega^2 + k_x^2) \\ c &= \frac{j}{\omega\epsilon_z}(-\mu\epsilon_z\omega^2 + k_y^2) \; ; \; \gamma = \frac{j}{\omega\mu}(\mu\epsilon_x\omega^2 - k_y^2) \\ d &= -\frac{j}{\omega\epsilon_z} k_x k_y \; ; \; \delta = \frac{j}{\omega\mu}(\mu\epsilon_{xy}\omega^2 + k_x k_y) . \end{aligned} \tag{3.14}$$

When $\tilde{\mathbf{J}}$ is a surface current confined to $z = z'$, i.e., $\tilde{\mathbf{J}} = \tilde{\mathbf{J}}_s\delta(z - z')$, then integration of (3.9) produces

$$\tilde{\mathbf{e}}^{(+)} - \tilde{\mathbf{e}}^{(-)} = \mathbf{U} \cdot \tilde{\mathbf{J}}_s . \tag{3.15}$$

The superscript $(+)$ denotes the limit as z approaches z' from above, and the superscript $(-)$ denotes the limit from below. Equation (3.15) is used to compute the Green's dyad for a layered workpiece.

Starting with these equations, Roberts [90] has developed a fairly complete theory of normal modes of biaxial anisotropic media. This work is based on, and extends, earlier work performed at Sabbagh Associates [92, 93]. From here on we specialize the theory developed in [90] to the case to be considered here, in which the media involved are transversely isotropic to the x-coordinate. The generalized electric permittivity tensor, in its principal-axis coordinate system, then takes the form

$$\boldsymbol{\epsilon} = \begin{bmatrix} \epsilon_x & 0 & 0 \\ 0 & \epsilon & 0 \\ 0 & 0 & \epsilon \end{bmatrix}, \tag{3.16}$$

which is typical of many graphite-epoxy composites.

The eigenmodes are solutions of (3.9) with the input currents set equal to zero, and are simply the eigenvalues of **S**. It is straightforward to compute these eigenvalues, $\pm\lambda_1$ and $\pm\lambda_3$:

$$\lambda_1 = \left[(\epsilon_x/\epsilon)k_x^2 + k_y^2 - \omega^2\mu_0\epsilon_x\right]^{1/2}, \quad \lambda_3 = \left[k_x^2 + k_y^2 - \omega^2\mu_0\epsilon\right]^{1/2}. \tag{3.17}$$

λ_1 corresponds to the extraordinary wave, and λ_3 to the ordinary wave. Clearly, when $\epsilon_x = \epsilon$, then $\lambda_1 = \lambda_3$, and the extraordinary wave becomes ordinary, which agrees with the results for an isotropic medium (such as free-space).

Corresponding to each eigenvalue is an eigenvector. We have some liberty in choosing the two independent equations that generate the eigenvectors; hence, there is some arbitrariness in choosing the eigenvectors. We choose the following because of their simple structure:

$$\mathbf{v}_1 = \begin{bmatrix} \alpha_1 \\ \alpha_2 \\ 0 \\ 1 \end{bmatrix}, \mathbf{v}_2 = \begin{bmatrix} -\alpha_1 \\ -\alpha_2 \\ 0 \\ 1 \end{bmatrix}, \mathbf{v}_3 = \begin{bmatrix} 0 \\ 1 \\ \gamma_1 \\ -\gamma_2 \end{bmatrix}, \mathbf{v}_4 = \begin{bmatrix} 0 \\ 1 \\ -\gamma_1 \\ \gamma_2 \end{bmatrix}, \tag{3.18}$$

where

$$\alpha_1 = S_{14}/\lambda_1,\ \alpha_2 = S_{24}/\lambda_1,\ \gamma_1 = S_{32}/\lambda_3,\ \gamma_2 = S_{31}/\lambda_3\ , \tag{3.19}$$

and the S_{ij} are defined in (3.13) and (3.14). $\mathbf{v}_1$ and $\mathbf{v}_2$ are associated with $+\lambda_1,\ -\lambda_1$, respectively, whereas $\mathbf{v}_3$ and $\mathbf{v}_4$ are associated with $+\lambda_3,\ -\lambda_3$, respectively. The second and fourth vectors are the two (+)-going modes, and the first and third are the two (−)-going modes, in the z-direction. Corresponding functions in free-space are designated by the subscript 0. Note that $\mathbf{v}_1,\ \mathbf{v}_2$ are transverse magnetic (TM) to x, and $\mathbf{v}_3,\ \mathbf{v}_4$ are transverse electric (TE) to x.

3.10 Computing a Green's Function for a Layered Workpiece

We now consider a layered workpiece of finite extent in the z-direction. We assume that the source is at $z' > 0$, the workpiece has a thickness, z_w, satisfies $-z_w < z < 0$, and is divided into, say, ten subregions, which are defined as follows:

$$\begin{aligned} \text{Layer1}: &\quad z_1 < z < z_0 = 0 \\ \text{Layer2}: &\quad z_2 < z < z_1 \\ \vdots &\quad\quad \vdots \\ \text{Layer10}: &\; -z_w = z_{10} < z < z_9 \end{aligned} \tag{3.20}$$

The spectral-domain Green's function is expanded in terms of the eigenvectors of (3.18) as

$$\widetilde{\mathbf{G}}(z, z') = \begin{cases} a\mathbf{v}_{20}e^{-\lambda_0(z-z')} + b\mathbf{v}_{40}e^{-\lambda_0(z-z')}, & z' < z \\ c\mathbf{v}_{10}e^{\lambda_0(z-z')} + d\mathbf{v}_{20}e^{-\lambda_0(z-z')} \\ \quad + e\mathbf{v}_{30}e^{\lambda_0(z-z')} + f\mathbf{v}_{40}e^{-\lambda_0(z-z')}, & 0 < z < z' \\ c^{(i)}\mathbf{v}_1^{(i)}e^{\lambda_1(z-z_{i-1})} + d^{(i)}\mathbf{v}_2^{(i)}e^{-\lambda_1(z-z_{i-1})} \\ \quad + e^{(i)}\mathbf{v}_3^{(i)}e^{\lambda_3(z-z_{i-1})} + f^{(i)}\mathbf{v}_4^{(i)}e^{-\lambda_3(z-z_{i-1})}, & z_i < z < z_{i-1},\ i = 1, \ldots, 10 \\ g\mathbf{v}_{10}e^{\lambda_0(z+z_w)} + h\mathbf{v}_{30}e^{\lambda_0(z+z_w)}, & z < -z_w \end{cases} \tag{3.21}$$

Using the boundary conditions and the conditions at the source, we can find the unknowns by solving (in the case of ten layers) a 44×44 system of equations for $d,\ f,\ c^{(1)},\ e^{(1)},\ d^{(1)},\ f^{(1)},\ \ldots,\ c^{(10)},\ e^{(10)},\ d^{(10)},\ f^{(10)},\ g,\ h$ and then find $c,\ a,\ e,\ b$. The form of the system is:

$$\mathscr{S}X = Y\,, \tag{3.22}$$

where

$$\mathscr{S} = \begin{bmatrix} R_1 & -S^{(1)} & 0_{4\times4} & 0_{4\times4} & \cdots & 0_{4\times4} & 0_{4\times4} & 0_{4\times2} \\ 0_{4\times2} & S^{(1)}E^{(1)} & -S^{(2)} & 0_{4\times4} & \cdots & 0_{4\times4} & 0_{4\times4} & 0_{4\times2} \\ 0_{4\times2} & 0_{4\times4} & S^{(2)}E^{(2)} & -S^{(3)} & \cdots & 0_{4\times4} & 0_{4\times4} & 0_{4\times2} \\ \vdots & \vdots & \vdots & \vdots & \ddots & -S^{(9)} & 0_{4\times4} & 0_{4\times2} \\ 0_{4\times2} & 0_{4\times4} & 0_{4\times4} & 0_{4\times4} & \cdots & S^{(9)}E^{(9)} & -S^{(10)} & 0_{4\times2} \\ 0_{4\times2} & 0_{4\times4} & 0_{4\times4} & 0_{4\times4} & \cdots & 0_{4\times4} & S^{(10)}E^{(10)} & -R_2 \end{bmatrix}, \tag{3.23}$$

with

$$
\begin{aligned}
R_1 &= (\mathbf{v}_{20}, \mathbf{v}_{40}) \\
R_2 &= (\mathbf{v}_{10}, \mathbf{v}_{30}) \\
S^{(i)} &= \left(\mathbf{v}_1^{(i)}, \mathbf{v}_3^{(i)}, \mathbf{v}_2^{(i)}, \mathbf{v}_4^{(i)}\right) \\
E^{(i)} &= \begin{bmatrix} e^{\lambda_1 h_i} & 0 & 0 & 0 \\ 0 & e^{\lambda_3 h_i} & 0 & 0 \\ 0 & 0 & e^{-\lambda_1 h_i} & 0 \\ 0 & 0 & 0 & e^{-\lambda_3 h_i} \end{bmatrix} \\
h_i &= z_i - z_{i-1} \\
X &= \left(\tilde{d}, \tilde{f}, c^{(1)}, e^{(1)}, d^{(1)}, f^{(1)}, \cdots, c^{(10)}, e^{(10)}, d^{(10)}, f^{(10)}, g, h\right)^T \\
\tilde{d} &= d e^{\lambda_0 z'} \\
\tilde{f} &= f e^{\lambda_0 z'} \\
Y &= \begin{pmatrix} -\mathbf{v}_{10} c - \mathbf{v}_{30} e \\ 0_{40\times 1} \end{pmatrix} e^{-\lambda_0 z'}
\end{aligned}
\tag{3.24}
$$

Up to this point, we have assumed that all layers are oriented in the same direction with respect to a global coordinate system. Let us now suppose that each layer can be rotated. Since eigenvalues are unchanged and eigenvectors are merely transformed by the rotation, it is straight forward to set up the equations for this situation. The rotation matrix in this case is

$$
\mathscr{T}(\theta) = \begin{pmatrix} \mathscr{R}(\theta) & 0_{2\times 2} \\ 0_{2\times 2} & \mathscr{R}(\theta) \end{pmatrix}, \tag{3.25}
$$

where

$$
\mathscr{R}(\theta) = \begin{pmatrix} \cos(\theta) & \sin(\theta) \\ -\sin(\theta) & \cos(\theta) \end{pmatrix}. \tag{3.26}
$$

If we let θ_i be the angle of rotation of the ith layer with respect to the global system, then the eigenvectors for this layer become

$$\mathbf{v}_1^{(i)'} = \mathscr{T}(\theta_i)\mathbf{v}_1^{(i)},\ \mathbf{v}_2^{(i)'} = \mathscr{T}(\theta_i)\mathbf{v}_2^{(i)},\ \mathbf{v}_3^{(i)'} = \mathscr{T}(\theta_i)\mathbf{v}_3^{(i)},\ \mathbf{v}_4^{(i)'} = \mathscr{T}(\theta_i)\mathbf{v}_4^{(i)}, \tag{3.27}$$

and the only change in (3.23) is the definition of $S^{(i)}$, which becomes

$$S^{(i)'} = \left(\mathbf{v}_1^{(i)'},\ \mathbf{v}_3^{(i)'},\ \mathbf{v}_2^{(i)'},\ \mathbf{v}_4^{(i)'}\right) \tag{3.28}$$

$$= \left(\mathscr{T}\mathbf{v}_1^{(i)},\ \mathscr{T}\mathbf{v}_3^{(i)},\ \mathscr{T}\mathbf{v}_2^{(i)},\ \mathscr{T}\mathbf{v}_4^{(i)}\right). \tag{3.29}$$

This is a general model for the computation of the Green's function and the computation of reflection coefficients that has been implemented in our computer codes. In our model, we have assumed that the conductivities of each layer are the same. The generalization to a model where each layer has its own conductivities could be easily accomplished, however, by computing the eigenvalues and eigenvectors of each layer in its own local coordinate system and then proceeding with the rotation stage above.

3.11 An Example of the Multilayer Model

The multilayer model that was developed above has been applied to a number of different configurations. Figure 3.15 shows one configuration that has been validated experimentally. The figure illustrates the computed and measured EMF induced into a probe coil due to a circular current loop above a graphite-epoxy workpiece, which consists of eighteen unidirectional layers in a $[0, 0, 0, 0, 0, 0, 0, 0, 90]_S$ lay-up. By this notation is meant that the first eight layers are aligned with each other, while the ninth is rotated 90°. The bottom nine layers form a mirror image of the top nine (the 'S' denotes a symmetrical configuration). Neither the conductivities of any of the layers, nor the conductivity of the bulk workpiece, were known, but were inferred by parameter-fitting. The peak computed value is 3.5 V, and the peak measured value is 4.4 V. See [91] for more details on the development of the model and [128] for the experimental validation.

3.12 A Bulk Model

A graphite epoxy panel is a layup of individual plies. We described in the preceding section a method for computing electromagnetic interactions *exactly* using a *multilayer* model. The multilayer model treats layers individually, and was used to produce the results shown in Fig. 3.15. One alternative to modeling layers individually is to model them in bulk with an equivalent single-layer slab. The bulk model is acceptable when wavelengths (or skin depths) are larger than the thickness

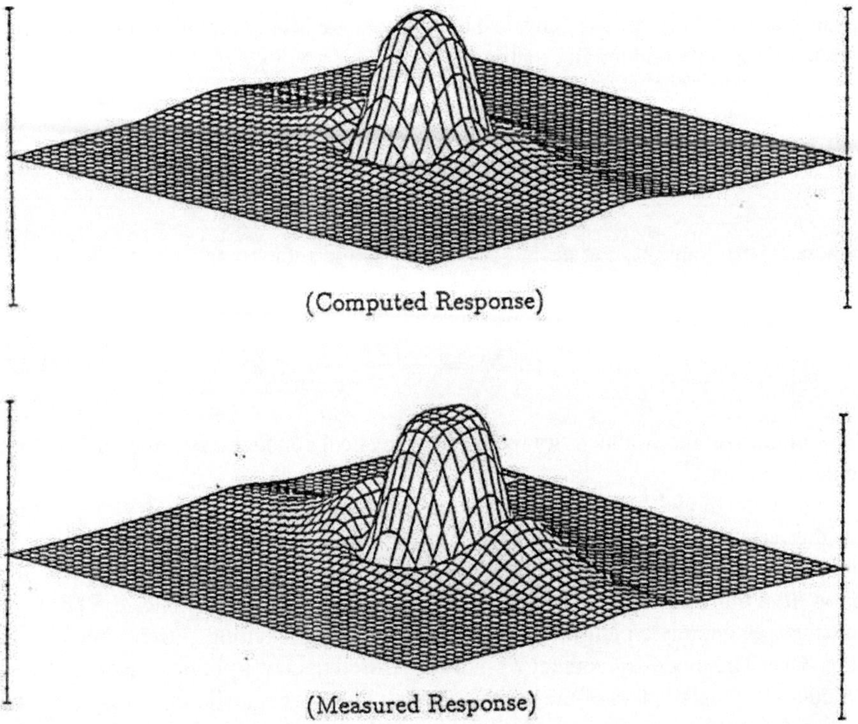

Fig. 3.15 EMF induced into a probe coil due to a circular current loop above a graphite-epoxy workpiece. (See the text for details)

of a few plies. It is computationally more efficient to use the bulk model, rather than the multilayer model, when the long-wavelength (long-skin depth) condition is satisfied, because we model, typically, *dozens* of layers with *just one* equivalent bulk layer.

Our bulk model computes interactions exactly for the most general stratified conductor–one with off-diagonal terms in the conductivity matrix:

$$\sigma = \begin{bmatrix} \sigma_{xx} & \sigma_{xy} & 0 \\ \sigma_{yx} & \sigma_{yy} & 0 \\ 0 & 0 & \sigma_{zz} \end{bmatrix} . \tag{3.30}$$

The z-axis is the axis of stratification; that is, the plies are xy-planes at various positions along the z-axis, which means that σ_{xz} and σ_{yz} are zero, because there is no way that a z-directed field can produce an x-directed current, and *vice versa*.

The bulk model has been set up for the most *general* case of a stratified conductor, namely one in which the conductivity elements can be positive, negative, or zero. Only symmetric conductivity matrices ($\sigma_{xy} = \sigma_{yx}$), with nonnegative diagonal elements, however, are associated with passive conductors, i.e., conductors that absorb energy. Furthermore, physical conductors are just one rotation removed from having *diagonal* conductivity matrices:

$$\sigma_{\mathbf{p}} = \begin{bmatrix} \sigma_x & 0 & 0 \\ 0 & \sigma_y & 0 \\ 0 & 0 & \sigma_{zz} \end{bmatrix}, \tag{3.31}$$

where σ_x, σ_y, and σ_{zz} are each ≥ 0, and ϕ_{p} is the rotation angle in the xy-plane, given by

$$\tan 2\phi_{\mathrm{p}} = \frac{2\sigma_{xy}}{\sigma_{xx} - \sigma_{yy}}. \tag{3.32}$$

To summarize: the conductivity matrix of a physical conductor is a diagonal matrix, plus a rotation.

The bulk model has been tested numerically and in the laboratory [100, 106]. In the tests reported there, a circular loop of current was placed above a $\pm 22.5°$ layup of graphite epoxy, and an EMF-sensing probe coil was scanned over the loop, in a transmit/receive configuration. The resulting plot illustrates the effect of *anisotropy*, namely an elliptical response to a circular exciting current, much as in Fig. 3.15. The model agreement with the measured data was not precise because the conductivity parameters of the graphite-epoxy layup were not known, but had to be estimated with about 50% uncertainty. Nevertheless, the utility of the bulk model was confirmed.

Further discussion of the relation between the multilayer and bulk models is given in [129], where a number of experimental verifications of each are given. Experimental data are also given for graphite-epoxy when it is configured in a 'satin-weave,' in which the fibers are 'woven,' in this case 'over four, under one.' This paper demonstrates that eddy-current measurements can give indications of fiber breakage, electrical conductivity, fiber density, layer thickness, and perhaps delaminations. The layer-by-layer detail of the material *is* important in modeling the electromagnetic field in the vicinity of the material.

Chapter 4
Application of the Set-Theoretic Algorithm to CFRP's

4.1 Background

A major milestone that we have demonstrated is that the set-theoretic algorithm works well with anisotropic media, such as CFRPs. Such problems require at least two variables to be reconstructed at each voxel, such as the longitudinal and transverse conductivities. The details are described in Sects. 4.3.1–4.3.2 in the context of measuring FAWT (Fiber Areal-Weight), which is an important parameter during the manufacture of CFRP prepregs. An interesting corollary is shown in Sect. 4.3.2, in which a single transmitting coil is used with the same receiver array as before, and very good results were achieved. This suggests that less extreme T/R-arrays can be used in the algorithm, with the possibility of using sparse-grid interpolation to fill in gaps in the measured data.

We begin our study of microstructure quantification problems in Sect. 4.4 with an investigation of detecting and measuring delaminations using the model shown in Fig. 4.7. The objective was to estimate the smallest delamination that is probably detectable. The results shown in Table 4.2 suggest that it may be difficult to distinguish a delamination 7 μm high from the background. There would appear to be a better chance of detecting one that is 14 μm tall, and probably a pretty good chance of detecting one that is 21 μm tall.

The problem suggested in Fig. 4.8 will be a major 'test-bed' for applying set-theoretic inversion to microstructure quantification. We have started the process during this quarter by analyzing the model of Fig. 4.9, which represents the structure of Fig. 4.8 without the voids. The host is as shown in Fig. 4.1, except that it is 0.4 mm thick. The receive-array is 21×21, as before in the FAWT studies, so that $N_s = 441$, and there are $11 \times 11 = 121$ transmit positions in the transmitter array. The excitation is at a frequency of 10GHz. Data for setting up the set-theoretic algorithm are being generated, and the actual inversion process will be completed during

H. A. Sabbagh et al., *Advanced Electromagnetic Models for Materials Characterization and Nondestructive Evaluation*, Scientific Computation,
https://doi.org/10.1007/978-3-030-67956-9_4

the next quarter. The parameters in the **VIC-3D®**-file for solving this problem are shown following Fig. 4.11.

The problem shown in Fig. 4.10 models Fig. 4.8 with a single void microcrack. This structure is embedded in an isotropic graphite-epoxy host whose conductivity is 20,000 S/m, and is 0.3 mm high. The objective of this study is to determine the feasibility of detecting and determining the width, W, of the microcrack when the structure is excited by the T probe shown in Fig. 4.7 as it is scanned past the microcrack from -1.6 mm to $+1.6$ mm at a frequency of 1 GHz.

To that end, we apply model-based inversion, starting with the blending functions that define the surrogate interpolation table. These are shown in Fig. 4.11 for the nodal values $W = (0.00, 0.025, 0.050)$ mm. The **VIC-3D®**-grid used to generate these functions was $N_x = 256$, $N_y = 8$, $N_z = 16$, which was sufficient to capture the variations in W. The test value of W was 0.030 mm.

The results of the inversion are shown in Table 4.3. Not only is the solution virtually identical to the test value, but all 500 initial starting points in NLSE converged to the same global minimum. The excellent quality of the inversion testifies to the fact that the blending functions of Fig. 4.11 are highly sensitive to W.

4.2 Statistical Analysis of the Feasible Set

One of the principal efforts of research in Set-Theoretic Estimation is to determine the number and nature of the experiments, $(Z(v), \mathbf{E}_{lmj}^{(i)}(v))$, and receiver scans, $\mathbf{E}_{LMJ}^{(n)}$, that produce a good feasible set for statistical analysis. As might be expected, this depends upon the complexity of the flaw that is to be reconstructed, and upon the resolution desired in the reconstruction.

After deducing the feasibility set, we are then faced with the task of assigning a single number for the conductivity of each cell; this is a problem of data analysis, which uses robust regression, as described in Chapter 13 of [111]. The aim of the statistical analysis of the feasible set is to fit a constant through the data, which are produced by the algorithm described in the preceding section, for all views (or experiments), v.

A beneficial feature of this algorithm is that the analysis of the data set for each cell is done independently of every other cell; i.e., a decision is made on a cell-by-cell basis. Since the decision to be made for each cell involves a nonlinear (robust) estimator, the computational burden is greatly reduced when compared to using a nonlinear estimator to solve for many cells jointly. (The bilinear conjugate-gradient algorithm is an exception.) Furthermore, this leads us to a constrained iterative ('layer-stripping') algorithm, that uses the known and accepted results for some cells to determine the results for others at a later stage of the iteration.

4.3 An Anisotropic Inverse Problem for Measuring FAWT

The model problem is shown in Fig. 4.1 excited at 40 MHz. First, a little background on the physical model: As shown in Fig. 4.2, the fiber-to-fiber contact is random, so that the two electrical parameters that define the contact, σ_y and σ_z, are essentially average values that will be taken to be equal; i.e., it is impossible to distinguish the direction of the contact between the fibers. Hence, we define only a single transverse conductivity, $\sigma_y = \sigma_z = \sigma_T = 100$ S/m as listed in Table 4.1, even though the model shows σ_y and σ_z separately. This is for the benefit of **VIC-3D®**, which requires three values for the anisotropic conductivity tensor. Furthermore, the most general expression for a biaxial conductivity tensor includes the off-diagonal terms, $\sigma_{xy} = \sigma_{yx}$, but these will be taken to be zero in the model calculation since we have no information as to what they might be (see Table 4.1).

4.3.1 First Set-Theoretic Result

The model consists of a 21×21 receiving array, so that $N_s = 441$, with an 11×11 transmitting array. This means that there will be 121 'experiments' with 242 outcomes, the real and imaginary parts of (E_x, J_x) and (E_y, J_y) for each conductivity (Figs. 4.3 and 4.4).

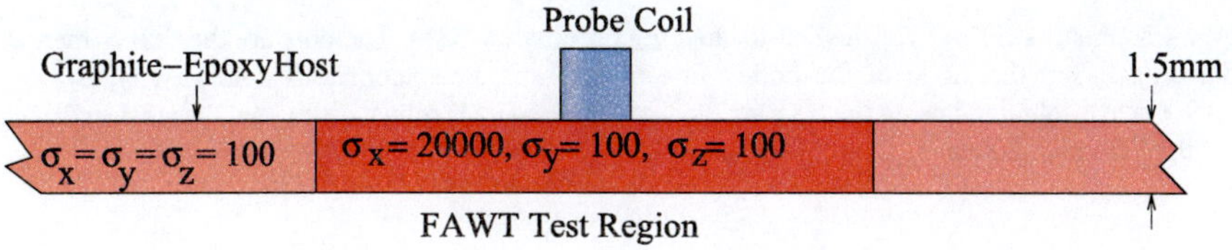

Fig. 4.1 Illustrating the model problem for analyzing FAWT. The host graphite-epoxy slab is isotropic, with the conductivities shown, whereas the FAWT region is anisotropic. The ratio of the longitudinal to the two transverse conductivities is typical for a FAWT = 60% for this particular sample of graphite-epoxy

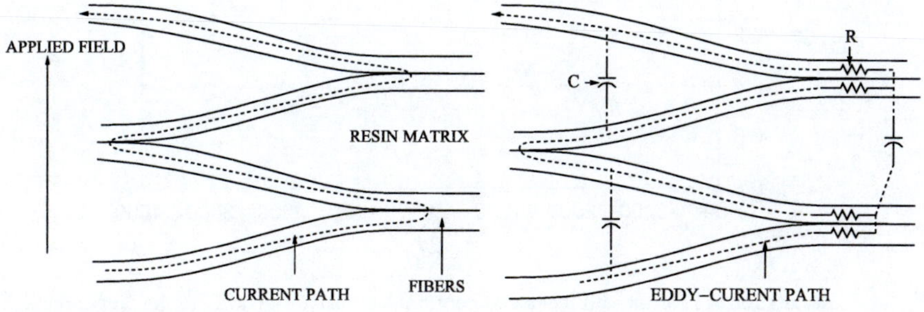

Fig. 4.2 How fiber-to-fiber contact allows transverse conduction (Left). A possible AC equivalent circuit for eddy-current flow (Right)

Table 4.1 Summary of electrical properties of some composites [5]

	Graphite-Epoxy	Boron-Epoxy	Kevlar
Permeability (μ_R)	1	1	1
Permittivity (ϵ_R)	Indeterminate	5.6	3.6
DC conductivity (S/m)			
Longitudinal (σ_L)	2×10^4	30	6×10^{-9}
Transverse (σ_T)	100	2×10^{-8}	6×10^{-9}
Anisotropy ratio (σ_L/σ_T)	200	1.5×10^9	1

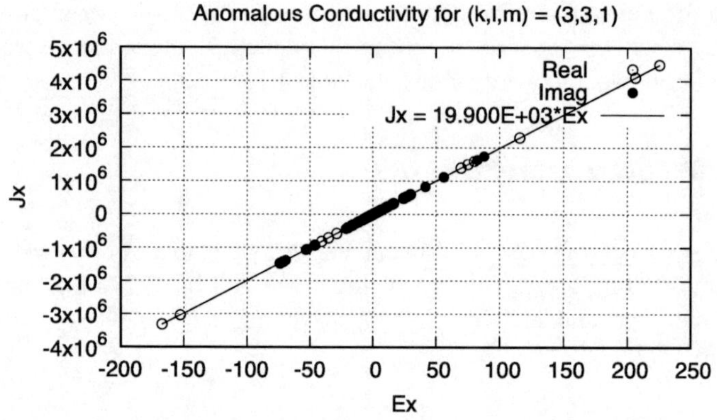

Fig. 4.3 Results for σ_{xx} for the cell located at coordinates (331). The dots are the 'experimental outcomes', and the slope of the dotted line is the estimated anomalous conductivity, $\sigma_{xx} = 19.900 \times 10^3$, obtained using the least-median-of-squares (LM) robust estimator. This is the correct value of the anomalous σ_{xx}

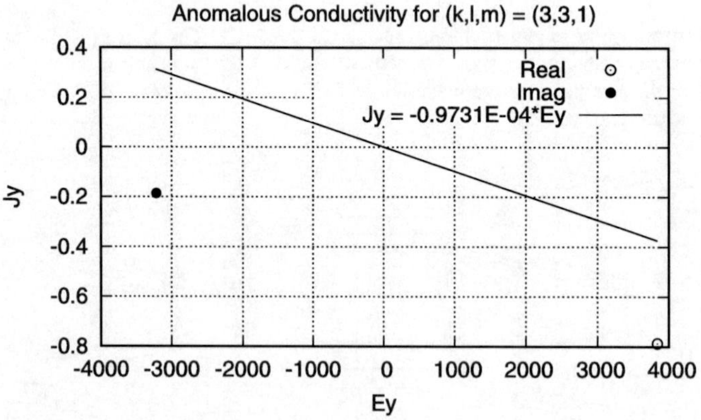

Fig. 4.4 Results for σ_{yy} for the cell located at coordinates (331). The dots are the 'experimental outcomes', and the slope of the dotted line is the estimated anomalous conductivity, $\sigma_{yy} = 0.2947 \times 10^{-4}$, obtained using the least-median-of-squares (LM) robust estimator. This is nine-orders of magnitude less than σ_{xx}, and is close to the correct value of zero for the anomalous σ_{yy}

FIRST SET-THEORETIC σ_{xx} FOR BOTTOM LAYER:SCALE OF 1000 S/m

19.49	20.12	20.22	20.13	19.84	20.30	20.63	19.19
19.05	20.04	19.94	19.88	19.69	19.67	18.91	19.02
20.05	19.98	19.90	19.91	19.93	19.92	20.18	19.96
19.37	20.30	19.91	19.69	19.83	19.97	19.10	19.37
19.13	20.10	19.87	19.93	20.05	19.82	20.38	19.87
19.89	19.99	19.89	19.90	19.87	19.90	19.77	19.66
19.41	19.46	20.29	19.75	19.83	19.48	19.64	19.81
19.35	19.92	20.99	20.39	19.80	20.43	20.11	19.61

FIRST SET-THEORETIC σ_{xx} FOR TOP LAYER OF ANOMALY: SCALE OF 1000 S/m

19.78	20.14	19.87	19.90	19.91	19.87	20.12	19.80
19.85	19.93	19.89	19.89	19.90	19.91	19.99	19.86
19.89	19.90	19.90	19.90	19.90	19.90	19.89	19.89
19.89	19.86	19.90	19.91	19.90	19.89	19.91	19.89
19.90	19.87	19.90	19.90	19.90	19.90	19.84	19.86
19.90	19.90	19.90	19.90	19.90	19.90	19.91	19.91
19.83	19.95	19.87	19.90	19.89	19.91	19.94	19.81
19.79	20.15	19.83	19.89	19.92	19.86	20.14	19.78

FIRST SET-THEORETIC σ_{yy} FOR BOTTOM LAYER OF ANOMALY: SCALE OF 0.001 S/m

-0.430	-6.317	-0.039	0.841	1.444	-3.513	-1.476	-3.174
5.454	-0.771	-0.190	-0.003	-0.290	0.148	3.636	-2.583
0.352	-1.499	-4.035	1.446	1.524	0.426	-0.964	1.483
1.820	-2.795	3.153	0.239	-0.155	2.163	0.019	0.887
-3.441	0.764	0.304	0.556	3.860	-0.871	-0.946	-0.239
-0.423	7.477	0.803	0.117	-1.061	-0.437	-5.926	-0.291
-3.240	2.590	-1.986	0.338	-0.165	-0.049	-0.385	-1.496
-4.172	0.368	1.350	0.503	-4.620	0.226	1.817	-0.150

FIRST SET-THEORETIC σ_{yy} FOR TOP LAYER OF ANOMALY: SCALE OF 0.001 S/m

-0.433	-1.824	0.085	1.838	2.779	-0.089	-2.072	-0.668
-0.516	0.085	0.102	0.649	-1.115	0.150	-1.332	-0.548
-0.814	-0.941	0.097	-0.100	0.092	-0.041	-0.628	-0.443
-0.216	0.237	0.030	-0.056	-0.100	-0.007	-0.501	-0.340
-0.384	-0.164	-0.012	-0.097	-0.100	-0.004	-0.194	0.085
-0.070	-0.961	0.013	-0.123	-0.030	-0.002	0.033	0.861
-1.132	-0.365	-0.106	-0.312	-2.665	-0.356	0.725	-0.860
-0.692	-0.270	0.139	2.411	2.685	-0.258	-0.234	-0.529

4.3.2 Second Set-Theoretic Result

Proceeding as before, except with a single transmitting coil. Hence, one experiment with two outcomes for each conductivity component (Figs. 4.5 and 4.6).

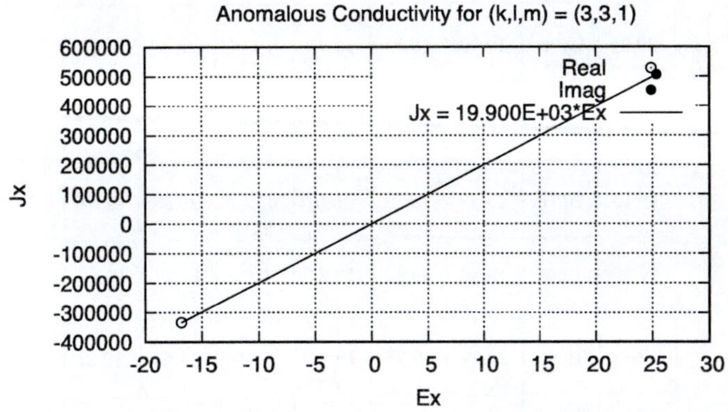

Fig. 4.5 Results for σ_{xx} for the cell located at coordinates (331). The dots are the 'experimental outcomes', and the slope of the dotted line is the estimated anomalous conductivity, $\sigma_{xx} = 19.900 \times 10^3$, obtained using the least-median-of-squares (LM) robust estimator. This is the correct value of the anomalous σ_{xx}

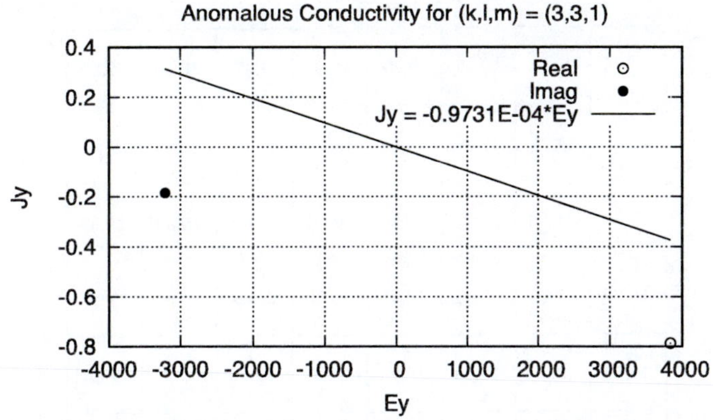

Fig. 4.6 Results for σ_{yy} for the cell located at coordinates (331). The dots are the 'experimental outcomes', and the slope of the dotted line is the estimated anomalous conductivity, $\sigma_{yy} = -0.9731 \times 10^{-4}$, obtained using the least-median-of-squares (LM) robust estimator. This is nine-orders of magnitude less than σ_{xx}, and is close to the correct value of zero for the anomalous σ_{yy}

SECOND SET-THEORETIC σ_{xx} FOR BOTTOM LAYER:SCALE OF 1000 S/m

19.49	20.12	20.22	20.13	19.84	20.30	20.63	19.19
19.05	20.04	19.94	19.88	19.69	19.67	18.91	19.02
20.05	19.98	19.90	19.91	19.93	19.92	20.18	19.96
19.37	20.30	19.91	19.69	19.83	19.97	19.10	19.37
19.13	20.10	19.87	19.93	20.05	19.82	20.38	19.87
19.89	19.99	19.89	19.90	19.87	19.90	19.77	19.66
19.41	19.46	20.29	19.75	19.83	19.48	19.64	19.81
19.35	19.92	20.99	20.39	19.80	20.43	20.11	19.61

SECOND SET-THEORETIC σ_{xx} FOR TOP LAYER OF ANOMALY: SCALE OF 1000 S/m

19.78	20.14	19.87	19.90	19.91	19.87	20.12	19.80
19.85	19.93	19.89	19.89	19.90	19.91	19.99	19.86
19.89	19.90	19.90	19.90	19.90	19.90	19.89	19.89
19.89	19.86	19.90	19.91	19.90	19.89	19.91	19.89
19.90	19.87	19.90	19.90	19.90	19.90	19.84	19.86
19.90	19.90	19.90	19.90	19.90	19.90	19.91	19.91
19.83	19.95	19.87	19.90	19.89	19.91	19.94	19.81
19.79	20.15	19.83	19.89	19.92	19.86	20.14	19.78

SECOND SET-THEORETIC σ_{yy} FOR BOTTOM LAYER OF ANOMALY: SCALE OF 0.001 S/m

-0.430	-6.317	-0.039	0.841	1.444	-3.513	-1.476	-3.174
5.454	-0.771	-0.190	-0.003	-0.290	0.148	3.636	-2.583
0.352	-1.499	-4.035	1.446	-1.524	0.426	-0.964	1.483
1.820	-2.795	3.153	0.239	-0.155	2.163	0.019	0.887
-3.441	0.764	0.304	0.556	3.860	-0.871	-0.946	-0.239
-0.423	7.477	0.803	0.117	-1.061	-0.437	-5.926	-0.291
-3.240	2.590	-1.986	0.338	-0.165	-0.049	-0.385	-1.496
-4.172	0.368	1.350	0.503	-4.620	0.226	1.817	-0.150

SECOND SET-THEORETIC σ_{yy} FOR TOP LAYER OF ANOMALY: SCALE OF 0.001 S/m

-0.433	-1.824	0.085	1.838	2.779	-0.089	-2.072	-0.668
-0.516	0.085	0.102	0.649	-1.115	0.150	-1.332	-0.548
-0.814	-0.941	0.097	-0.100	0.092	-0.041	-0.628	-0.443
-0.216	0.237	0.030	-0.056	-0.100	-0.007	-0.501	-0.340
-0.384	-0.164	-0.012	-0.097	-0.100	-0.004	-0.194	0.085
-0.070	-0.961	0.013	-0.123	-0.030	-0.002	0.033	0.861
-1.132	-0.365	-0.106	-0.312	-2.665	-0.356	0.725	-0.860
-0.692	-0.270	0.139	2.411	2.685	-0.258	-0.234	-0.529

4.3.3 Comment

In practice, we would not put receive sensors on the dense 21×21 grid suggested in the model problems. They would be placed on a much sparser grid, and then the dense data required would be generated by interpolation. This would also hold for placement of the transmit sensors when they are placed over the entire anomalous region. The idea is to use 'sparse-grid' methods to defeat the 'curse of dimensionality' as described in Chap. 9.

4.4 Modeling Microstructure Quantification Problems

4.4.1 Delaminations

We begin the study of microstructure quantification with an investigation of detecting and measuring delaminations. The model problem is shown in Fig. 4.7. Our interest is in determining the minimum height, h, of the delamination that is likely to be detectable. To that end, we perform a **VIC-3D®**-model with a grid of $16 \times 16 \times$

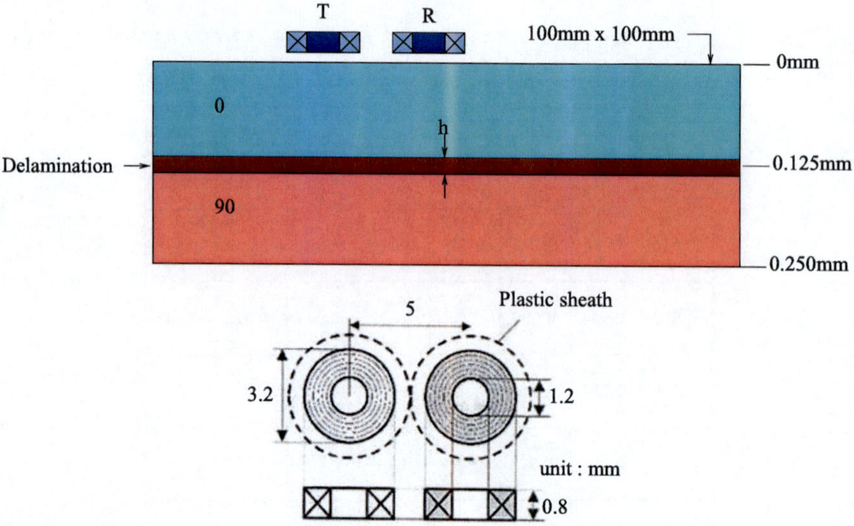

Fig. 4.7 Illustrating a model of a delamination located between two layers of a composite structure. The label, 0, indicates a ply oriented in the 0° fiber direction, and 90 indicates a cross-ply oriented 90° relative to the first. The delamination is modeled as a void whose conductivity tensor is diagonal with zero entries. The system is excited by a T/R-probe, shown in the bottom of the figure [144], at 10 MHz. The graphite-epoxy host is identical to that in Fig. 4.1, except that it is 0.250 mm thick

Table 4.2 Results of the model problem of Fig. 4.7

h, mm	R, Ohms	X, Ohms	‖Z‖, Ohms	Phase, Deg.	L, mH
0	0.043101	−0.013585	0.045191	−17.495	−2.1622E−7
0.007	0.042193	−0.012985	0.044146	−17.106	−2.0667E−7
0.014	0.041288	−0.012412	0.043113	−16.731	−1.9754E−7
0.021	0.04032	−0.011821	0.042017	−16.34	−1.8814E−7

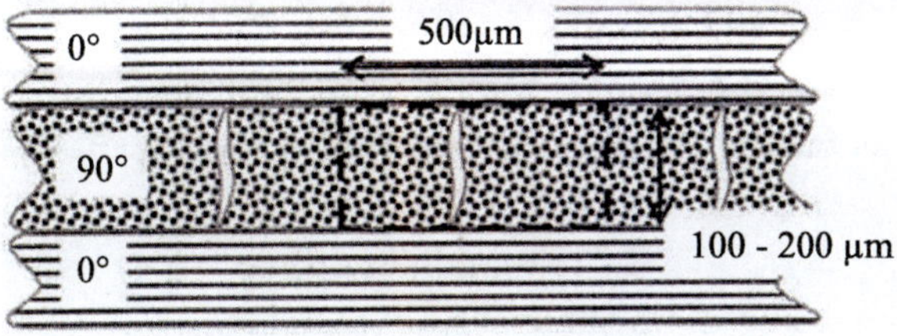

Fig. 4.8 Typical dimensions of the volume of a transverse ply in a laminated engineering composite within which stress is modified when a single transverse microcrack forms. The composite is loaded in tension along the 0° fiber direction. From [35]

256 cells and allow h to range over the values (0.0, 0.007, 0.014, 0.021) mm. The results are shown in Table 4.2. They indicate that it will be difficult to distinguish a delamination of $h = 7\,\mu$m from the background in which $h = 0\,\mu$m. There would appear to be a better chance of detecting a delamination that is 14 μm tall, and probably a pretty good chance of detecting one that is 21 μm tall, especially if we use reactance, X, as the measured data. For the latter case, the change in reactance is about 15%. For comparison, 1 mil = 25.4 μm.

4.4.2 *Transverse Ply with Microcrack*

The problem suggested in Fig. 4.8 is a major 'test-bed' for applying set-theoretic inversion to microstructure quantification. We start the process by analyzing the model of Fig. 4.9, which represents the structure of Fig. 4.8 without the voids. The host is as shown in Fig. 4.1, except that it is 0.4 mm thick. The receive-array is 21×21, as before in the FAWT studies, so that $N_s = 441$, and there are $11 \times 11 = 121$ transmit positions in the transmitter array. The excitation is at a frequency of 1 GHz.

The problem shown in Fig. 4.10 models Fig. 4.8 with a single void microcrack. This structure is embedded in an isotropic graphite-epoxy host whose conductivity is 20,000 S/m, and is 0.3 mm high. The objective of this study is to determine the feasibility of detecting and determining the width, W, of the microcrack when

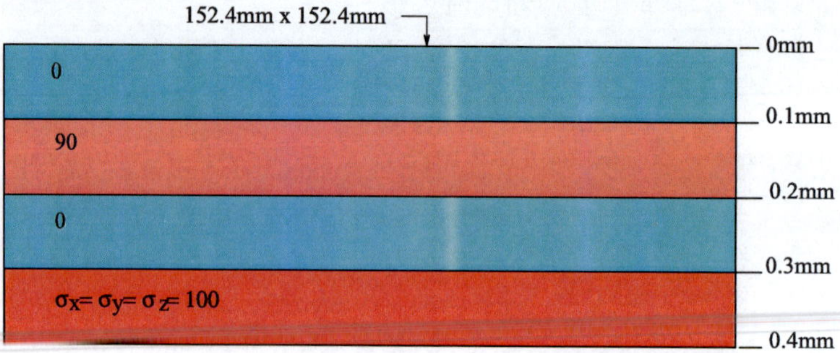

Fig. 4.9 Model for studying the laminated engineering composite structure shown in Fig. 4.8 without the voids

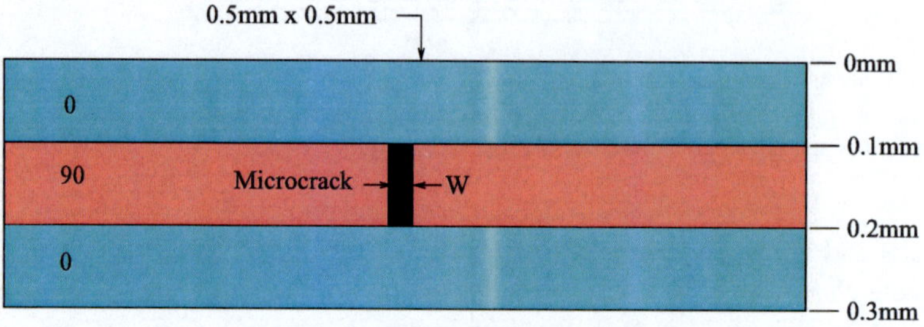

Fig. 4.10 Model for studying the laminated engineering composite structure shown in Fig. 4.8 with a single void microcrack. The width of the void is W, and the depth into the figure is 0.5 mm. This structure is embedded in an isotropic graphite-epoxy host whose conductivity is 20,000 S/m, and is 0.3 mm high

the structure is excited by the T probe shown in Fig. 4.7 as it is scanned past the microcrack from -1.6 mm to $+1.6$ mm at a frequency of 1 GHz.

To that end, we apply model-based inversion, starting with the blending functions that define the surrogate interpolation table. These are shown in Fig. 4.11 for the nodal values $W = (0.00, 0.025, 0.050)$ mm. The **VIC-3D®**-grid used to generate these functions was $N_x = 256$, $N_y = 8$, $N_z = 16$, which was sufficient to capture the variations in W. The test value of W was 0.030 mm.

The result of the inversion is shown in Table 4.3. Not only is the solution virtually identical to the test value, but all 500 initial starting points in NLSE converged to the same global minimum. The excellent quality of the inversion testifies to the fact that the blending functions of Fig. 4.11 are highly sensitive to W.

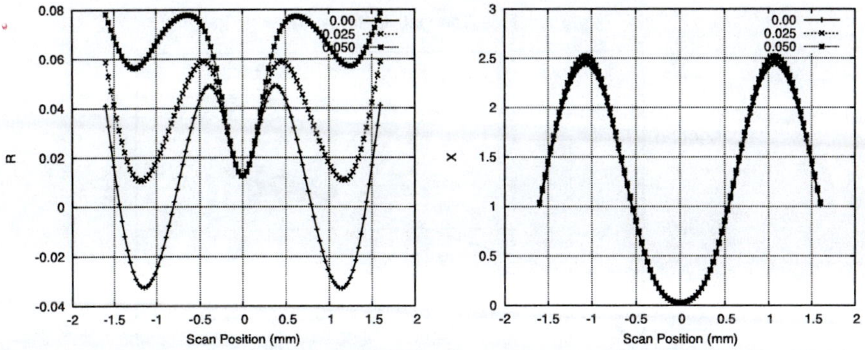

Fig. 4.11 Blending functions for the model-based inversion of the microcrack shown in Fig. 4.10. The nodal values are (0, 0.025, 0.050) mm

Table 4.3 Result of inversion to determine W in Fig. 4.10

Φ	W/sensit	No. pts.
0.322(−2)	0.0294/0.550(−3)	500

4.5 Layer-Stripping for Anisotropic Flaws

We want to extend the layer stripping algorithm used in conjuction with the set-theoretic inversion algorithm to flaws with anisotropic conductivity. In this layer stripping algorithm, we use our knowledge of the conductivity of some flaw cells to construct constraint equations to be added to the data equation

$$\sum_{KLM} E^{(i)(x)}_{KLM}[\nu]J^x_{KLM} + \sum_{KLM} E^{(i)(y)}_{KLM}[\nu]J^y_{KLM} = Z[\nu]. \tag{4.1}$$

where ν labels the 'views' produced by different incident fields, $\mathbf{E}^{(i)}[\nu]$, of the receiver coil. Solving the data equation with these constraints gives a feasible $(\mathbf{J}, \mathbf{E})$ pair that is consistent with the known conductivities used to construct the constraint equations. That is, it forces the set-theoretic algorithm to produce feasible $(\mathbf{J}, \mathbf{E})$ pairs that give the correct conductivity for these cells.

4.6 Advanced Features for Set-Theoretic Microstructure Quantification

We continue our application of the set-theoretic voxel-based inversion algorithm to the study of microstructure quantification of CFRPs, in particular to the problem shown in Figs. 4.8 and 4.9.

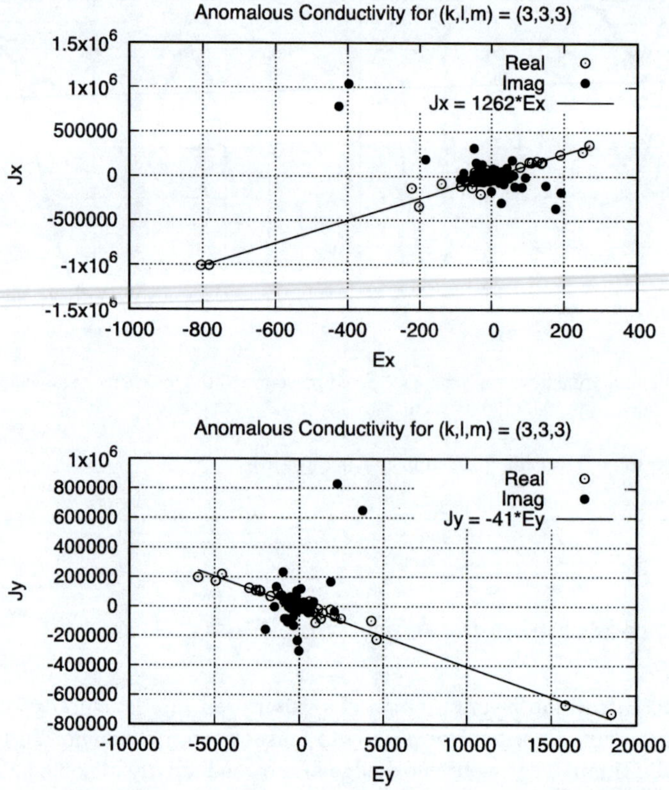

Fig. 4.12 Results for cell $(k, l, m) = (3, 3, 3)$ after pinning the bottom three layers to zero, and allowing only the top layer to be free. The lines are an eyeball fit to the real values of the 'experimental outcomes', ignoring the imaginary values. Top: σ_{xx}. Bottom: σ_{yy}

Central to the study is an understanding of the role played by constraints on the inversion process. It is clear, for example, that the model of Fig. 4.9 should produce zeros for the anomalous conductivities in the bottom layer, since that layer is host. Figures 4.12, 4.13, 4.14, and 4.15 illustrate the effect on cell (k,l,m) = (3,3,3) of 'pinning' various layers to zero, i.e., on placing constraints on these layers. Figures 4.16, 4.17, 4.18, 4.19, 4.20, 4.21, 4.22, and 4.23 illustrate the reconstruction, using the LMS(least-median-of-squares)-estimator, of the entire structure when the bottom layer is constrained to zero in both σ_{xx} and σ_{yy}.

Figures 4.24, 4.25, and 4.26 introduce a key feature of this, namely the application of 'inverse-quality metrics' to the inversion process. This metric indicates the quality of the inversion by showing the convergence to the minimum. In particular, we see the advantage in establishing constraints where they are appropriate. This

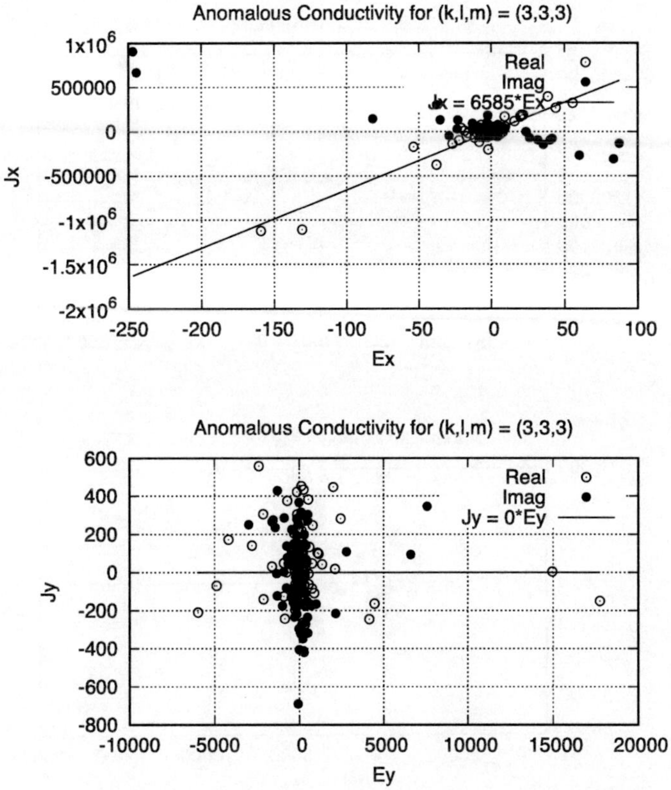

Fig. 4.13 Results for cell $(k, l, m) = (3, 3, 3)$ after pinning the bottom two layers to zero, and allowing only the top two layers to be free. The lines are an eyeball fit to the real values of the 'experimental outcomes', ignoring the imaginary values. Top: σ_{xx}. Bottom: σ_{yy}

reduces the number of unknowns, and stabilizes the inversion procedure. These figures are typical of the convergence of a robust estimator, such as LMS, and serve the same purpose as exploring configuration space for the minimum of a least-squares problem. Typically, the sharper and deeper the minimum, the better the solution.

So that raises the question: 'How can we tell where the constraints lie?' This is a problem of classification theory, in which we want to assign regions of space to the host material or to the anomalous region. The host, of course, carries zero anomalous current, which results in a null anomalous conductivity. The result of classifying the problem space is the creation of a 'zero-cutoff threshold.' Solutions that are smaller than the threshold are assumed to be host material, and are 'pinned' to zero. We have developed a heuristic iterative scheme to classify the solution,

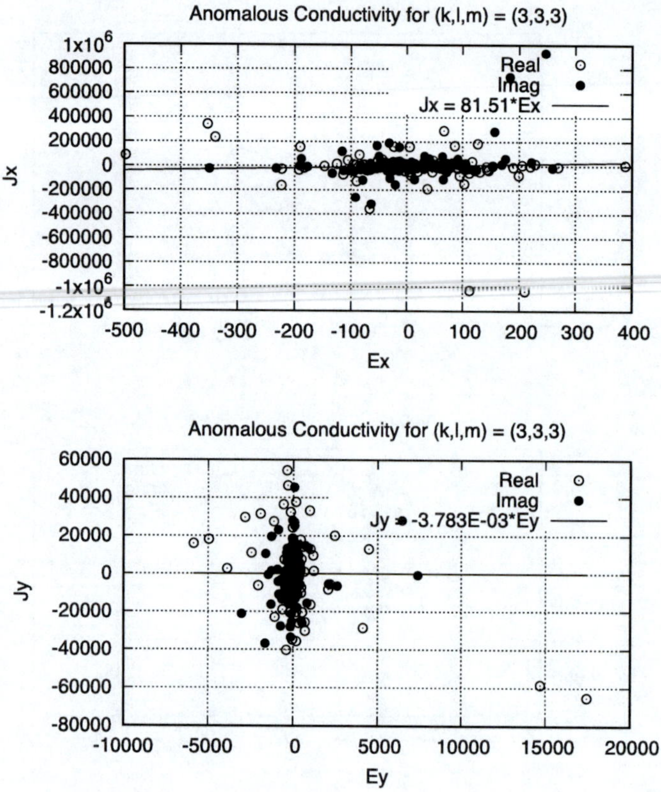

Fig. 4.14 Results for cell $(k, l, m) = (3, 3, 3)$ after removing all pins and allowing all currents to be nonzero. The lines are the set-theoretic estimates, in which all 'experimental outcomes' are used. Top: σ_{xx}. Bottom: σ_{yy}

and it is explained in the discussion before Fig. 4.27. Figures 4.27, 4.28, 4.29, 4.30, 4.31, and 4.32 illustrate the result of applying the heuristic scheme, starting with the zeroth iteration and ending with the fourth. The results are quite good, and give us confidence in the method. Nevertheless, we believe that it can be sharpened, and no longer be 'heuristic', by applying formal statistical decision theory.

The development of the layer-stripping algorithm that was described in Sect. 4.5, together with the inverse-quality metric and the heuristic classifier are three extremely important results of our research, as they strengthen the set-theoretic, voxel-based inversion algorithm.

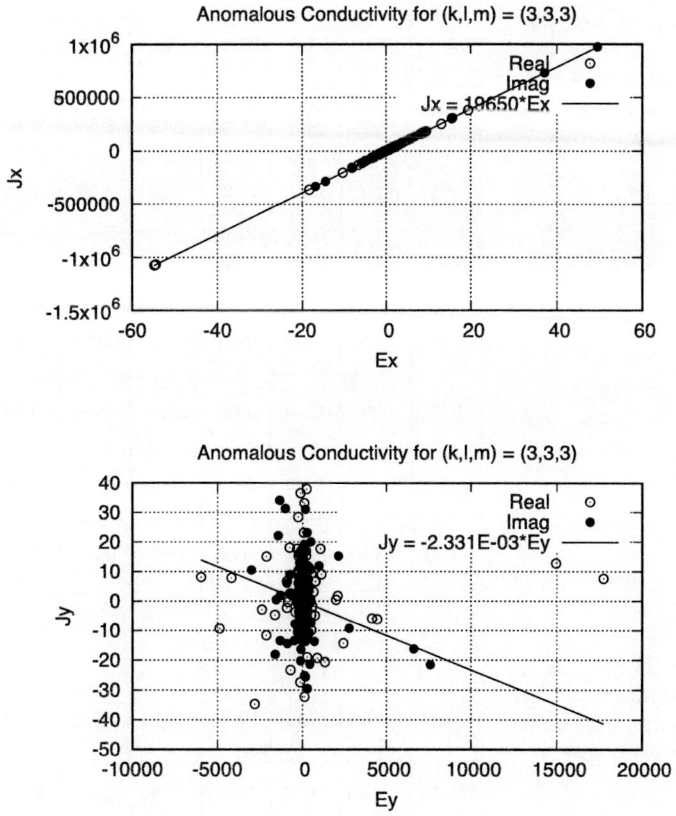

Fig. 4.15 Results for cell $(k, l, m) = (3, 3, 3)$ after pinning only the bottom layer and allowing the top three layers to be free. The lines are the set-theoretic estimates, in which all 'experimental outcomes' are used. Top: σ_{xx}. Bottom: σ_{yy}

Fig. 4.16 Results for σ_{xx} in the bottom layer. The zeros are forced as a constraint because the bottom layer is host material

0.00	0.00	0.00	0.00	0.00	0.00	0.00	0.00
0.00	0.00	0.00	0.00	0.00	0.00	0.00	0.00
0.00	0.00	0.00	0.00	0.00	0.00	0.00	0.00
0.00	0.00	0.00	0.00	0.00	0.00	0.00	0.00
0.00	0.00	0.00	0.00	0.00	0.00	0.00	0.00
0.00	0.00	0.00	0.00	0.00	0.00	0.00	0.00
0.00	0.00	0.00	0.00	0.00	0.00	0.00	0.00
0.00	0.00	0.00	0.00	0.00	0.00	0.00	0.00

Fig. 4.17 Results for σ_{xx} in the first layer, obtained using the LMS-estimator:scale of 1000 S/m

19.63	19.85	19.86	19.56	19.56	19.52	20.24	14.10
19.46	19.86	19.72	19.79	19.99	20.05	19.83	19.10
19.79	19.07	20.77	19.72	19.35	20.01	19.70	19.95
19.87	19.84	19.11	19.23	19.92	20.04	19.97	20.25
19.95	19.83	19.85	19.43	20.11	20.19	20.07	19.61
19.91	20.09	19.13	19.79	19.31	19.95	18.94	19.94
20.27	19.22	19.85	19.54	19.92	19.23	19.91	19.95
19.73	19.74	19.81	19.84	19.09	19.76	20.00	17.37

Fig. 4.18 Results for σ_{xx} in the second layer, obtained using the LMS-estimator:scale of 0.001 S/m

-71.67	378.9	-91.44	124.9	144.7	79.01	-82.49	-191.6
59.34	-86.55	-250.5	171.0	44.64	-43.43	-102.2	-711.4
26.80	411.2	36.45	531.8	-322.7	-199.6	-500.7	214.0
-131.8	-43.14	238.5	123.4	294.2	162.0	51.63	89.20
133.6	49.88	-72.30	328.5	-856.8	-153.8	6.767	183.7
150.2	47.33	144.7	62.56	4.828	183.3	1677.0	77.70
-70.02	-42.38	-63.05	-368.8	-49.5	340.4	-54.87	-108.0
94.2	400.4	-131.8	-80.11	-29.33	-137.0	-78.11	-13.48

Fig. 4.19 Results for σ_{xx} in the third layer, obtained using the LMS-estimator:scale of 1000 S/m

19.71	19.61	19.63	19.70	19.58	19.62	19.73	19.98
19.60	19.64	19.69	19.58	19.66	19.68	19.65	19.64
19.63	19.57	19.64	19.68	19.74	19.62	19.68	19.65
19.68	19.55	19.59	19.58	19.65	19.54	19.60	19.67
19.67	19.63	19.66	19.65	19.68	19.66	19.66	19.66
19.65	19.54	19.61	19.67	19.43	19.54	19.62	19.68
19.63	19.62	19.61	19.62	19.63	19.64	19.62	19.63
19.7	19.64	19.68	19.69	19.68	19.70	19.63	19.56

Fig. 4.20 Results for σ_{yy} in the bottom layer. The zeros are forced as a constraint because the bottom layer is host material

0.00	0.00	0.00	0.00	0.00	0.00	0.00	0.00
0.00	0.00	0.00	0.00	0.00	0.00	0.00	0.00
0.00	0.00	0.00	0.00	0.00	0.00	0.00	0.00
0.00	0.00	0.00	0.00	0.00	0.00	0.00	0.00
0.00	0.00	0.00	0.00	0.00	0.00	0.00	0.00
0.00	0.00	0.00	0.00	0.00	0.00	0.00	0.00
0.00	0.00	0.00	0.00	0.00	0.00	0.00	0.00
0.00	0.00	0.00	0.00	0.00	0.00	0.00	0.00

Fig. 4.21 Results for σ_{yy} in the first layer, obtained using the LMS-estimator:scale of 0.001 S/m

175.0	67.42	−122.8	−101.4	184.2	−921.1	4378.0	−17.18
−538.3	67.78	600.9	−506.0	−2071.	2376.0	292.3	266.5
−318.0	18.58	−27.35	1271.0	1033.0	−129.4	−125.3	631.4
168.3	−598.2	−305.2	−1590.	2237.0	−742.3	−243.1	169.9
584.4	−711.0	1087.0	293.5	−18.72	−2929.	−678.9	437.4
−141.6	586.4	1052.0	−921.0	2506.0	1034.0	−6.045	−212.3
42.95	−510.6	1637.0	419.9	753.6	−707.7	128.7	−1114.
−435.6	182.3	−547.0	−472.1	−3118.	105.6	297.7	430.5

Fig. 4.22 Results for σ_{yy} in the second layer, obtained using the LMS-estimator:scale of 1000 S/m

19.91	19.90	19.89	19.89	19.90	19.91	19.91	19.89
19.91	19.91	19.91	19.91	19.90	19.92	19.92	19.88
19.91	19.91	19.91	19.98	19.94	19.93	19.91	19.91
19.81	19.86	19.90	19.87	19.91	19.90	19.83	19.91
19.92	19.89	19.90	19.92	19.90	19.92	19.90	19.96
19.92	19.93	19.91	19.88	19.90	19.94	19.91	19.93
19.90	19.89	19.90	19.92	19.90	19.90	19.90	19.90
19.86	19.92	19.80	19.90	19.87	19.90	19.90	19.92

Fig. 4.23 Results for σ_{yy} in the third layer, obtained using the LMS-estimator:scale of 0.001 S/m

0.3218	1.739	0.3090	0.7071	0.7306	−0.91	1.454	0.7928
1.474	−1.856	−1.691	12.57	0.5385	−0.975	1.723	2.027
−1.354	−0.994	1.433	0.7844	2.361	−2.831	0.0248	6.357
−0.254	0.7373	1.149	13.27	1.036	0.2012	−0.526	−2.090
−0.634	4.818	0.8086	−2.331	4.495	0.9460	−0.332	−3.042
−0.574	0.720	−0.305	−2.447	2.548	−1.427	−0.377	−2.816
−2.952	1.106	−1.051	3.656	1.087	−0.827	−0.292	−1.172
−3.139	−0.166	−0.382	−2.141	−0.007	−1.408	1.047	−0.167

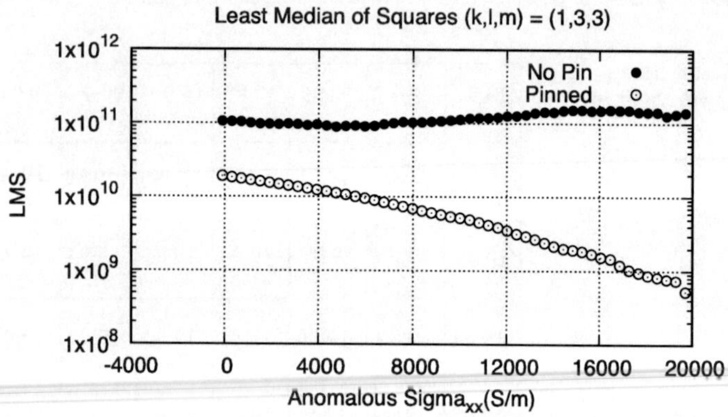

Fig. 4.24 Comparing the inverse-quality metric for the LMS-estimator when applied to cell (1,3,3). The curve labeled 'No Pin' is for the condition in which the currents in the zeroth layer are not forced to zero, and the other curve is for the condition in which the currents are forced to zero

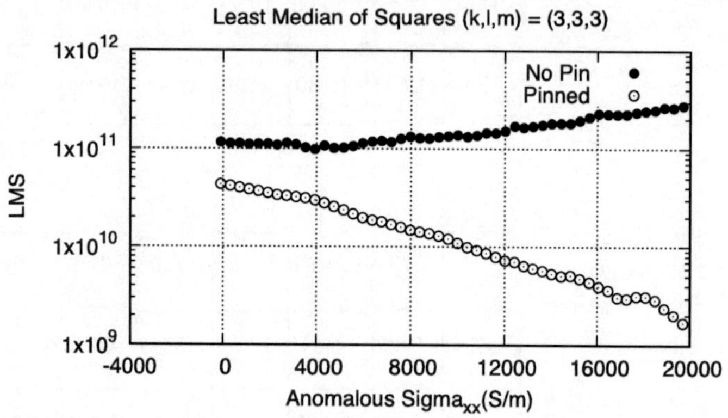

Fig. 4.25 Comparing the inverse-quality metric for the LMS-estimator when applied to cell (3,3,3). The curve labeled 'No Pin' is for the condition in which the currents in the zeroth layer are not forced to zero, and the other curve is for the condition in which the currents are forced to zero

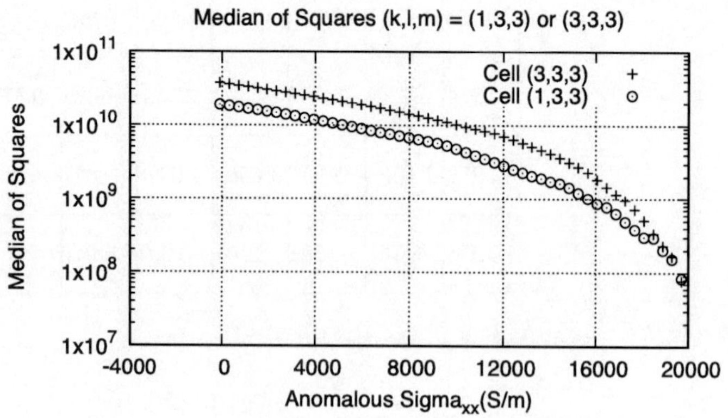

Fig. 4.26 The inverse-quality metric for the LMS-estimator when applied to cells (3,3,3) and (1,3,3) under the conditions that we pin all x currents in layers 0 and 2 and all y currents in layers 0, 1 and 3 to zero

−0.123	−0.201	−0.083	−0.139	0.0494	−0.063	−0.013	0.1813	20.51	22.16	25.30	21.33	19.95	19.86	22.60	21.53
0.0056	0.0637	−0.043	−0.158	−0.034	−0.118	−0.163	−0.141	22.59	20.43	14.18	24.96	14.01	20.49	25.69	24.20
−0.090	−0.214	−0.184	−0.689	−0.516	−0.482	−0.131	−0.055	21.40	17.39	8.053	4.977	5.288	2.619	23.85	23.64
−0.197	−0.338	−0.140	0.4589	−0.069	−0.333	−0.537	−0.352	23.29	25.49	25.45	8.144	5.896	4.179	19.82	24.37
0.0408	0.0015	−0.329	−0.075	0.0345	0.0478	−0.309	−0.092	20.52	19.61	5.158	4.357	5.027	4.963	20.09	15.40
−0.139	−0.258	−0.293	−0.167	−0.151	−0.492	−0.186	−0.088	23.95	25.98	28.69	6.695	26.14	6.873	25.12	21.51
−0.142	−0.155	−0.221	−0.122	−0.019	−0.144	0.013	−0.116	24.53	23.41	28.93	26.24	29.59	26.89	19.93	23.64
0.089	−0.073	−0.024	−0.107	−0.01	−0.250	−0.148	0.039	22.30	23.48	21.79	23.86	24.88	21.17	24.64	23.96
0.4681	0.3842	0.3720	0.1792	−0.103	−0.066	0.2046	−0.219	20.07	19.80	19.42	19.91	19.96	20.03	19.71	19.90
0.0542	−0.318	0.3366	0.6483	0.4430	0.6739	0.4089	0.5058	19.45	19.61	19.87	19.91	19.98	20.00	19.58	19.75
0.3494	0.2488	0.1930	1.015	1.145	1.396	0.5795	0.5740	20.08	20.30	18.95	19.28	19.54	19.45	18.77	20.04
0.7296	0.6614	0.7888	−0.196	0.7157	1.270	1.036	0.8415	19.75	19.75	19.96	19.76	19.22	19.98	20.46	20.14
0.0168	−0.055	2.063	2.112	0.7167	1.452	0.4696	0.9739	19.51	19.91	19.59	19.42	20.03	19.79	20.22	19.45
0.4847	1.132	0.7018	0.9622	0.5395	0.8357	0.5964	0.6732	20.16	20.00	19.33	19.69	19.32	19.39	19.73	20.39
0.2018	0.6856	0.6293	0.4301	0.3059	0.6592	0.1015	0.1763	19.83	19.89	19.56	19.56	19.50	19.49	19.91	20.16
0.2133	0.5076	0.0770	0.1300	0.1795	0.2290	0.5569	0.1343	19.95	19.89	19.88	19.63	19.73	20.04	19.99	19.89

Fig. 4.27 Results for σ_{xx} in the bottom (top left) to third layer (bottom right) after the zeroth iteration, obtained using the LMS-estimator at 10 GHz. The signed-zero, +0.000, indicates a number that is very small but is above the zero-constraint threshold. The scale of these results is 1000 S/m

0.019	−0.047	−0.037	−0.009	−0.085	−0.022	−0.023	0.016	−0.042	0.1501	−0.03	0.0404	0.1237	0.0754	0.0504	−0.04
−0.065	−0.034	−0.071	−0.069	−0.091	−0.055	−0.005	−0.022	0.2473	0.1437	0.2877	0.3552	0.0883	0.0903	0.0657	0.0713
−0.044	−0.080	−0.001	−0.051	−0.043	−0.096	−0.067	−0.011	0.1563	0.3673	−0.125	0.1672	0.5166	0.2958	0.2000	0.0152
−0.011	−0.030	−0.059	−0.025	−0.008	−0.077	−0.025	0.000	0.0445	0.1623	0.4734	0.2547	0.0768	0.2760	0.1638	−0.029
0.000	−0.075	−0.091	0.01	0.068	−0.098	−0.034	−0.005	0.0219	0.4151	0.3945	0.0002	0.1009	0.4074	0.1107	0.000
0.0399	0.129	−0.160	−0.083	−0.005	−0.112	−0.087	−0.012	0.1287	0.5344	0.5760	0.4123	−0.148	0.4688	0.3584	0.0542
−0.034	−0.042	−0.060	−0.057	−0.074	−0.059	−0.029	−0.022	0.1491	0.1460	0.3060	0.1474	0.3060	0.1756	0.1620	0.1097
−0.004	−0.035	−0.000	−0.017	−0.049	−0.058	−0.02	0.049	0.0414	0.1713	0.0886	0.1892	0.1434	0.0575	0.1400	−0.025
18.69	16.42	18.31	16.47	16.99	15.97	16.57	17.82	−0.014	−0.021	0.002	−0.002	0.004	−0.003	−0.026	−0.004
19.14	16.88	18.42	16.73	15.62	17.02	17.54	16.87	−0.012	−0.043	0.023	−0.020	−0.012	0.041	−0.023	−0.021
18.55	16.54	13.95	11.89	11.65	14.35	14.92	19.37	−0.034	−0.047	−0.014	−0.100	−0.069	−0.058	−0.017	0.033
18.96	15.27	14.66	12.39	9.460	16.31	15.79	18.36	−0.008	−0.080	−0.002	−0.077	−0.125	−0.009	−0.014	−0.002
18.05	13.59	13.64	13.20	11.98	11.41	12.14	18.35	0.000	−0.093	−0.060	−0.074	−0.011	−0.077	−0.002	0.018
18.74	15.26	11.96	11.04	11.76	16.77	13.73	19.02	0.027	−0.078	−0.094	−0.081	0.018	−0.067	−0.068	0.000
17.91	16.97	18.15	17.20	17.36	17.42	16.24	17.45	−0.036	−0.041	−0.056	−0.050	−0.025	−0.019	−0.036	−0.013
17.66	13.27	13.82	15.28	16.97	16.07	16.38	16.76	−0.015	−0.049	−0.052	−0.031	−0.045	−0.055	−0.060	−0.035

Fig. 4.28 Results for σ_{yy} in the bottom (top left) to third layer (bottom right) after the zeroth iteration, obtained using the LMS-estimator at 10 GHz. The signed-zeros, ± 0.000, indicate numbers that are very small but are above the zero-constraint threshold. The scale of these results is 1000 S/m

0.000	0.000	0.000	0.000	0.000	0.000	0.000	0.000	19.91	19.91	19.92	19.91	19.87	19.91	19.91	19.92
−0.001	0.000	0.000	0.000	0.000	0.000	0.000	0.000	19.93	19.90	19.93	19.87	19.88	19.89	19.89	19.90
0.000	0.000	0.000	0.000	0.000	0.000	0.000	0.000	19.91	19.91	19.90	19.89	19.90	19.93	19.88	19.92
0.000	0.000	0.000	0.000	0.000	0.000	0.000	0.000	19.89	19.89	19.92	19.89	19.89	19.93	19.87	19.91
−0.001	−0.001	0.000	0.000	0.003	0.002	0.000	0.000	19.94	19.92	19.92	19.88	19.87	19.91	19.88	19.92
0.000	0.000	0.000	0.000	0.000	0.000	0.000	0.000	19.91	19.89	19.91	19.91	19.89	19.87	19.89	19.91
0.000	0.000	0.000	0.000	0.000	0.000	0.000	0.000	19.89	19.88	19.89	19.88	19.88	19.91	19.90	19.91
0.000	0.000	0.000	0.000	0.000	0.000	0.000	0.000	19.93	19.90	19.90	19.87	19.89	19.91	19.90	19.83
−0.004	−0.001	0.000	0.0006	0.000	0.000	0.000	0.000	19.91	19.89	19.90	19.90	19.90	19.90	19.90	19.90
0.000	0.000	−0.004	0.000	0.0023	0.0008	0.000	0.0007	19.90	19.89	19.90	19.90	19.90	19.91	19.90	19.90
0.0010	−0.003	0.0018	0.000	0.000	−0.002	0.0005	0.000	19.90	19.90	19.90	19.90	19.90	19.90	19.91	19.90
0.0011	0.0019	0.0011	0.000	−0.003	−0.001	0.0052	0.0052	19.91	19.90	19.91	19.90	19.89	19.90	19.91	19.91
0.000	0.000	0.0041	0.0044	−0.008	0.0022	0.0017	0.000	19.90	19.90	19.90	19.90	19.90	19.89	19.90	19.90
0.0045	0.0038	−0.002	−0.006	0.0031	0.0049	0.0025	0.000	19.91	19.90	19.91	19.89	19.90	19.92	19.91	19.90
0.0035	0.0016	0.0031	0.0051	0.0015	0.000	−0.001	−0.002	19.91	19.91	19.90	19.90	19.91	19.89	19.90	19.90
0.0006	0.0011	0.0004	0.0017	−0.005	0.000	0.0009	0.000	19.90	19.90	19.90	19.90	19.90	19.90	19.90	19.89

Fig. 4.29 Results for σ_{xx} in the bottom (top left) to third layer (bottom right) after the first iteration, obtained using the LMS-estimator at 10 GHz. The signed-zero, +0.000, indicates a number that is very small but is above the zero-constraint threshold. The scale of these results is 1000 S/m

0.000	0.000	0.000	0.000	0.000	0.000	0.000	0.000	0.000	−0.000	0.000	−0.000	−0.000	−0.000	+0.000	0.000
0.000	0.000	0.000	0.000	0.000	0.000	0.000	0.000	−0.000	+0.000	0.000	+0.000	−0.000	+0.000	+0.000	+0.000
0.000	0.000	0.000	0.000	0.000	0.000	0.000	0.000	−0.000	−0.000	0.000	−0.000	+0.000	−0.000	+0.000	+0.000
0.000	0.000	0.000	0.000	0.000	0.000	0.000	0.000	−0.000	+0.000	−0.000	+0.000	+0.000	+0.000	−0.000	0.000
0.000	0.000	0.000	0.000	0.000	0.000	0.000	0.000	−0.000	−0.000	+0.000	−0.000	+0.000	+0.000	−0.000	0.000
0.000	0.000	0.000	0.000	0.000	0.000	0.000	0.000	−0.000	−0.000	+0.000	+0.000	0.000	+0.000	+0.000	−0.000
0.000	0.000	0.000	0.000	0.000	0.000	0.000	0.000	−0.000	+0.000	−0.000	−0.000	+0.000	−0.000	−0.000	−0.000
0.000	0.000	0.000	0.000	0.000	0.000	0.000	0.000	−0.000	−0.000	+0.000	−0.000	+0.000	−0.000	−0.000	0.000

19.90	19.90	19.90	19.90	19.90	19.90	19.90	19.90	0.000	0.000	−0.000	0.000	0.000	0.000	0.000	0.000
19.91	19.90	19.91	19.91	19.90	19.91	19.90	19.91	0.000	0.000	+0.000	0.000	0.000	0.000	0.000	0.000
19.91	19.91	19.90	19.91	19.90	19.90	19.91	19.91	0.000	0.000	0.000	0.000	0.000	0.000	0.000	0.000
19.91	19.90	19.90	19.90	19.89	19.90	19.90	19.91	0.000	0.000	0.000	0.000	0.000	0.000	0.000	0.000
19.91	19.90	19.90	19.89	19.90	19.91	19.90	19.91	0.000	0.000	0.000	0.000	0.000	0.000	0.000	0.000
19.91	19.91	19.90	19.90	19.90	19.90	19.91	19.91	0.000	0.000	0.000	0.000	0.000	0.000	0.000	0.000
19.91	19.90	19.91	19.91	19.91	19.91	19.90	19.91	0.000	0.000	0.000	0.000	0.000	0.000	0.000	0.000
19.90	19.90	19.90	19.90	19.91	19.90	19.90	19.90	0.000	0.000	0.000	0.000	0.000	0.000	0.000	0.000

Fig. 4.30 Results for σ_{yy} in the bottom (top left) to third layer (bottom right) after the first iteration, obtained using the LMS-estimator at 10 GHz. The signed-zeros, ±0.000, indicate numbers that are very small but are above the zero-constraint threshold. The scale of these results is 1000 S/m

0.000	0.000	0.000	0.000	0.000	0.000	0.000	0.000
0.000	0.000	0.000	0.000	0.000	0.000	0.000	0.000
0.000	0.000	0.000	0.000	0.000	0.000	0.000	0.000
0.000	0.000	0.000	0.000	0.000	0.000	0.000	0.000
0.000	0.000	0.000	0.000	0.000	0.000	0.000	0.000
0.000	0.000	0.000	0.000	0.000	0.000	0.000	0.000
0.000	0.000	0.000	0.000	0.000	0.000	0.000	0.000
0.000	0.000	0.000	0.000	0.000	0.000	0.000	0.000

19.92	19.91	19.91	19.89	19.87	19.91	19.91	19.92
19.90	19.90	19.92	19.89	19.88	19.90	19.89	19.90
19.91	19.90	19.93	19.89	19.89	19.93	19.89	19.91
19.89	19.90	19.93	19.89	19.88	19.91	19.89	19.89
19.91	19.90	19.92	19.89	19.88	19.91	19.90	19.91
19.90	19.89	19.89	19.88	19.89	19.90	19.90	19.91
19.90	19.90	19.93	19.88	19.89	19.92	19.90	19.90
19.92	19.91	19.93	19.89	19.87	19.91	19.91	19.92

0.000	0.000	0.000	0.000	0.000	0.000	0.000	0.000
0.000	0.000	0.000	0.000	+0.000	0.001	0.000	0.000
0.000	0.000	0.000	0.000	0.000	0.000	0.000	0.000
0.000	0.000	0.000	0.000	0.000	0.000	0.000	0.000
0.000	0.000	0.000	0.000	0.000	0.000	0.000	0.000
0.003	0.002	0.000	0.000	0.000	0.000	0.000	0.000
0.000	0.000	0.000	0.000	0.000	0.000	0.000	0.000
0.000	0.000	0.000	0.000	0.000	0.000	0.000	0.000

19.90	19.90	19.90	19.90	19.90	19.90	19.90	19.90
19.90	19.90	19.90	19.90	19.90	19.90	19.90	19.90
19.90	19.90	19.90	19.90	19.90	19.90	19.90	19.91
19.90	19.90	19.91	19.90	19.90	19.91	19.90	19.90
19.91	19.90	19.90	19.90	19.90	19.91	19.90	19.91
19.91	19.90	19.90	19.90	19.90	19.90	19.90	19.90
19.90	19.90	19.90	19.90	19.90	19.90	19.90	19.90
19.90	19.90	19.90	19.90	19.90	19.90	19.90	19.90

Fig. 4.31 Results for σ_{xx} in the bottom (top left) to third layer (bottom right) after the fourth iteration, obtained using the LMS-estimator at 10 GHz. The signed-zero, +0.000, indicates a number that is very small but is above the zero-constraint threshold. The scale of these results is 1000 S/m

0.000	0.000	0.000	0.000	0.000	0.000	0.000	0.000	0.000	0.000	0.000	0.000	0.000	0.000	+0.000	0.000
0.000	0.000	0.000	0.000	0.000	0.000	0.000	0.000	0.000	0.000	0.000	0.000	0.000	0.000	+0.000	0.000
0.000	0.000	0.000	0.000	0.000	0.000	0.000	0.000	0.000	0.000	0.000	0.000	0.000	0.000	0.000	0.000
0.000	0.000	0.000	0.000	0.000	0.000	0.000	0.000	0.000	0.000	0.000	0.000	0.000	0.000	0.000	0.000
0.000	0.000	0.000	0.000	0.000	0.000	0.000	0.000	0.000	0.000	0.000	0.000	0.000	0.000	0.000	0.000
0.000	0.000	0.000	0.000	0.000	0.000	0.000	0.000	0.000	0.000	0.000	0.000	0.000	0.000	0.000	0.000
0.000	0.000	0.000	0.000	0.000	0.000	0.000	0.000	0.000	0.000	0.000	0.000	+0.000	0.000	0.000	0.000
0.000	0.000	0.000	0.000	0.000	0.000	0.000	0.000	0.000	0.000	0.000	0.000	+0.000	0.000	0.000	0.000
19.90	19.90	19.90	19.90	19.90	19.90	19.90	19.90	0.000	0.000	0.000	0.000	0.000	0.000	0.000	0.000
19.91	19.90	19.91	19.91	19.90	19.90	19.90	19.91	0.000	0.000	0.000	0.000	0.000	0.000	0.000	0.000
19.91	19.91	19.90	19.90	19.90	19.90	19.91	19.91	0.000	0.000	0.000	0.000	0.000	0.000	0.000	0.000
19.91	19.90	19.90	19.89	19.89	19.90	19.90	19.91	0.000	0.000	0.000	0.000	0.000	0.000	0.000	0.000
19.91	19.90	19.90	19.90	19.90	19.90	19.90	19.91	0.000	0.000	0.000	0.000	0.000	0.000	0.000	0.000
19.91	19.91	19.90	19.90	19.90	19.90	19.91	19.91	0.000	0.000	0.000	0.000	0.000	0.000	0.000	0.000
19.91	19.90	19.91	19.91	19.91	19.91	19.90	19.91	0.000	0.000	0.000	0.000	0.000	0.000	0.000	0.000
19.90	19.90	19.90	19.90	19.91	19.90	19.90	19.90	0.000	0.000	0.000	0.000	0.000	0.000	0.000	0.000

Fig. 4.32 Results for σ_{yy} in the bottom (top left) to third layer (bottom right) after the fourth iteration, obtained using the LMS-estimator at 10 GHz. The signed-zero, +0.000, indicates a number that is very small but is above the zero-constraint threshold. The scale of these results is 1000 S/m

4.6.1 *A Heuristic Iterative Scheme to Determine a Zero-Cutoff Threshold*

```
        Our problems are largely due to errant
small values of J and E, so what follows is an approach to resolve
this problem,  all
using the LMS estimator on the 10GHz data.

1) Run the calculation throwing out all (E,J) pairs for which
E is less than 20\% of the maximum size for each cell and
each experiment. We'll call the outputs 'sigmaxx_0.txt'
and 'sigmayy_0.txt'.

2) Rerun the calculation throwing out (E,J) pairs with E less
than the 20\% threshold, and also pinning to zero all currents
associated with conductivity values in 'sigmaxx_0.txt' and
'sigmayy_0.txt' that are less than or equal to zero. We'll
call the outputs 'sigmaxx_1.txt' and 'sigmayy_1.txt'

3) Repeat 2) substituting 'sigmaxx_1.txt' for 'sigmaxx_0.txt'
   and 'sigmayy_1.txt' for 'sigmayy_0.txt' to obtain
   'sigmaxx_2.txt' and 'sigmayy_2.txt'

4) Continue in this manner until there are no negative
   conductivity values in the output.

This converged after 4 iterations, the outputs for which are
attached. You can see that the first step (zeroth iteration)
gets us into the ball park (it's much better than with no
threshold on the size of E). The results of the first iteration
are pretty good. The fourth iteration is very good.

This, of course, assumes that we know that there are no negative
anomalous conductivities, which will not always be the case,
so this is not a general algorithm. It also assumes that we
can throw out the small (E,J) pairs at a 20\% threshold, which
may also not be the case in general, so again it is not general
algorithm. But it may be that for any given problem a threshold
for (E,J) size and conductivity value can be found that will work.
If our threshold on conductivity is nonzero, we cannot pin the
currents to zero, but will have to rely on a constraint equation,
which is in development.
```

4.7 Progress in Modeling Microstructure Quantification

The model shown in Fig. 4.33 is our 'test-bed' for microstructure quantification. The model consists of this three-layer structure (with just one crack) embedded in an isotropic host whose conductivity is 100 S/m (Fig. 4.34).

Figures 4.35, 4.36, 4.37, and 4.38 show the results of inverting the impedances obtained for a transverse ply sample with 4 layers ($m = 0, 1, 2, 3$) and a crack with zero conductivity running through the center of the upper middle layer ($m = 2$) that is 2 cells wide and 8 cells long. The x and y components of the anomalous conductivity were fixed to zero for the bottom layer ($m = 0$), which is host material. The only cells of the anomalous region that were not perfectly reconstructed are some of the crack cells, which should all have anomalous conductivity of −100 S/m since the host conductivity is 100 S/m. Figures 4.39, 4.40, 4.41, and 4.42 show plots of **J** versus **E** for a cell near the middle of the top ($klm = 333$) and the cracked ($klm = 332$) layers.

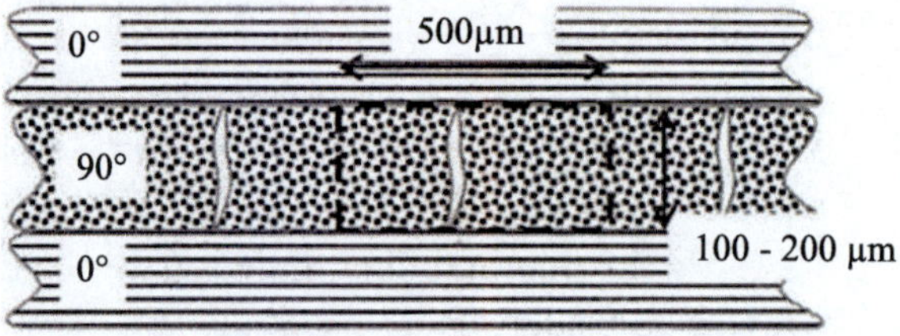

Fig. 4.33 Typical dimensions of the volume of a transverse ply in a laminated engineering composite within which stress is modified when a single transverse microcrack forms. The composite is loaded in tension along the 0° fiber direction. From [35]

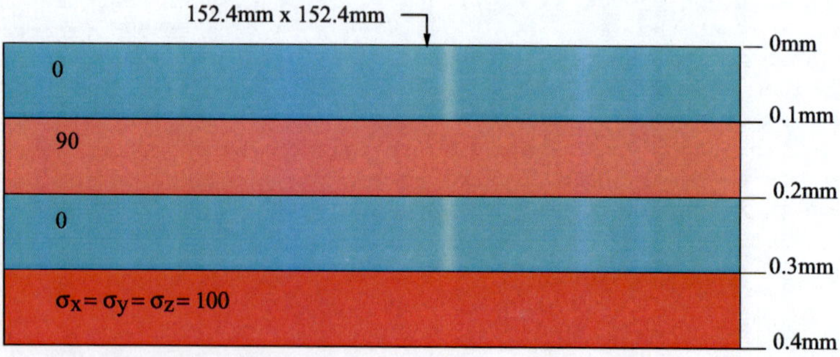

Fig. 4.34 Model for studying the laminated engineering composite structure shown in Fig. 4.33 without the voids

Fig. 4.35 Anomalous σ_{xx} for $m = 1$ or $m = 3$

19900	19900	19900	19900	19900	19900	19900	19900
19900	19900	19900	19900	19900	19900	19900	19900
19900	19900	19900	19900	19900	19900	19900	19900
19900	19900	19900	19900	19900	19900	19900	19900
19900	19900	19900	19900	19900	19900	19900	19900
19900	19900	19900	19900	19900	19900	19900	19900
19900	19900	19900	19900	19900	19900	19900	19900
19900	19900	19900	19900	19900	19900	19900	19900

Fig. 4.36 Anomalous σ_{xx} for $m = 2$

0.000	0.000	0.000	-92.90	-92.90	0.000	0.000	0.000
0.000	0.000	0.000	-100.5	-100.5	0.000	0.000	0.000
0.000	0.000	0.000	-100.5	-100.5	0.000	0.000	0.000
0.000	0.000	0.000	-85.25	-85.25	0.000	0.000	0.000
0.000	0.000	0.000	-85.25	-85.25	0.000	0.000	0.000
0.000	0.000	0.000	-100.5	-100.5	0.000	0.000	0.000
0.000	0.000	0.000	-100.5	-100.5	0.000	0.000	0.000
0.000	0.000	0.000	-92.90	-92.90	0.000	0.000	0.000

Fig. 4.37 Anomalous σ_{yy} for $m = 1$ or $m = 3$

0.000	0.000	0.000	0.000	0.000	0.000	0.000	0.000
0.000	0.000	0.000	0.000	0.000	0.000	0.000	0.000
0.000	0.000	0.000	0.000	0.000	0.000	0.000	0.000
0.000	0.000	0.000	0.000	0.000	0.000	0.000	0.000
0.000	0.000	0.000	0.000	0.000	0.000	0.000	0.000
0.000	0.000	0.000	0.000	0.000	0.000	0.000	0.000
0.000	0.000	0.000	0.000	0.000	0.000	0.000	0.000
0.000	0.000	0.000	0.000	0.000	0.000	0.000	0.000

Fig. 4.38 Anomalous σ_{yy} for $m = 2$

19900	19900	19900	-100.0	-100.0	19900	19900	19900
19900	19900	19900	-100.0	-100.0	19900	19900	19900
19900	19900	19900	-100.0	-100.0	19900	19900	19900
19900	19900	19900	-100.0	-100.0	19900	19900	19900
19900	19900	19900	-100.0	-100.0	19900	19900	19900
19900	19900	19900	-100.0	-100.0	19900	19900	19900
19900	19900	19900	-100.0	-100.0	19900	19900	19900
19900	19900	19900	-100.0	-100.0	19900	19900	19900

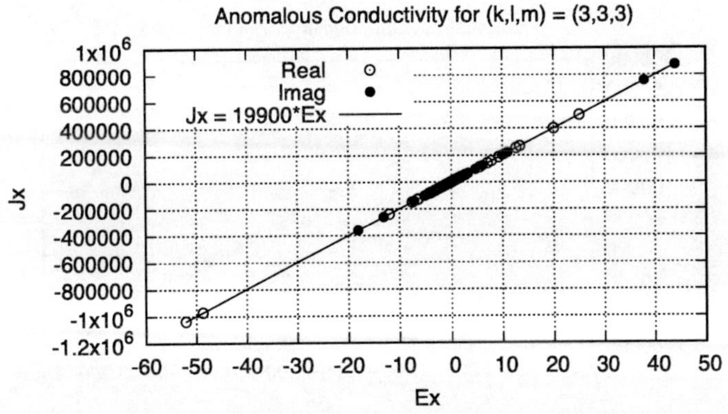

Fig. 4.39 Showing J_x vs. E_x for a cell near the middle of the top ($klm = 333$) layer

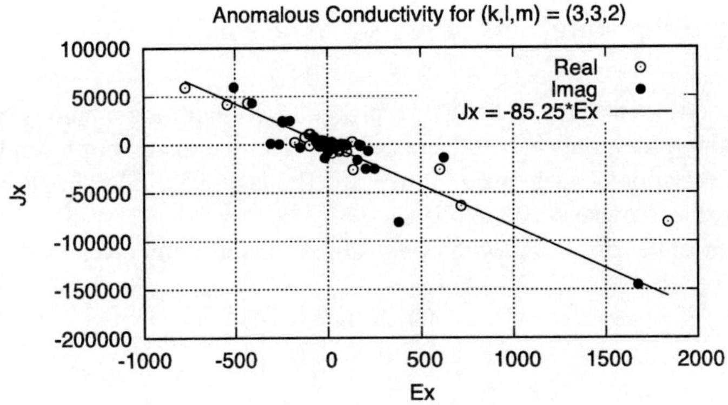

Fig. 4.40 Showing J_x vs. E_x for a cell near the middle of the cracked ($klm = 332$) layer

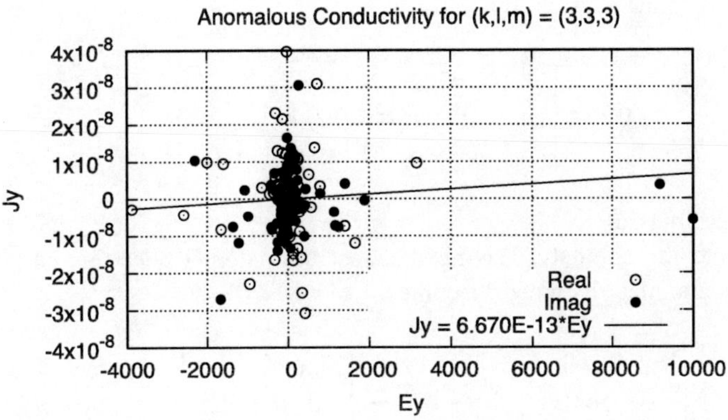

Fig. 4.41 Showing J_y vs. E_y for a cell near the middle of the top ($klm = 333$) layer

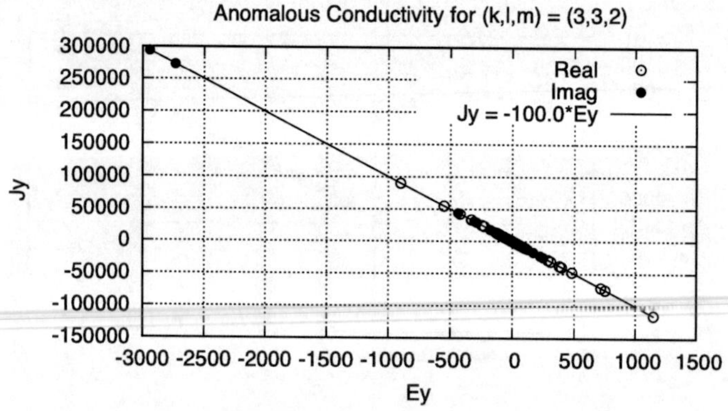

Fig. 4.42 Showing J_y vs.E_y for a cell near the middle of the cracked ($klm = 332$) layer

4.8 Handling Rotations of Anisotropic Media

In this section, we develop an analytical procedure for inferring rotations in the set-theoretic algorithm. It appears to be feasible to infer the rotation of a single ply, or to infer the rotation of each voxel in the grid. The latter possibility might be useful in studying the 'waviness' of the fibers.

Let the host be anisotropic with principal-axis conductivity given by the usual

$$\boldsymbol{\sigma}_h = \begin{bmatrix} \sigma_{h1} & 0 & 0 \\ 0 & \sigma_{h2} & 0 \\ 0 & 0 & \sigma_{h3} \end{bmatrix}, \tag{4.2}$$

and the anomalous region have a biaxial conductivity tensor with distinct eigenvalues in its principal-axis system. Then the anomalous conductivity tensor is given in the same coordinate system by

$$\boldsymbol{\sigma}_a(\mathbf{r}) = \begin{bmatrix} \sigma_1(\mathbf{r}) - \sigma_{h1} & 0 & 0 \\ 0 & \sigma_2(\mathbf{r}) - \sigma_{h2} & 0 \\ 0 & 0 & \sigma_3(\mathbf{r}) - \sigma_{h3} \end{bmatrix}, \tag{4.3}$$

which is also biaxial.

We rotate the anomaly, leaving the host unrotated. Under this condition, the anomalous conductivity tensor becomes

$$\boldsymbol{\sigma}_a(\mathbf{r}) = \mathbf{M} \begin{bmatrix} \sigma_1(\mathbf{r}) & 0 & 0 \\ 0 & \sigma_2(\mathbf{r}) & 0 \\ 0 & 0 & \sigma_3(\mathbf{r}) \end{bmatrix} \mathbf{M}^T - \begin{bmatrix} \sigma_{h1} & 0 & 0 \\ 0 & \sigma_{h2} & 0 \\ 0 & 0 & \sigma_{h3} \end{bmatrix} = \begin{bmatrix} \Sigma_{xx} & \Sigma_{xy} & \Sigma_{xz} \\ \Sigma_{yx} & \Sigma_{yy} & \Sigma_{yz} \\ \Sigma_{zx} & \Sigma_{zy} & \Sigma_{zz} \end{bmatrix}, \tag{4.4}$$

where

$$\begin{aligned}
\Sigma_{xx} &= \sigma_{1a} + m_{12}^2(\sigma_2 - \sigma_1) + m_{13}^2(\sigma_3 - \sigma_1) \\
\Sigma_{xy} &= m_{12}m_{22}(\sigma_2 - \sigma_1) + m_{13}m_{23}(\sigma_3 - \sigma_1) \\
\Sigma_{xz} &= m_{12}m_{32}(\sigma_2 - \sigma_1) + m_{13}m_{33}(\sigma_3 - \sigma_1) \\
\Sigma_{yx} &= m_{21}m_{11}(\sigma_1 - \sigma_2) + m_{23}m_{13}(\sigma_3 - \sigma_2) \\
\Sigma_{yy} &= \sigma_{2a} + m_{21}^2(\sigma_1 - \sigma_2) + m_{23}^2(\sigma_3 - \sigma_2) \\
\Sigma_{yz} &= m_{21}m_{31}(\sigma_1 - \sigma_2) + m_{23}m_{33}(\sigma_3 - \sigma_2) \\
\Sigma_{zx} &= m_{31}m_{11}(\sigma_1 - \sigma_3) + m_{32}m_{12}(\sigma_2 - \sigma_3) \\
\Sigma_{zy} &= m_{31}m_{21}(\sigma_1 - \sigma_3) + m_{32}m_{22}(\sigma_2 - \sigma_3) \\
\Sigma_{zz} &= \sigma_{3a} + m_{31}^2(\sigma_1 - \sigma_3) + m_{32}^2(\sigma_2 - \sigma_3)\,,
\end{aligned} \tag{4.5}$$

and $\sigma_{1a}(\mathbf{r}) = \sigma_1(\mathbf{r}) - \sigma_{h1}$, $\sigma_{2a}(\mathbf{r}) = \sigma_2(\mathbf{r}) - \sigma_{h2}$, and $\sigma_{3a}(\mathbf{r}) = \sigma_3(\mathbf{r}) - \sigma_{h3}$. Note that this is a symmetric tensor in the rotated coordinate system, as can be shown by making use of the orthonormality of the columns of **M**.

The m_{ij} are functions of the three Euler angles that define a three-dimensional rotation of coordinate systems. We are only interested in rotations, ϕ, about the z−axis, which simplifies the results of (4.5) considerably, as many of the terms vanish. The result is

$$\begin{aligned}
\Sigma_{xx} &= \sigma_{1a} + \sin^2(\phi)(\sigma_2 - \sigma_1) \\
\Sigma_{xy} &= -\sin(\phi)\cos(\phi)(\sigma_2 - \sigma_1) \\
\Sigma_{xz} &= 0 \\
\Sigma_{yx} &= \sin(\phi)\cos(\phi)(\sigma_1 - \sigma_2) \\
\Sigma_{yy} &= \sigma_{2a} + \sin^2(\phi)(\sigma_1 - \sigma_2) \\
\Sigma_{yz} &= 0 \\
\Sigma_{zx} &= 0 \\
\Sigma_{zy} &= 0 \\
\Sigma_{zz} &= \sigma_{3a}\,.
\end{aligned} \tag{4.6}$$

Therefore, in the rotated coordinate system we have

$$\begin{aligned}
J_x &= \Sigma_{xx}E_x + \Sigma_{xy}E_y \\
J_y &= \Sigma_{yx}E_x + \Sigma_{yy}E_y \\
J_z &= \sigma_{3a}E_z\,.
\end{aligned} \tag{4.7}$$

The anomalous currents, J_i, are measured by the T/R-array, and the E_i are computed in the usual manner for the set-theoretic algorithm, given the J_i. Thus, the only unknown is the rotation angle, ϕ, which must be determined by solving (4.7), given the expressions in (4.6).

Chapter 5
An Electromagnetic Model for Anisotropic Media: Green's Dyad for Plane-Layered Media

5.1 Theory

The Green's dyad, which is the electric-field response to a delta-function vector current source, plays a principal role in volume-integral equations. In this chapter we sketch the theory of the Green's dyad for plane-parallel layered media as it is applied to titanium and titanium-like alloys that possess 6 mm symmetry.

Eigenmodes of Anisotropic Media

We will consider plane-parallel bodies of infinite extent in the (x, y) plane, which are made up of layers of homogeneous, anisotropic material. To be specific, we consider host materials that are characterized by the following biaxial generalized electrical permittivity matrix:

$$\boldsymbol{\epsilon}_h = \begin{bmatrix} \epsilon_x & \epsilon_{xy} & 0 \\ \epsilon_{yx} & \epsilon_y & 0 \\ 0 & 0 & \epsilon_z \end{bmatrix}, \tag{5.1}$$

where the entries are generalized permittivities $\epsilon + \sigma/j\omega$.

Maxwell's equations for an electrically anisotropic body are

$$\begin{aligned} \nabla \times \mathbf{E} &= -j\omega\mu_h\mathbf{H} - j\omega(\mu(\mathbf{r}) - \mu_h)\mathbf{H} \\ &= -j\omega\mu_h\mathbf{H} + \mathbf{J}_m \\ \nabla \times \mathbf{H} &= j\omega\boldsymbol{\epsilon}_h \cdot \mathbf{E} + j\omega(\boldsymbol{\epsilon}(\mathbf{r}) - \boldsymbol{\epsilon}_h) \cdot \mathbf{E} \\ &= j\omega\boldsymbol{\epsilon}_h \cdot \mathbf{E} + \mathbf{J}_e, \end{aligned} \tag{5.2}$$

H. A. Sabbagh et al., *Advanced Electromagnetic Models for Materials Characterization and Nondestructive Evaluation*, Scientific Computation,
https://doi.org/10.1007/978-3-030-67956-9_5

where $\mathbf{J}_m$ and $\mathbf{J}_e$ are anomalous magnetic and electric currents that account for the presence of flaws, or anomalies, in the otherwise-uniform host material. From here on we drop the subscript h on the generalized host permittivity and permeability.

Because of the material anisotropy, it is convenient to work with a matrix formulation of these equations that has been useful in crystal optics, plasmas and microwave devices [7–9, 17, 59–61, 115, 125]. If the body is homogeneous with respect to (x, y), then Maxwell's equations can be Fourier transformed with respect to (x, y), and written as the following four-vector matrix differential equation in the spectral domain:

$$\frac{d\tilde{\mathbf{e}}}{dz} = \mathbf{S} \cdot \tilde{\mathbf{e}} + \mathbf{U} \cdot \tilde{\mathbf{J}} \tag{5.3}$$

$$\tilde{E}_z = \frac{k_y}{\epsilon_z \omega} \tilde{H}_x - \frac{k_x}{\epsilon_z \omega} \tilde{H}_y + \frac{j}{\epsilon_z \omega} \tilde{J}_{ez} \tag{5.4}$$

$$\tilde{H}_z = \frac{-k_y}{\mu \omega} \tilde{E}_x + \frac{k_x}{\mu \omega} \tilde{E}_y - \frac{j}{\mu \omega} \tilde{J}_{mz}, \tag{5.5}$$

where the tilde denotes a function defined in the transform domain (k_x, k_y), and

$$\tilde{\mathbf{e}} = \begin{bmatrix} \tilde{E}_x \\ \tilde{E}_y \\ \tilde{H}_x \\ \tilde{H}_y \end{bmatrix} ; \tilde{\mathbf{J}} = \begin{bmatrix} \tilde{J}_{ex} \\ \tilde{J}_{ey} \\ \tilde{J}_{ez} \\ \tilde{J}_{mx} \\ \tilde{J}_{my} \\ \tilde{J}_{mz} \end{bmatrix} . \tag{5.6}$$

The subscript e denotes an electric current, and m denotes a magnetic current. The matrices in (5.3) are given by

$$\mathbf{S} = - \begin{bmatrix} 0 & 0 & a & b \\ 0 & 0 & c & d \\ \alpha & \beta & 0 & 0 \\ \gamma & \delta & 0 & 0 \end{bmatrix} ; \mathbf{U} = \begin{bmatrix} 0 & 0 & k_x/\omega\epsilon_z & 0 & 1 & 0 \\ 0 & 0 & k_y/\omega\epsilon_z & -1 & 0 & 0 \\ 0 & 1 & 0 & 0 & 0 & -k_x/\omega\mu \\ -1 & 0 & 0 & 0 & 0 & -k_y/\omega\mu \end{bmatrix} . \tag{5.7}$$

The entries of $\mathbf{S}$ are given in terms of the entries of (5.1) by

$$a = \frac{j}{\omega \epsilon_z} k_x k_y \; ; \; \alpha = \frac{j}{\omega \mu} (-\mu \epsilon_{yx} \omega^2 - k_x k_y)$$

$$b = \frac{j}{\omega \epsilon_z} (\mu \epsilon_z \omega^2 - k_x^2) \; ; \; \beta = \frac{j}{\omega \mu} (-\mu \epsilon_y \omega^2 + k_x^2)$$

$$c = \frac{j}{\omega \epsilon_z} (-\mu \epsilon_z \omega^2 + k_y^2) \; ; \; \gamma = \frac{j}{\omega \mu} (\mu \epsilon_x \omega^2 - k_y^2)$$

$$d = -\frac{j}{\omega\epsilon_z} k_x k_y \; ; \; \delta = \frac{j}{\omega\mu}(\mu\epsilon_{xy}\omega^2 + k_x k_y). \tag{5.8}$$

When $\widetilde{\mathbf{J}}$ is a surface current confined to $z = z'$, i.e., $\widetilde{\mathbf{J}} = \widetilde{\mathbf{J}}_s \delta(z - z')$, then integration of (5.3) produces

$$\widetilde{\mathbf{e}}^{(+)} - \widetilde{\mathbf{e}}^{(-)} = \mathbf{U} \cdot \widetilde{\mathbf{J}}_s \, . \tag{5.9}$$

The superscript $(+)$ denotes the limit as z approaches z' from above, and the superscript $(-)$ denotes the limit from below. Equation (5.9) is used to compute the Green's dyad for a layered workpiece.

Starting with these equations, Roberts [90] has developed a fairly complete theory of normal modes of biaxial anisotropic media. This work is based on, and extends, earlier work performed at Sabbagh Associates [92, 93]. From here on we specialize the theory developed in [90] to the case to be considered here, in which the media involved are transversely isotropic to the z-coordinate. The generalized electric permittivity tensor, in its principal-axis coordinate system, then takes the form

$$\boldsymbol{\epsilon} = \begin{bmatrix} \epsilon_t & 0 & 0 \\ 0 & \epsilon_t & 0 \\ 0 & 0 & \epsilon_z \end{bmatrix} , \tag{5.10}$$

where, for pure titanium $\epsilon_t = \epsilon_0 - j2.205 \times 10^6/\omega \approx -j2.205 \times 10^6/\omega$, $\epsilon_z = \epsilon_0 - j2.083 \times 10^6/\omega \approx -j2.083 \times 10^6/\omega$.

The entries in $\mathbf{S}$ now become

$$\begin{aligned} a &= \frac{j}{\omega\epsilon_z} k_x k_y \; ; \; \alpha = \frac{j}{\omega\mu}(-k_x k_y) \\ b &= \frac{j}{\omega\epsilon_z}(\mu\epsilon_z\omega^2 - k_x^2) \; ; \; \beta = \frac{j}{\omega\mu}(-\mu\epsilon_t\omega^2 + k_x^2) \\ c &= \frac{j}{\omega\epsilon_z}(-\mu\epsilon_z\omega^2 + k_y^2) \; ; \; \gamma = \frac{j}{\omega\mu}(\mu\epsilon_t\omega^2 - k_y^2) \\ d &= -\frac{j}{\omega\epsilon_z} k_x k_y \; ; \; \delta = \frac{j}{\omega\mu}(k_x k_y). \end{aligned} \tag{5.11}$$

Let's introduce some notation: $k_x^2 + k_y^2 = k_t^2$, $\omega^2\mu\epsilon_t = \Omega_t^2$, $\omega^2\mu\epsilon_z = \Omega_z^2$, $\epsilon = \epsilon_t/\epsilon_z$. Then the eigenvalues of $\mathbf{S}$ are

$$\lambda_1 = \sqrt{k_t^2 - \Omega_t^2} \; \lambda_2 = -\lambda_1 \; \lambda_3 = \sqrt{\epsilon}\sqrt{k_t^2 - \Omega_z^2} \; \lambda_4 = -\lambda_3 \; . \tag{5.12}$$

The linearly-independent eigenvectors that correspond to these eigenvalues are:

$$\mathbf{v}_1 = \begin{bmatrix} -j\omega\mu_0 k_y \\ j\omega\mu_0 k_x \\ \lambda_1 k_x \\ \lambda_1 k_y \end{bmatrix} \quad \mathbf{v}_2 = \begin{bmatrix} -j\omega\mu_0 k_y \\ j\omega\mu_0 k_x \\ -\lambda_1 k_x \\ -\lambda_1 k_y \end{bmatrix} \quad \mathbf{v}_3 = \begin{bmatrix} \lambda_3 k_x \\ \lambda_3 k_y \\ j\omega\epsilon_t k_y \\ -j\omega\epsilon_t k_x \end{bmatrix} \quad \mathbf{v}_4 = \begin{bmatrix} \lambda_3 k_x \\ \lambda_3 k_y \\ -j\omega\epsilon_t k_y \\ j\omega\epsilon_t k_x \end{bmatrix} . \tag{5.13}$$

When $k_x = k_y = 0$, the following are linearly-independent eigenvectors:

$$\mathbf{v}_1 = \begin{bmatrix} 1 \\ 0 \\ 0 \\ -\sqrt{\dfrac{\epsilon_t}{\mu_0}} \end{bmatrix} \quad \mathbf{v}_2 = \begin{bmatrix} 1 \\ 0 \\ 0 \\ \sqrt{\dfrac{\epsilon_t}{\mu_0}} \end{bmatrix} \quad \mathbf{v}_3 = \begin{bmatrix} 0 \\ \sqrt{\dfrac{\mu_0}{\epsilon_t}} \\ 1 \\ 0 \end{bmatrix} \quad \mathbf{v}_4 = \begin{bmatrix} 0 \\ -\sqrt{\dfrac{\mu_0}{\epsilon_t}} \\ 1 \\ 0 \end{bmatrix} . \tag{5.14}$$

When we substitute $\mathbf{v}_1, \mathbf{v}_2$ of (5.13) into (5.4), with the source currents set to zero, we find that $\widetilde{E}_z = 0$; hence, $\mathbf{v}_1, \mathbf{v}_2$ are transverse electric (TE) modes, with respect to z. Similarly, $\mathbf{v}_3, \mathbf{v}_4$ are transverse magnetic (TM) modes. Note that the TE modes are orthogonal to the TM modes. This will facilitate the computation of the Green's dyadic. $\mathbf{v}_1$ and $\mathbf{v}_3$ are downward-traveling waves in the z-direction; i.e., they represent waves that travel in the negative z-direction. $\mathbf{v}_2, \mathbf{v}_4$ are upward-traveling waves (in the positive z-direction). We see from (5.4) and (5.5) that all modes are TEM (transverse electric and magnetic) with respect to z for $k_x = k_y = 0$, which is the condition for infinite plane-waves traveling in the $z-$direction. Furthermore, under this condition $\lambda_1 = \lambda_3$, which means that the anisotropy does not manifest itself. The effects of anisotropy are most pronounced on the TE and TM modes when the transverse wave-number, $k_t >> \Omega_t$ and Ω_z.

We can justify the interpretation of upward- and downward-traveling waves for the various eigenvectors by returning to the fundamental differential equation (5.3), which has the solution when $\widetilde{\mathbf{J}} = 0$:

$$\begin{aligned} \widetilde{\mathbf{e}}(z) &= \exp(z\mathbf{S}) \cdot \widetilde{\mathbf{e}}_0 \\ &= \mathbf{I} \cdot \widetilde{\mathbf{e}}_0 + z\mathbf{S} \cdot \widetilde{\mathbf{e}}_0 + \frac{z^2\mathbf{S}^2}{2} \cdot \widetilde{\mathbf{e}}_0 + \cdots , \end{aligned} \tag{5.15}$$

where $\widetilde{\mathbf{e}}_0$ is the solution at $z = 0$, and the second line defines the matrix exponential operator, $\exp(z\mathbf{S})$.

Let $\widetilde{\mathbf{e}}_0 = \mathbf{v}_1$; then

$$\widetilde{\mathbf{e}}(z) = e^{z\lambda_1}\mathbf{v}_1 , \tag{5.16}$$

which follows upon substituting $\mathbf{v}_1$ into the second line of (5.15), and then making use of the fact that $\mathbf{v}_1$ is an eigenvector of $\mathbf{S}$ with eigenvalue λ_1. This result is in the form of a wave propagating in the negative z-direction, and justifies our calling $\mathbf{v}_1$ a downward-traveling TE wave. A similar analysis holds for the other three eigenvectors.

The formal solution of (5.3) defines the Green's dyadic, $\mathbf{G}(z, z')$:

$$\widetilde{\mathbf{e}}(z) = \int \mathbf{G}(z, z') \cdot \mathbf{J}(z') dz' , \tag{5.17}$$

where we are still working in the transverse Fourier-transform domain. The computation of $\mathbf{G}(z, z')$ makes use of the eigenvectors $\mathbf{v}_1, \cdots, \mathbf{v}_4$, together with the discontinuity condition, (5.9), as described in the references cited in this appendix. Once we have $\mathbf{G}$, we transform (5.17) back into the spatial domain, and work with the spatial volume-integral equation, as is currently done in **VIC-3D®**.

Modeling Anisotropic Anomalies

The integro-differential equation to which we will apply the method-of-moments is simply gotten by equating the total electric field, $\boldsymbol{\sigma}_a^{-1}(\mathbf{r}) \cdot \mathbf{J}^{(e)}(\mathbf{r})$, to the sum of the incident field, due to the coil, and the infinite-space and layered-space scattered fields:

$$\mathbf{E}^{(i)}(\mathbf{r}) = \boldsymbol{\sigma}_a^{-1}(\mathbf{r}) \cdot \mathbf{J}^{(e)}(\mathbf{r}) - \mathbf{E}^{(0)}(\mathbf{r})[\mathbf{J}^{(e)}] - \mathbf{E}^{(s)}(\mathbf{r})[\mathbf{J}^{(e)}] , \tag{5.18}$$

where $\boldsymbol{\sigma}_a(\mathbf{r}) = j\omega(\boldsymbol{\epsilon}(\mathbf{r}) - \epsilon_h)$ is the anomalous conductivity tensor.

Now, let's talk about **Euler Angles**. Let a rotation about O carry the orthogonal triad $(\mathbf{I}, \mathbf{J}, \mathbf{K})$ into $(\mathbf{i}, \mathbf{j}, \mathbf{k})$. We break this rotation into three rotations. First, rotate about $\mathbf{K}$ so as to make the new position of the plane $(\mathbf{I}, \mathbf{K})$ contain $\mathbf{k}$, say through an angle ϕ; this gives a transformation

$$(\mathbf{I}, \mathbf{J}, \mathbf{K}) \rightarrow (\mathbf{I}_1, \mathbf{J}_1, \mathbf{K}_1) \left\{ \begin{array}{l} \mathbf{I}_1 = \mathbf{I}\cos\phi + \mathbf{J}\sin\phi \\ \mathbf{J}_1 = -\mathbf{I}\sin\phi + \mathbf{J}\cos\phi \\ \mathbf{K}_1 = \mathbf{K} \end{array} \right\} . \tag{5.19}$$

Secondly, rotate about $\mathbf{J}_1$ to bring $\mathbf{K}_1$ to $\mathbf{k}$, say through an angle θ; this gives a transformation

$$(\mathbf{I}_1, \mathbf{J}_1, \mathbf{K}_1) \rightarrow (\mathbf{I}_2, \mathbf{J}_2, \mathbf{k}) \left\{ \begin{array}{l} \mathbf{I}_2 = \mathbf{I}_1\cos\theta - \mathbf{K}_1\sin\theta \\ \mathbf{J}_2 = \mathbf{J}_1 \\ \mathbf{k} = \mathbf{I}_1\sin\theta + \mathbf{K}_1\cos\theta \end{array} \right\} . \tag{5.20}$$

Finally, rotate about $\mathbf{k}$ to bring $\mathbf{I}_2$ to $\mathbf{i}$ and $\mathbf{J}_2$ to $\mathbf{j}$, say through an angle ψ; this gives the transformation

$$(\mathbf{I}_2, \mathbf{J}_2, \mathbf{k}) \rightarrow (\mathbf{i}, \mathbf{j}, \mathbf{k}) \left\{ \begin{array}{l} \mathbf{i} = \mathbf{I}_2\cos\psi + \mathbf{J}_2\sin\psi \\ \mathbf{j} = -\mathbf{I}_2\sin\psi + \mathbf{J}_2\cos\psi \\ \mathbf{k} = \mathbf{k} \end{array} \right\} . \tag{5.21}$$

The angles (θ, ϕ, ψ) are the Euler angles. Their values determine the position of the triad $(\mathbf{i}, \mathbf{j}, \mathbf{k})$ relative to $(\mathbf{I}, \mathbf{J}, \mathbf{K})$. The angles range over the following values:

$$\begin{aligned} 0 &\le \theta \le \pi \\ 0 &\le \phi < 2\pi \quad . \\ 0 &\le \psi < 2\pi \end{aligned} \tag{5.22}$$

From the above equations of transformation, we can obtain the matrix $\mathbf{M}$, that defines orthogonal rotations. We use the notation, $c = \cos$, $s = \sin$ and let the subscripts, 1, 2, 3, refer to θ, ϕ, ψ, respectively:

(5.23)

	i	**j**	**k**
I	$c_1c_2c_3 - s_2s_3$	$-c_1c_2s_3 - s_2c_3$	s_1c_2
J	$c_1s_2c_3 + c_2s_3$	$-c_1s_2s_3 + c_2c_3$	s_1s_2
K	$-s_1c_3$	s_1s_3	c_1

Now, we'll apply this to the problem at hand. Let the host be transversely anisotropic about the z-axis, with a diagonal conductivity tensor, $\boldsymbol{\sigma}_h = j\omega\boldsymbol{\epsilon}_h$, and define $\boldsymbol{\sigma}(\mathbf{r}) = j\omega\boldsymbol{\epsilon}(\mathbf{r})$. In its rotated coordinate system, this conductivity tensor becomes

$$\begin{aligned} \boldsymbol{\sigma}(\mathbf{r}) &= \mathbf{M}\begin{bmatrix} \sigma_1 & 0 & 0 \\ 0 & \sigma_1 & 0 \\ 0 & 0 & \sigma_2 \end{bmatrix}\mathbf{M}^T \\ &= \begin{bmatrix} \sigma_1 + m_{13}^2(\sigma_2 - \sigma_1) & (\sigma_2 - \sigma_1)m_{13}m_{23} & (\sigma_2 - \sigma_1)m_{13}m_{33} \\ (\sigma_2 - \sigma_1)m_{23}m_{13} & \sigma_1 + (\sigma_2 - \sigma_1)m_{23}^2 & (\sigma_2 - \sigma_1)m_{23}m_{33} \\ (\sigma_2 - \sigma_1)m_{13}m_{33} & (\sigma_2 - \sigma_1)m_{23}m_{33} & \sigma_1 + (\sigma_2 - \sigma_1)m_{33}^2 \end{bmatrix} \end{aligned} \tag{5.24}$$

We will restrict $\mathbf{M}$ to those Euler angles that leave $\boldsymbol{\sigma}(\mathbf{r})$ in its principal axes. These are $\theta = 90°,\ \phi = 0°,\ \psi = 0°$, for which

$$\boldsymbol{\sigma}(\mathbf{r}) = \begin{bmatrix} \sigma_2 & 0 & 0 \\ 0 & \sigma_1 & 0 \\ 0 & 0 & \sigma_1 \end{bmatrix}, \tag{5.25}$$

and $\theta = 90°,\ \phi = 90°,\ \psi = 90°$, for which

$$\boldsymbol{\sigma}(\mathbf{r}) = \begin{bmatrix} \sigma_1 & 0 & 0 \\ 0 & \sigma_2 & 0 \\ 0 & 0 & \sigma_1 \end{bmatrix} . \tag{5.26}$$

Actually, all we need is (5.25) because we can always orient the flaw or probe scan appropriately to simulate (5.26).

Because the host and variable conductivity tensors are both diagonal, we can write the anomalous conductivity as

$$\begin{aligned} \boldsymbol{\sigma}_a(\mathbf{r}) &= \begin{bmatrix} \sigma_1 - \sigma_x(\mathbf{r}) & 0 & 0 \\ 0 & \sigma_1 - \sigma_y(\mathbf{r}) & 0 \\ 0 & 0 & \sigma_2 - \sigma_z(\mathbf{r}) \end{bmatrix} \\ &= \begin{bmatrix} \sigma_a^x & 0 & 0 \\ 0 & \sigma_a^y & 0 \\ 0 & 0 & \sigma_a^z \end{bmatrix} . \end{aligned} \tag{5.27}$$

The reciprocal of this tensor is simply the reciprocal of the diagonal entries:

$$\boldsymbol{\sigma}_a^{-1} = \begin{bmatrix} 1/\sigma_a^x(\mathbf{r}) & 0 & 0 \\ 0 & 1/\sigma_a^y(\mathbf{r}) & 0 \\ 0 & 0 & 1/\sigma_a^z(\mathbf{r}) \end{bmatrix} . \tag{5.28}$$

Now, we are at the same place as in the original formulation of the **Q** matrices, except that we replace the scalar σ with $\sigma_i,\ i = x,\ y,\ z$. This yields the following replacements for the original **Q**s:

$$Q_{kK}^{(xx)} = \frac{\delta x \delta y \delta z}{6} \begin{cases} \dfrac{1}{\sigma_{klm}^x} & \text{if } K = k - 1 \\ \dfrac{2}{\sigma_{klm}^x} + \dfrac{2}{\sigma_{k+1,lm}^x} & \text{if } K = k \\ \dfrac{1}{\sigma_{k+1,lm}^x} & \text{if } K = k + 1 \end{cases}$$

$$Q_{lL}^{(yy)} = \frac{\delta x \delta y \delta z}{6} \begin{cases} \dfrac{1}{\sigma_{klm}^y} & \text{if } L = l - 1 \\ \dfrac{2}{\sigma_{klm}^y} + \dfrac{2}{\sigma_{k,l+1,m}^y} & \text{if } L = l \\ \dfrac{1}{\sigma_{k,l+1,m}^y} & \text{if } L = l + 1 \end{cases}$$

$$Q_{mM}^{(zz)} = \frac{\delta x \delta y \delta z}{6} \begin{cases} \dfrac{1}{\sigma_{klm}^z} & \text{if } M = m-1 \\ \dfrac{2}{\sigma_{klm}^z} + \dfrac{2}{\sigma_{klm+1}^z} & \text{if } M = m \\ \dfrac{1}{\sigma_{klm+1}^z} & \text{if } M = m+1 \end{cases} \tag{5.29}$$

The first entry in each of these matrices is the lower diagonal, the second the main diagonal, and the third the upper diagonal.

The **Q**'s contain the reciprocals of the anomalous conductivities, which may vanish when a flaw cell contains only host material. This condition forces the corresponding unknowns to vanish, but we are still left with an indeterminate 'zero over zero' expression to evaluate. In order to overcome this problem we form the reciprocals of the diagonal entries of the **Q**'s of (5.29),

$$\nu_{klm}^{(xx)} = \frac{6}{\delta x \delta y \delta z} \frac{\sigma_{klm}^x \sigma_{k+1,lm}^x}{2(\sigma_{klm}^x + \sigma_{k+1,lm}^x)}$$

$$\nu_{klm}^{(yy)} = \frac{6}{\delta x \delta y \delta z} \frac{\sigma_{klm}^y \sigma_{k,l+1,m}^y}{2(\sigma_{klm}^y + \sigma_{k,l+1,m}^y)}$$

$$\nu_{klm}^{(zz)} = \frac{6}{\delta x \delta y \delta z} \frac{\sigma_{klm}^z \sigma_{kl,m+1}^z}{2(\sigma_{klm}^z + \sigma_{kl,m+1}^z)}, \tag{5.30}$$

and multiply both sides of (5.18) by them. The resulting $\mathbf{Q}'$ matrices are also tri-diagonal, with the following non-zero entries:

$$Q_{kK}^{\prime(xx)} = \begin{cases} \dfrac{\sigma_{k+1,lm}^x}{2(\sigma_{klm}^x + \sigma_{k+1,lm}^x)} & \text{if } K = k-1 \\ 1 & \text{if } K = k \\ \dfrac{\sigma_{klm}^x}{2(\sigma_{klm}^x + \sigma_{k+1,lm}^x)} & \text{if } K = k+1 \end{cases}$$

$$Q_{lL}^{\prime(yy)} = \begin{cases} \dfrac{\sigma_{k,l+1,m}^y}{2(\sigma_{klm}^y + \sigma_{k,l+1,m}^y)} & \text{if } L = l-1 \\ 1 & \text{if } L = l \\ \dfrac{\sigma_{klm}^y}{2(\sigma_{klm}^y + \sigma_{k,l+1,m}^y)} & \text{if } L = l+1 \end{cases}$$

$$Q'^{(zz)}_{mM} = \begin{cases} \dfrac{\sigma^z_{klm+1}}{2(\sigma^z_{klm} + \sigma^z_{klm+1})} & \text{if } M = m - 1 \\ 1 & \text{if } M = m \\ \dfrac{\sigma^z_{klm}}{2(\sigma^z_{klm} + \sigma^z_{klm+1})} & \text{if } M = m + 1 \end{cases} \tag{5.31}$$

Note that if the anomalous conductivity in two adjacent cells vanishes, then the indeterminate terms in the Q''s are equal to $1/4$. Also, if the conductivity of either cell (or both) vanishes, then the corresponding ν_{klm} of (5.30) also vanishes. This follows because the numerators of the ν's vanish to a higher-order than the denominators. The result is that the corresponding $\mathbf{J}_{KLM}$ vanishes.

A Test Problem

Figure 5.1 shows the results for a problem in which a circular coil is placed on a halfspace having $\sigma_{11} = \sigma_{22} = \sigma_{33} = 6.04 \times 10^5$ S/m and containing a $12.7 \times 12.7 \times 0.508$ mm surface-breaking flaw with $\sigma_{11} = 5.90 \times 10^5$, $\sigma_{22} = \sigma_{33} = 6.04 \times 10^5$ S/m, and $\epsilon_{11} = \epsilon_{22} = \epsilon_{33} = 1$. Figure 5.2 illustrates the setup. The angle Φ in Fig. 5.2 goes from 0 to 360° as the center of the coil traces a circle around the center of the flaw. The model uses a single grid of volume fractions, but in order to simulate a region with randomly oriented (along x, y, or z) conductivities, we need 3 grids of volume fractions, one for each direction.

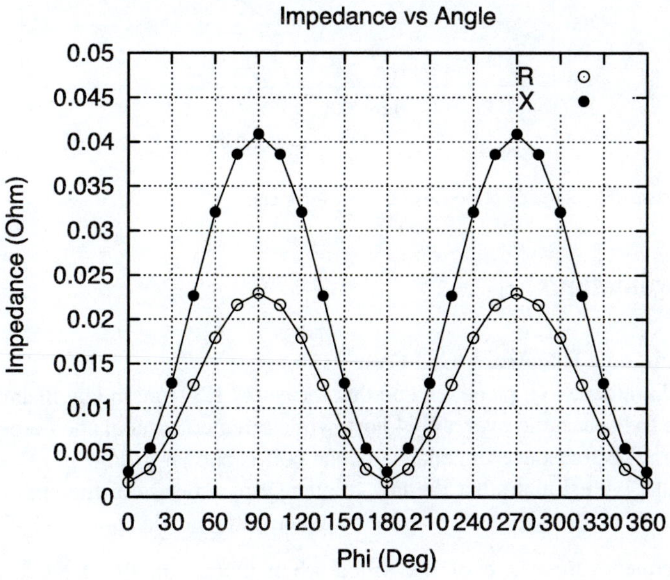

Fig. 5.1 Results for a problem with a circular coil excitation

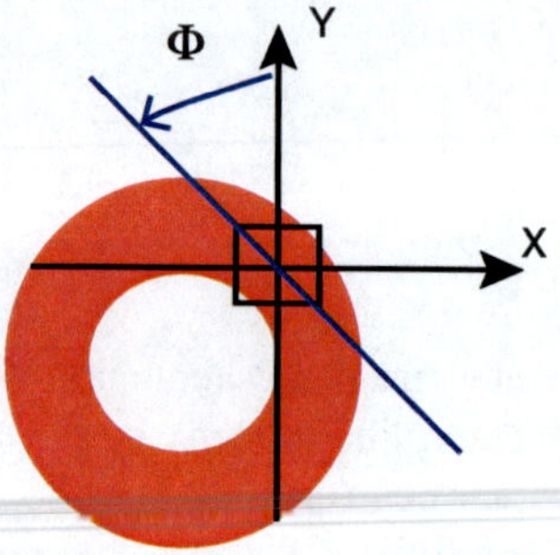

Fig. 5.2 Illustrating the setup that produced the data of Fig. 5.1

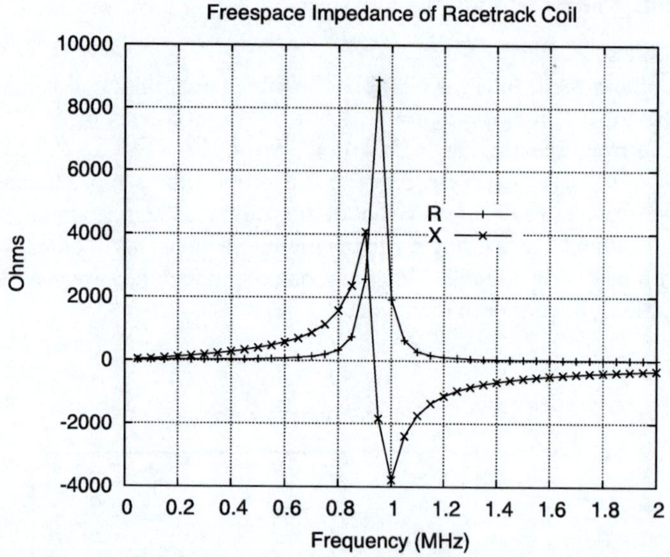

Fig. 5.3 Freespace frequency response of the racetrack coil

5.2 Applications

Characterizing a Racetrack Coil Over Ti64

We have characterized a racetrack coil that measures 12.9 mm in length and 3.5 mm in width in freespace and over a Ti64 host, while oriented parallel and perpendicular to the host. The freespace impedance of the coil is shown in Fig. 5.3. Based upon these results, we estimate that the low-frequency parameters of the coil are $L_0 = 90\,\mu\text{H}$ and $R_0 = 12.6\Omega$. The freespace resonant frequency is about 0.9345 MHz.

The frequency response of the probe when placed on the Ti64 host in the 'parallel' orientation is shown in Fig. 5.4 and in the 'perpendicular' orientation in Fig. 5.5. The reference orientation is a very light machining of the host.

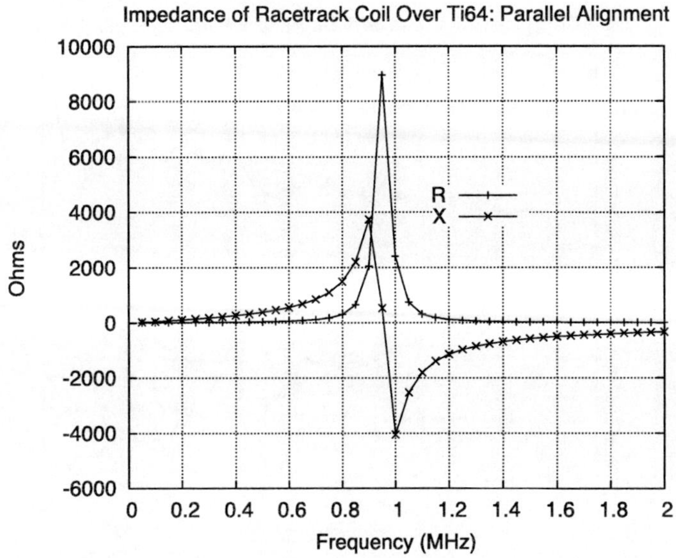

Fig. 5.4 Frequency response of the racetrack coil when oriented in the 'parallel' direction on the Ti64 workpiece

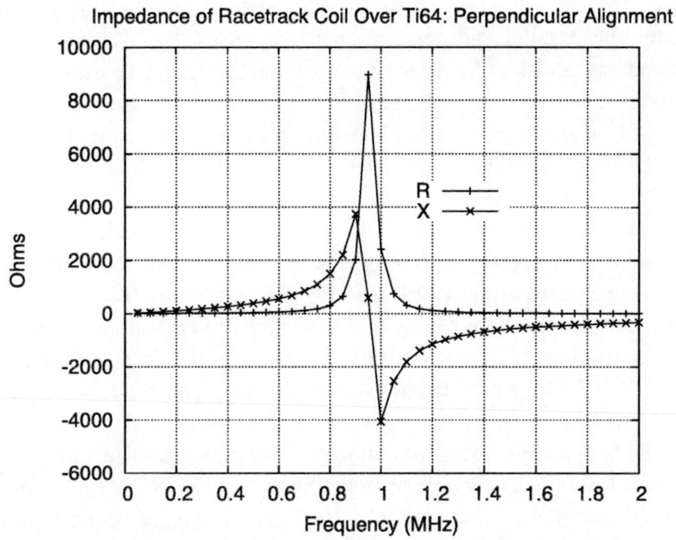

Fig. 5.5 Frequency response of the racetrack coil when oriented in the 'perpendicular' direction on the Ti64 workpiece

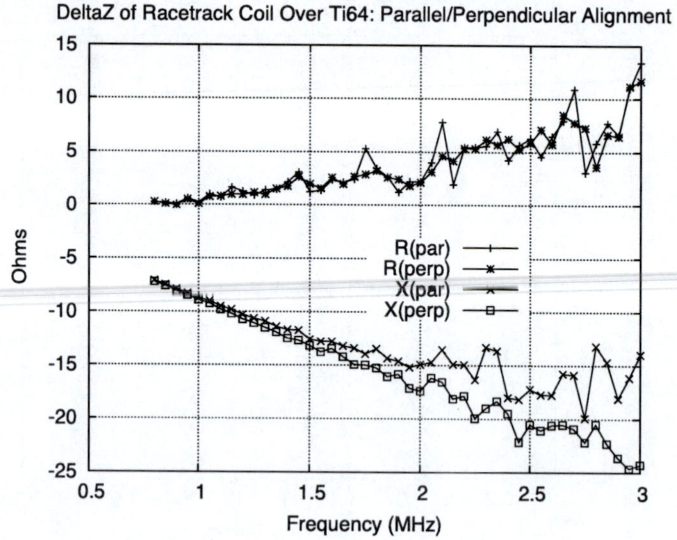

Fig. 5.6 Frequency response of δZ when the racetrack coil is oriented in the parallel and perpendicular directions

It appears from Figs. 5.4 and 5.5 that there is virtually no distinguishable feature that separates the parallel and perpendicular responses, but when we remove the parasitic elements and the freespace coil response, we get the results shown in Fig. 5.6 for the change in impedance, δZ for the two orientations as a function of frequency. Two things are quite clear in this figure: the results are quite noisy because of the poor coupling of the racetrack coil to the workpiece, and starting at about 1.5 MHz there is a clear distinction in δX between the two orientations.

The frequency response of a **VIC-3D®** model of the probe with a ferrite core over the Ti64 sample is shown in Fig. 5.7. Notice that the difference in the parallel and perpendicular responses around 3 MHz in Fig. 5.7 is comparable to that in Fig. 5.6, and suggests that the response of Fig. 5.6 may be due to crystalline 'texturing,' after all. Keep in mind, however, that with the ferrite core, the freespace inductance of the probe is 787.33 μH, which is almost nine times greater than the 90 μH that we estimated earlier.

Figure 5.8 [108] shows a tangent coil over a flawed workpiece, corresponding to the benchmark test of [20]. The 0° response shown in Fig. 5.7 is reminiscent of the response of this tangent coil, as shown in Fig. 5.9. The magnetic-moment vector of the tangent coil is along the long axis of the slot, which means that the 'effective conductivity' of the slot within its host will be smaller than when the coil is rotated 90°. This corresponds to the condition in Fig. 5.7 at 0°. When the reactance at this orientation in Fig. 5.7 is normalized to the freespace reactance of the coil, we get an effective inductance that will be reasonably constant over most of the frequency range, which is in agreement with the effective inductance shown in Fig. 5.9.

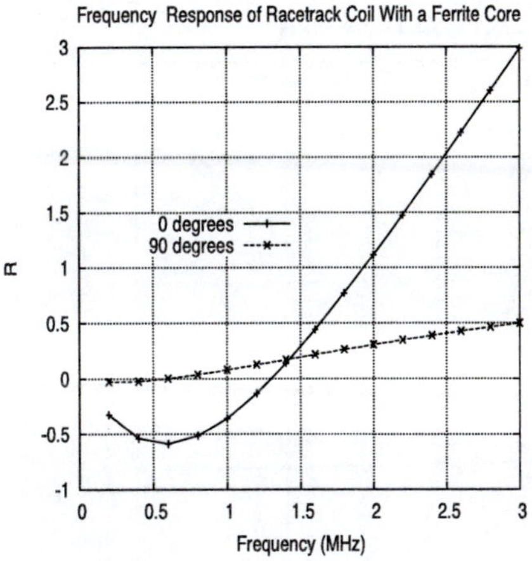

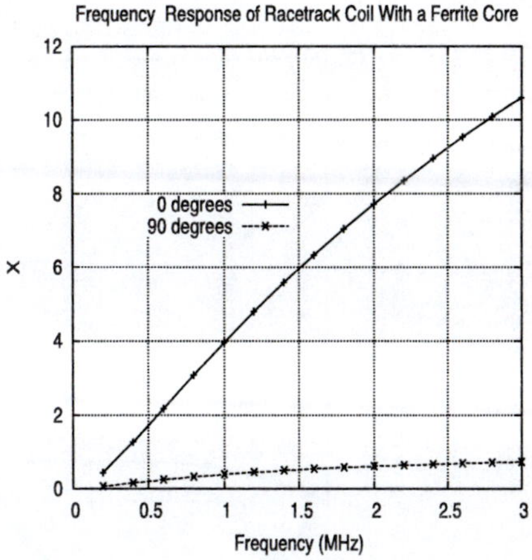

Fig. 5.7 Frequency response of the racetrack coil with a ferrite core when oriented at 0° and 90° over the anisotropic patch of Ti64

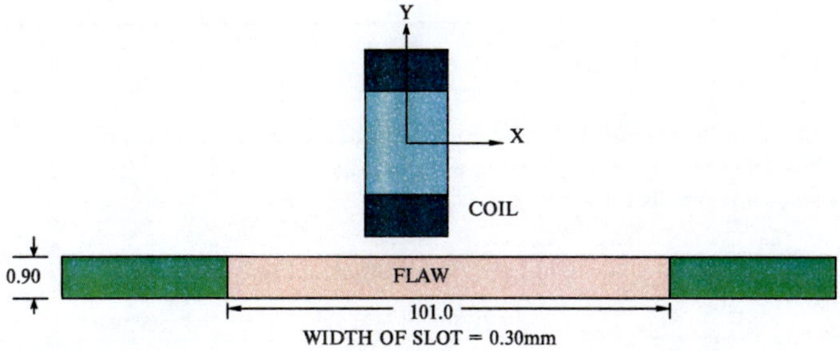

Fig. 5.8 Illustrating a tangent coil over a flawed workpiece, corresponding to the benchmark test of [20]

5.3 Some Inverse Problems with Random Anisotropies

We use the configuration shown in Fig. 5.10 of a host, anisotropic patch, and racetrack coil to develop the model problems in this section. There are two models for generating random anisotropies in **VIC-3D®**: (1) via random permutations of the nonrandom principal conductivities, which in this case will be $\sigma_{11} = 5.9 \times 10^5$ S/m, $\sigma_{22} = \sigma_{33} = 6.04 \times 10^5$ S/m, and (2) via random values assigned to the principal axes, generated with either a uniform or Gaussian probability density function (pdf).

Figure 5.11 illustrates an ensemble of ten random functions produced with the 'random permutations' process. These are the inputs to a Monte Carlo run with NLSE, with the intention of determining effective (nonrandom) values for σ_{11} and σ_{22}.

To begin the inversion process, we generate a 2×2 interpolation grid for σ_{11} and σ_{22} with nodal values of 5.5×10^5 S/m and 6.5×10^5 S/m. The blending

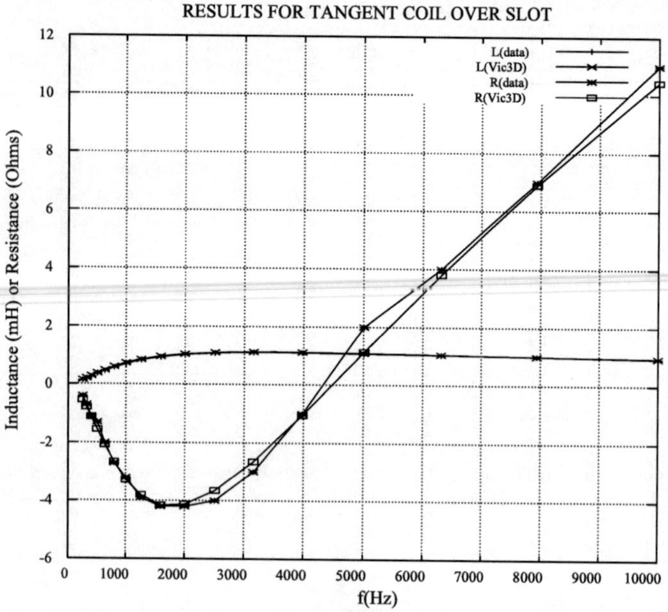

Fig. 5.9 Comparison of **VIC-3D®** calculations with the frequency-scan benchmark data of [20]. The results shown here were obtained using a modest flaw grid of 128 × 2 × 4 cells, and a very coarse grid of 2 × 4 × 4 cells for the probe

Table 5.1 Result of NLSE inversion of the ten functions in Fig. 5.11

Function	Φ	σ_{11}/sensit	σ_{22}/sensit
1	0.3529(−1)	601488/1.653(−2)	603655/1.911(−2)
2	0.3626(−1)	601484/1.698(−2)	603888/1.963(−2)
3	0.3808(−1)	601426/1.784(−2)	603703/2.062(−2)
4	0.3632(−1)	601389/1.701(−2)	603866/1.966(−2)
5	0.3643(−1)	601480/1.706(−2)	603807/1.973(−2)
6	0.3792(−1)	601344/1.776(−2)	603788/2.053(−2)
7	0.3637(−1)	601561/1.703(−2)	603681/1.969(−2)
8	0.3646(−1)	601523/1.707(−2)	603861/1.974(−2)
9	0.3531(−1)	601346/1.654(−2)	603800/1.912(−2)
10	0.3729(−1)	601248/1.747(−2)	604021/2.019(−2)
AVG	0.3657(−1)	601429	603807
STD	0.008(−1)	289.6	330.1

(interpolation) functions at these nodes are shown in Fig. 5.12. Note that curves A and D correspond to isotropic conditions, but the impedances vary because we are seeing the edge effects of the anomalous patch. The result of passing each of the ten functions in Fig. 5.11 through NLSE is shown in Table 5.1.

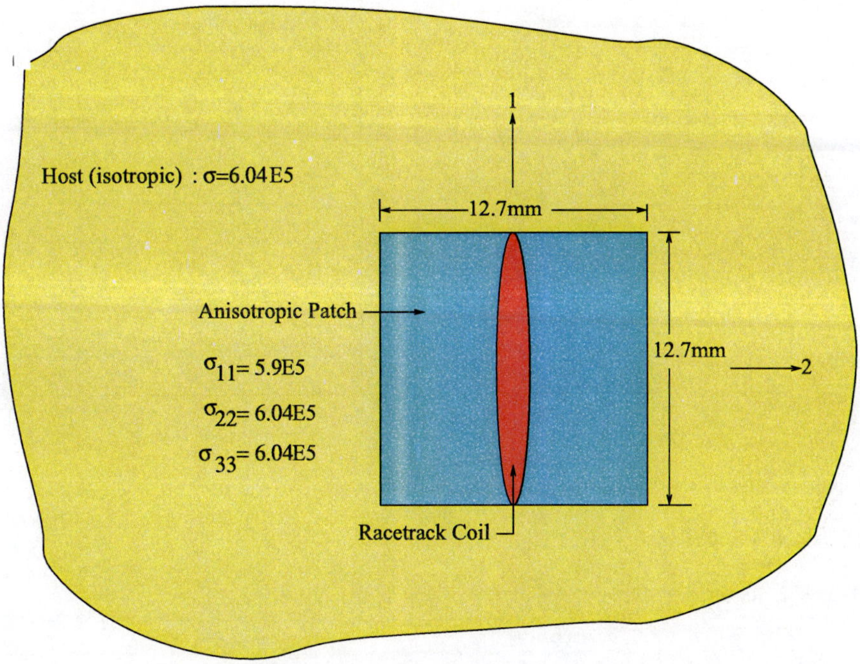

Fig. 5.10 Illustrating the host, anisotropic block and racetrack coil configuration

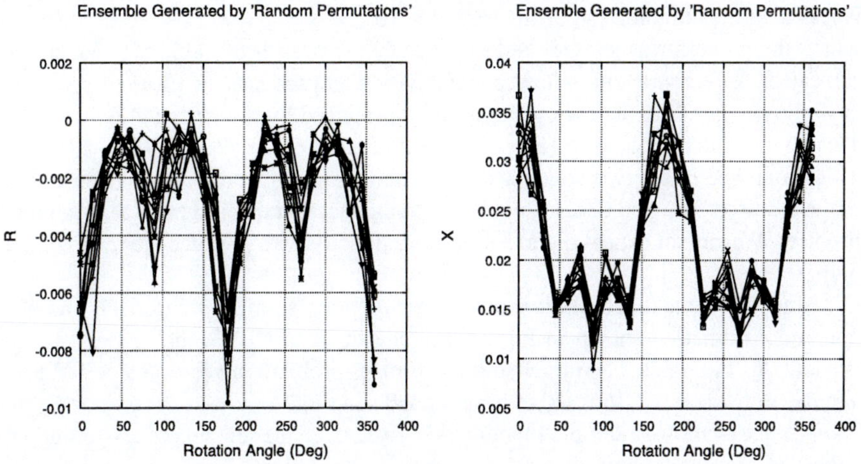

Fig. 5.11 An ensemble of ten random functions produced with the 'random permutations' process

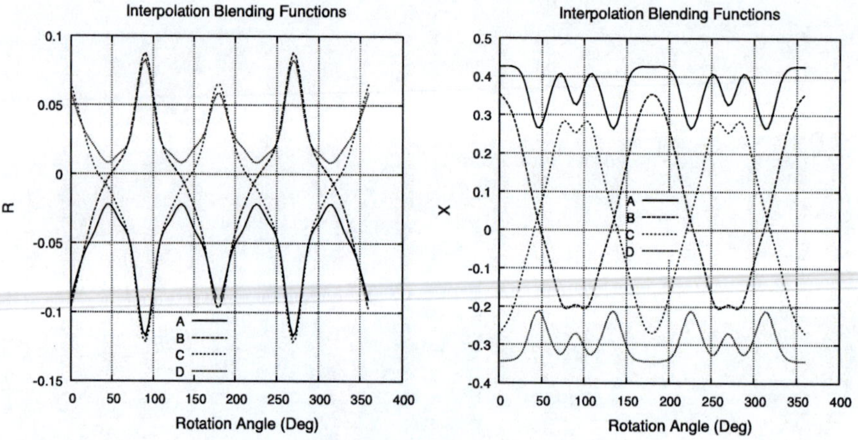

Fig. 5.12 Blending functions for the interpolation grid. A:$\sigma_{11} = 5.5 \times 10^5$ S/m, $\sigma_{22} = 5.5 \times 10^5$ S/m; B:$\sigma_{11} = 5.5 \times 10^5$ S/m, $\sigma_{22} = 6.5 \times 10^5$ S/m; C:$\sigma_{11} = 6.5 \times 10^5$ S/m, $\sigma_{22} = 5.5 \times 10^5$ S/m; D:$\sigma_{11} = 6.5 \times 10^5$ S/m, $\sigma_{22} = 6.5 \times 10^5$ S/m

Table 5.2 Result of NLSE inversion of the nonrandom function in Fig. 5.13

Φ	σ_{11}/sensit	σ_{22}/sensit
0.3424(−1)	591717/1.603(-2)	605798/1.856(−2)

Note that the degree of anisotropy induced by the random permutation process is less than the combination of $\sigma_{11} = 5.9 \times 10^5$, $\sigma_{22} = \sigma_{33} = 6.04 \times 10^5$ S/m by which the process was started. Note further that the result of Table 5.1, which we can call an 'effective (nonrandom) conductivity' that produces the mean impedance, is slightly larger than the mean conductivity produced by the permutation process. This conductivity is equal to $(5.9 \times 10^5 + 2 \times 6.04 \times 10^5)/3 = 5.99 \times 10^5$ S/m. This difference in the two conductivities is familiar to us from previous work with random anisotropic grain noise, and is due to the fact that this is a nonlinear random problem. We cannot expect the mean of the conductivity to produce the mean of the output.

Using the same interpolation blending functions shown in Fig. 5.12 and the nonrandom function shown in Fig. 5.13 as the input to NLSE, the inverted result is shown in Table 5.2. The target solution for this problem is $\sigma_{11} = 5.9 \times 10^5$ S/m and $\sigma_{22} = 6.04 \times 10^5$ S/m, so we have a good solution.

Figure 5.14 shows a sample function generated from the second process, 'random values of the conductivity assigned to the principal axes.' In the example given here, we assume a uniform pdf, with mean values of $\sigma_{11} = 5.9 \times 10^5$ S/m, and $\sigma_{22} = \sigma_{33} = 6.04 \times 10^5$ S/m, with a standard deviation of 5200 S/m for all three conductivities. The result of applying NLSE with this function as the input, and using the same blending functions as before, is shown in Table 5.3. The values of the function in Fig. 5.14 are much larger than those in the ten-sample ensemble shown in Fig. 5.11, so the larger value for the norm of the residuals, Φ, in Table 5.3,

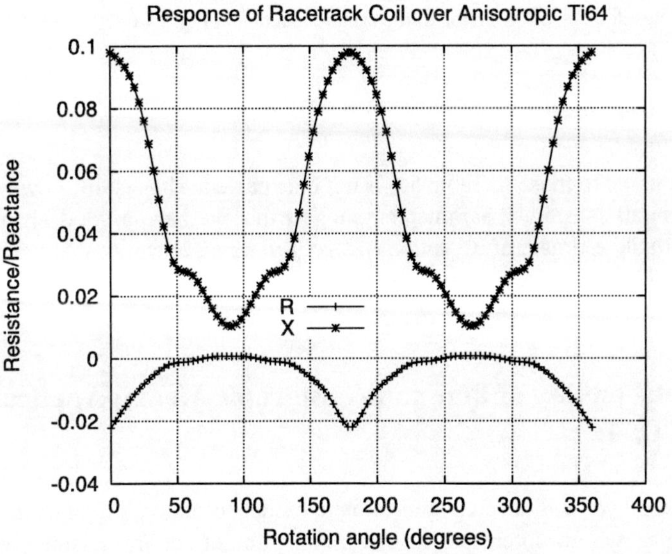

Fig. 5.13 Response of a rotated racetrack coil at 1 MHz over an anisotropic patch of Ti64

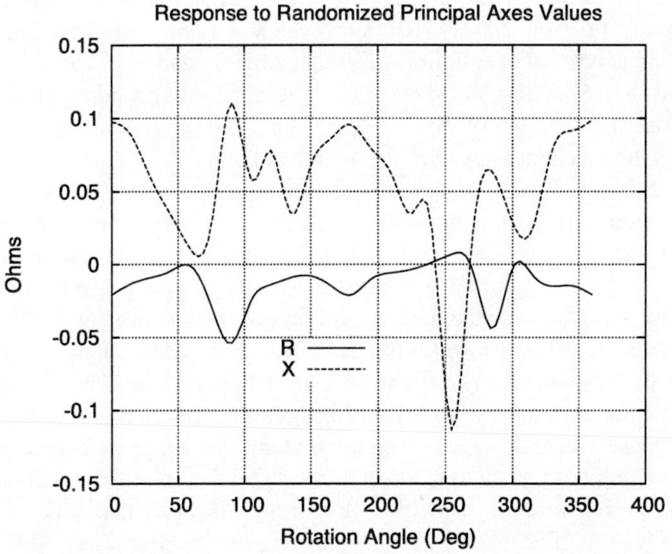

Fig. 5.14 Sample function produced by assigning random values of conductivity to the principal axes with a uniform pdf. Parameter values are given in the text

Table 5.3 Result of NLSE inversion of the sample function in Fig. 5.14

Φ	σ_{11}/sensit	σ_{22}/sensit
0.2301	598203/5.375(−2)	608706/6.222(−2)

when compared to those in Table 5.1 is not unexpected. The results, however, along with the small sensitivity parameters, suggest that we have a good inversion that gives a reliable estimate of σ_{11} and σ_{22}. (Keep in mind, however, that we have only used a single sample function to draw this conclusion.)

5.4 Detectability of Flaws in Anisotropic Media: Application to Ti64

In this section we will develop some simple models of notches in anisotropic Ti64. In particular, we are interested in determining the effects of anisotropies on the impedance response of a coil when scanned over a notch, and then using these results as a basis for determining the effects that anisotropies can have on the detectability of flaws.

Figure 5.15 illustrates the first model calculations. The host is isotropic with a conductivity of $\sigma = 6.04 \times 10^5$ S/m, which is typical of Ti64, and the two anisotropic patches differ only in the exchange of σ_{11} and σ_{22}. The values shown correspond to Ti64 with 6 mm symmetry. The notch is aligned along the 2-axis, which is also the direction of the coil scan. The patch dimensions are $0.1 \times 0.1 \times 0.02$in, and the notch measures $0.01 \times 0.06 \times 0.02$in.

Figure 5.16 shows the impedance-plane response at 2 MHz when the notch is omitted from Fig. 5.15. This will be referred to as the 'patch only' response. The labels 'Top' and 'Bottom' in Fig. 5.16 refer to the top and bottom models in Fig. 5.15. The anisotropy of the patch makes a significant impact in the responses.

When we introduce the notch back into the two models, we get the impedance-plane response at 2 MHz shown in Fig. 5.17. It is clear, when comparing Figs. 5.16 and 5.17, that the 'anisotropy' induced by the 0.01 in-wide flaw dominates the crystalline anisotropy of the host Ti64, in the sense that there is little difference in the response when the notch is oriented along the 1-axis or 2-axis. When we reduce the width of the flaw to 0.005 in, however, and redo the calculations, we get the responses shown in Fig. 5.18. It is clear from this figure that the anisotropy of the host plays a significant role in distinguishing the two responses. This suggests that we will need to do a sensitivity study to determine the minimum detectable flaw size when in the environment of a random crystalline anisotropy.

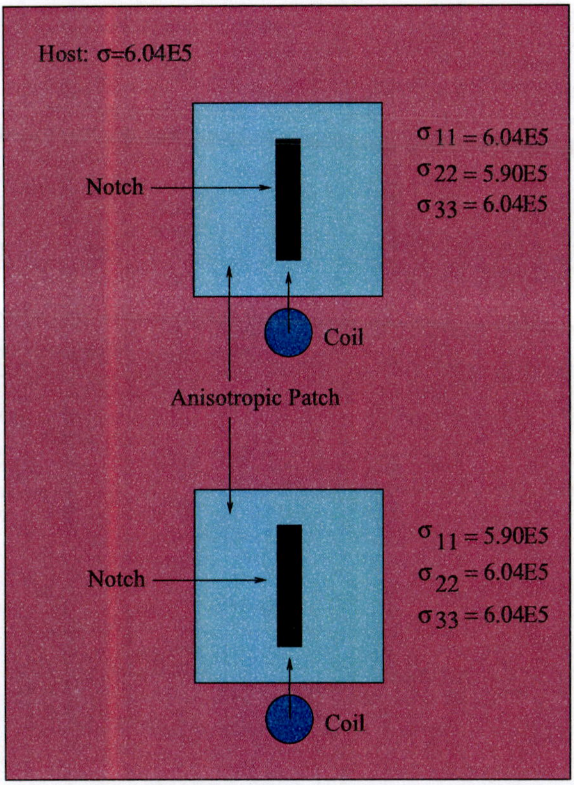

Fig. 5.15 A model of a notch in Ti64. The host is isotropic with a conductivity of $\sigma = 6.04 \times 10^5$ S/m, which is typical of Ti64. The two anisotropic patches differ only in the exchange of σ_{11} and σ_{22}. These values are those for Ti64 with 6 mm symmetry

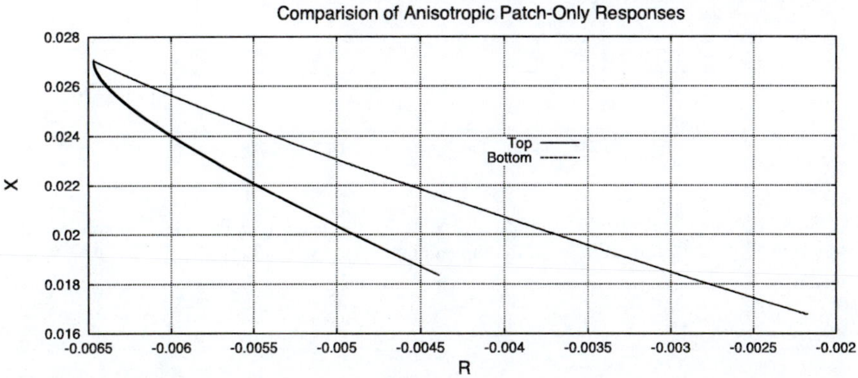

Fig. 5.16 The impedance-plane response at 2 MHz for the 'patch only' configuration, in which the notch is omitted in Fig. 5.15. The labels 'Top' and 'Bottom' refer to the top and bottom models in Fig. 5.15

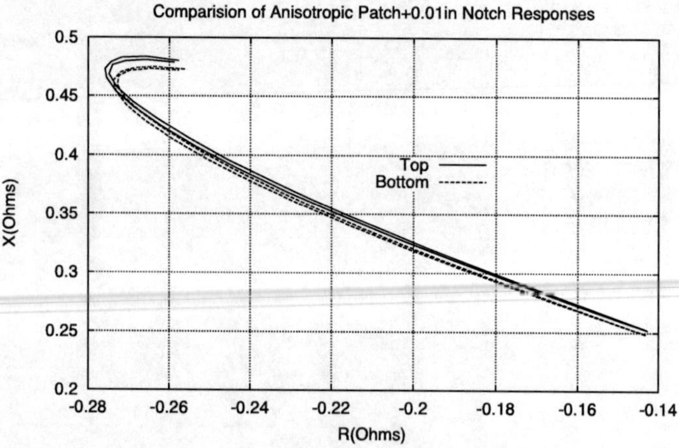

Fig. 5.17 The impedance-plane response at 2 MHz for the patch plus the 0.01" notch models. The labels 'Top' and 'Bottom' refer to the top and bottom models in Fig. 5.15

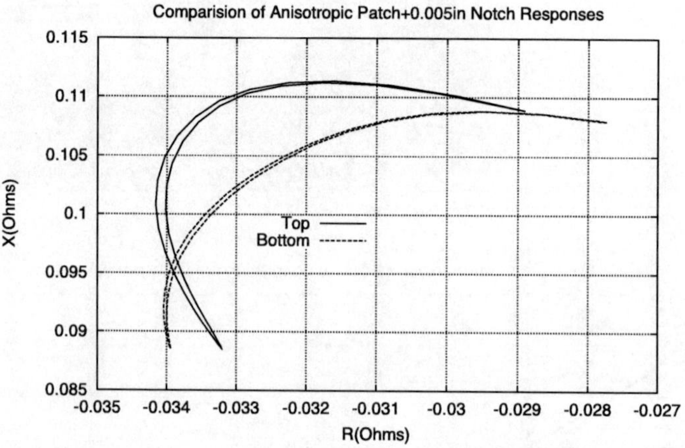

Fig. 5.18 The impedance-plane response at 2 MHz for the patch plus the 0.005" notch models. The labels 'Top' and 'Bottom' refer to the top and bottom models in Fig. 5.15

Chapter 6
Stochastic Inverse Problems: Models and Metrics

6.1 Introducing the Problem

Over the past 2 years, we have been developing a theory of uncertainty quantification and propagation that is computationally feasible with large numbers of unknowns. We have applied it to a problem of characterizing the eddy-current response of a shot-peened surface, where the surface is modeled as a one-dimensional random conductivity field with a known covariance function. We are currently extending the model to more general materials characterization problems, such as modeling two-dimensional random anisotropic grain noise in titanium alloys. In this case, we assume the existence of a (two-dimensional) covariance function for the random distribution of Euler angles that define the orientation of each crystallite within the material.

With this background, we want to develop a theory of stochastic inverse problems for more traditional eddy-current NDE flaw characterization and sizing. Instead of a random material, we assume that the flaw can be characterized as a random process. That this is a reasonable approach is suggested by reference to Fig. 6.1, which shows the typical shape of fatigue-crack growth progression in cold-worked fastener holes. Clearly, the ensemble of cracks cannot be modeled by a simple canonical shape with three parameters, length, width, height, so we will need to invoke a stochastic model for analyzing such cracks.

With such a stochastic model, we can draw parallels between 'probability of detection' (POD) and 'likelihood of inversion' (LOI). In the former, we are given a flaw, and ask ourselves, 'Can we detect it, and what are the metrics that measure our success?' In the latter, we are given data, and ask ourselves, 'Can we associate a flaw with them, and what are the metrics that measure our success?'

H. A. Sabbagh et al., *Advanced Electromagnetic Models for Materials Characterization and Nondestructive Evaluation*, Scientific Computation,
https://doi.org/10.1007/978-3-030-67956-9_6

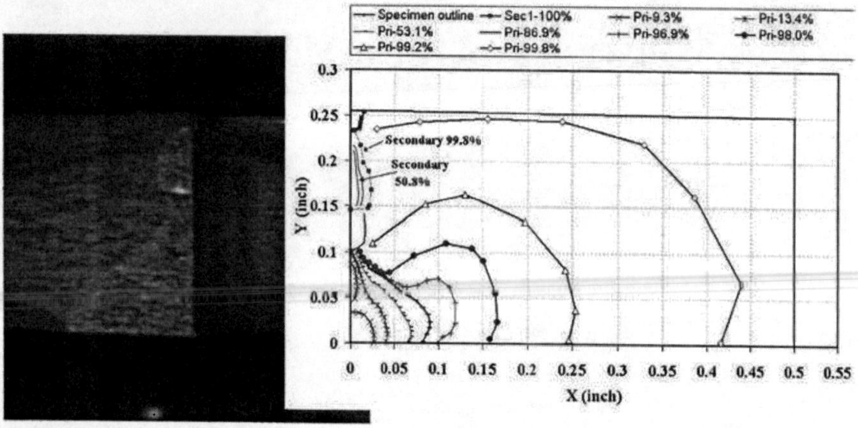

Fig. 6.1 Typical shape of fatigue-crack growth progression in cold-worked fastener holes. Images courtesty of D. Forsyth, TRI/Austin

The stochastic model will be described later, but first we will develop some background tools that are currently resident in **VIC-3D®**, and will be the basis of our stochastic computational model.

6.2 NLSE: Nonlinear Least-Squares Parameter Estimation

Let

$$Z = g(p_1, \ldots, p_N, f) , \tag{6.1}$$

where $p_1, \ldots, p_N$ are the N parameters of interest, and f is a control parameter at which the impedance, Z, is measured. f can be frequency, scan-position, lift-off, etc. It is, of course, known; it is not one of the parameters to be determined. To be explicit during our initial discussion of the theory, we will call f 'frequency.'

In order to determine $p_1, \ldots, p_N$, we measure Z at M frequencies, $f_1, \ldots, f_M$, where $M > N$:

$$\begin{aligned} Z_1 &= g(p_1, \ldots, p_N, f_1) \\ &\vdots \\ Z_M &= g(p_1, \ldots, p_N, f_M) . \end{aligned} \tag{6.2}$$

The right-hand side of (6.2) is computed by applying the volume-integral code to a model of the problem, usually at a discrete number of values of the vector, **p**, forming a multidimensional interpolation grid.

Because the problem is nonlinear, we use a Gauss-Newton iteration scheme to perform the inversion. First, we decompose (6.2) into its real and imaginary parts, thereby doubling the number of equations (we assume the $p_1, \ldots, p_N$ are real). Then we use the linear approximation to the resistance, R_i, and reactance, X_i, at the ith frequency:

$$\begin{bmatrix} R_1 \\ X_1 \\ \vdots \\ R_M \\ X_M \end{bmatrix} \approx \begin{bmatrix} R_1(p_1^{(q)}, \ldots, p_N^{(q)}) \\ X_1(p_1^{(q)}, \ldots, p_N^{(q)}) \\ \vdots \\ R_M(p_1^{(q)}, \ldots, p_N^{(q)}) \\ X_M(p_1^{(q)}, \ldots, p_N^{(q)}) \end{bmatrix} + \begin{bmatrix} \frac{\partial R_1}{\partial p_1} & \cdots & \frac{\partial R_1}{\partial p_N} \\ \frac{\partial X_1}{\partial p_1} & \cdots & \frac{\partial X_1}{\partial p_N} \\ & \vdots & \\ \frac{\partial R_M}{\partial p_1} & \cdots & \frac{\partial R_M}{\partial p_N} \\ \frac{\partial X_M}{\partial p_1} & \cdots & \frac{\partial X_M}{\partial p_N} \end{bmatrix}_{(p_1^{(q)}, \ldots, p_N^{(q)})} \begin{bmatrix} p_1 - p_1^{(q)} \\ \vdots \\ p_N - p_N^{(q)} \end{bmatrix}, \tag{6.3}$$

where the superscript (q) denotes the qth iteration, and the partial derivatives are computed numerically by the software. The left side of (6.3) is taken to be the measured values of resistance and reactance. We rewrite (6.3) as

$$0 \approx r + Jp\,, \tag{6.4}$$

where r is the $2M$-vector of residuals, J is the $2M \times N$ Jacobian matrix of derivatives, and p is the N-dimensional correction vector. Equation (6.4) is solved in a least-squares manner starting with an initial value, $(x_1^{(0)}, \ldots, x_N^{(0)})$, for the vector of unknowns, and then continuing by replacing the initial vector with the updated vector $(x_1^{(q)}, \ldots, x_N^{(q)})$ that is obtained from (6.3), until convergence occurs.

We are interested in determining a bound for the sensitivity of the residual norm to changes in some linear combination of the parameters. Given an $\epsilon > 0$ and a unit vector, v, the problem is to determine a sensitivity (upper) bound, σ, such that

$$\|r(x^* + \sigma v)\| \le (1+\epsilon)\|r(x^*)\|\,. \tag{6.5}$$

We will derive an estimate of σ. Equation (6.5) is equivalent to

$$\|r(x^* + \sigma v)\| - \|r(x^*)\| \le \epsilon\|r(x^*)\|\,. \tag{6.6}$$

The left-hand side of (6.6) can be approximated to the second order in σ by the second-order Taylor expansion:

$$\|r(x^* + \sigma v)\| - \|r(x^*)\| \approx \sigma v \cdot \nabla\|r(x^*)\| + \frac{\sigma^2}{2}\sum_{i,j} \frac{\partial^2\|r(x)\|}{\partial x_j \partial x_i}|_{x^*} v_i v_j \;, \tag{6.7}$$

where ∇ is the gradient operator in $N-$dimensional space. Even though the gradient vanishes at the minimum point, we will compute it to get the algebra started:

$$\nabla\|r(x)\| = \nabla\left[f_1^2(x) + f_2^2(x) + \cdots + f_{2M}^2(x)\right]^{1/2}$$

$$= \frac{1}{\|r(x)\|}\begin{bmatrix} f_1\frac{\partial f_1}{\partial x_1} + \cdots + f_{2M}\frac{\partial f_{2M}}{\partial x_1} \\ \vdots \\ f_1\frac{\partial f_1}{\partial x_N} + \cdots + f_{2M}\frac{\partial f_{2M}}{\partial x_N} \end{bmatrix}^T$$

$$= \frac{r(x)^T}{\|r(x)\|}\begin{bmatrix} \frac{\partial f_1}{\partial x_1} & \cdots & \frac{\partial f_1}{\partial x_N} \\ & \vdots & \\ \frac{\partial f_{2M}}{\partial x_1} & \cdots & \frac{\partial f_{2M}}{\partial x_N} \end{bmatrix}$$

$$= e^T(x) \cdot J \;, \tag{6.8}$$

where the superscript T denotes the transpose of a matrix (or vector), and $e(x) = r(x)/\|r(x)\|$ is a unit vector.

The second derivative that we want is the gradient of (6.8):

$$\nabla\nabla\|r(x)\| = -\frac{\nabla\|r(x)\|}{\|r(x)\|^2}\begin{bmatrix} f_1\frac{\partial f_1}{\partial x_1} + \cdots + f_{2M}\frac{\partial f_{2M}}{\partial x_1} \\ \vdots \\ f_1\frac{\partial f_1}{\partial x_N} + \cdots + f_{2M}\frac{\partial f_{2M}}{\partial x_N} \end{bmatrix}^T$$

$$+ \frac{1}{\|r(x)\|}\nabla\begin{bmatrix} f_1\frac{\partial f_1}{\partial x_1} + \cdots + f_{2M}\frac{\partial f_{2M}}{\partial x_1} \\ \vdots \\ f_1\frac{\partial f_1}{\partial x_N} + \cdots + f_{2M}\frac{\partial f_{2M}}{\partial x_N} \end{bmatrix}^T . \tag{6.9}$$

Before going further, we can immediately drop the first term in (6.9) because the gradient of the norm vanishes at the solution x^*. Thus, (6.9) becomes, using index notation,

$$
\begin{aligned}
\nabla\nabla\|r(x)\| &= \frac{1}{\|r(x)\|}\frac{\partial}{\partial x_j}\left[f_1\frac{\partial f_1}{\partial x_i}+\cdots+f_{2M}\frac{\partial f_{2M}}{\partial x_i}\right], \quad i,j=1,\ldots,N \\
&= \frac{1}{\|r(x)\|}\sum_\alpha\left[\frac{\partial f_\alpha}{\partial x_j}\frac{\partial f_\alpha}{\partial x_i}+f_\alpha\frac{\partial^2 f_\alpha}{\partial x_j\partial x_i}\right], \quad \alpha=1,\ldots,2M\,. \qquad (6.10)
\end{aligned}
$$

Following [88, page 523], we discard the second-derivative term in (6.10) by arguing that the residual vector for a good model fit should be small, which would make the second derivative term small. Furthermore, it is likely that the residual vector should have terms that are uncorrelated with each other and with the model, thus tending to cancel the second derivative terms when summed over α. We will call (6.10) the first-order curvature tensor, Γ_{ij}, of the mapping (or deformation) of the parameter space, $\{x_i\}$, into the model-measurement space. If we call the ith column of the Jacobian matrix, c_i, then it follows from (6.10) that

$$
\Gamma_{ij}(x^*) = \frac{c_i(x^*)\cdot c_j(x^*)}{\|r(x^*)\|}\,, \qquad (6.11)
$$

where we are ignoring the second-derivative term in (6.10).

Digression on Computing $\boldsymbol{\Gamma}_{ij}(x^*)$ We can use the MINPACK code that is already in NLSE to compute $c_i(x^*)\cdot c_j(x^*)$. The computation of the diagonal elements is already available as the 'self sensitivities,' so that leaves the off-diagonal elements. Consider $\|c_i(x^*)/\sqrt{2}+c_j(x^*)/\sqrt{2}\|^2 = \frac{\|c_i(x^*)\|^2}{2}+c_i(x^*)\cdot c_j(x^*)+\frac{\|c_j(x^*)\|^2}{2}$. Hence, it follows that $c_i(x^*)\cdot c_j(x^*) = \|c_i(x^*)/\sqrt{2}+c_j(x^*)/\sqrt{2}\|^2 - \frac{\|c_i(x^*)\|^2+\|c_j(x^*)\|^2}{2}$, where the right-hand side is already calculable using MINPACK in NLSE.

Substituting this result into (6.7) yields an upper bound for the quadratic term:

$$
\begin{aligned}
\sigma^2\sum_{i,j}\frac{\partial^2\|r(x)\|}{\partial x_j\partial x_i}\Big|_{x^*}v_iv_j &= \frac{\sigma^2}{\|r(x^*)\|}\sum_\alpha\left[\sum_{i,j}\frac{\partial f_\alpha}{\partial x_i}v_i\frac{\partial f_\alpha}{\partial x_j}v_j\right]_{x^*} \\
&= \frac{\sigma^2}{\|r(x^*)\|}(J(x^*)\cdot v)\cdot(J(x^*)\cdot v) \\
&= \frac{\sigma^2}{\|r(x^*)\|}\|J(x^*)\cdot v\|^2\,, \qquad (6.12)
\end{aligned}
$$

and if we equate this to the right-hand side of (6.6), we get the final result

$$
\sigma_v = \epsilon^{1/2}\left(\frac{\|r(x^*)\|}{\|J(x^*)\cdot v\|}\right). \qquad (6.13)
$$

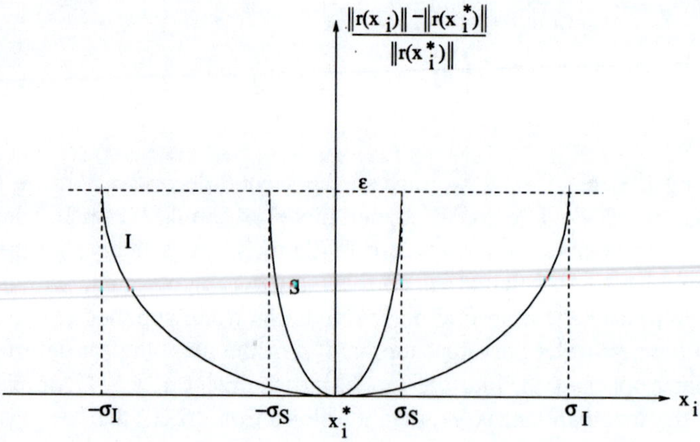

Fig. 6.2 Showing sensitivity parameters for two system responses to x_i. Response S is sensitive to x_i at x_i^*, whereas response I is not

We will call this the 'first-order' approximation, in the sense that we have truncated the Taylor series expansion with the first nonzero term, and have ignored the second-derivative terms in (6.10). This is the expression that is stated, but not derived, in [77].

Note that if $\|J(x^*) \cdot v\|$ is small compared to $\|r(x^*)\|$, then σ is large and the residual norm is insensitive to changes in the linear combination of the parameters specified by v. If $v = e_i$, the ith column of the $N \times N$ identity matrix, then (6.13) produces σ_i, the sensitivity bound for the ith parameter. Since σ_i will vary in size with the magnitude of x_i^*, it is better to compare the ratios σ_i/x_i^* for $i = 1, \ldots, N$ before drawing conclusions about the fitness of a solution.

The importance of these results is that we now have metrics for the inversion process: $\Phi = \|r(x^*)\|$, the norm of the residual vector at the solution, tells us how good the fit is between the model data and measured data. The smaller this number the better, of course, but the 'smallness' depends upon the experimental setup and the accuracy of the model to fit the experiment. Heuristic judgement based on experience will help in determining the quality of the solution for a given Φ.

The sensitivity coefficient, σ, is more subtle, but just as important. It, too, should be small, but, again, the quality of the 'smallness' will be determined by heuristics based upon the problem. If σ is large in some sense, it suggests that the solution is relatively independent of that parameter, so that we cannot reasonably accept the value assigned to that parameter as being meaningful, as suggested in Fig. 6.2, which shows a system, S, for which the system is sensitive to variable, x_i, at the solution point, x_i^*, and another system, I, for which the system is insensitive to x_i.

An example occurs when one uses a high-frequency excitation, with its attendant small skin depth, to interrogate a deep-seated flaw. The flaw will be relatively invisible to the probe at this frequency, and whatever value is given for its parameters will be highly suspect. When this occurs we will either choose a new parameter to characterize the flaw, or acquire data at a lower frequency.

These metrics are not available to us in the current inspection method, in which analog instruments acquire data that are then interpreted by humans using hardware standards. The opportunity to use these metrics is a significant advantage to the model-based inversion paradigm that we propose in this paper.

6.3 Confidence Levels: Stochastic Global Optimization

We can extend the previous results to obtain a statistical measure of confidence in the solution. Referring to Fig. 6.2, we have the probability relation

$$\mathrm{Prob}[x_i^* - \sigma_v v \le x_i \le x_i^* + \sigma_v v] = \mathrm{Prob}\left[\frac{\|r(x_i)\| - \|r(x_i^*)\|}{\|r(x_i^*)\|} \le \epsilon\right]. \tag{6.14}$$

Arguing that $\dfrac{\|r(x_i)\| - \|r(x_i^*)\|}{\|r(x_i^*)\|}$ is a random variable allows us to transform the inverse methods of [111] into the realm of 'stochastic inverse problems.'

This approach is based on the current 'Multi-Level Single Linkage' algorithm that is used in NLSE to reach the global minimum with probability one [21, 78, 89, 94], and also fits our concept of 'stochastic inversion.' Furthermore, it allows us to use prior knowledge of the unknown parameters. Let the model parameters, $\{x_n\}$, be a set of independent random variables, each uniformly distributed over its known range of values. We'll sample the parameter space by choosing, say, 500 points randomly, in accordance with the distribution function of each parameter, and compute the norm of the residual vector at each of the points, as in the first step of NLSE. In NLSE, these points are trial initial points for the minimization algorithm, (6.3), and the lowest of the resulting 500 minima is guaranteed to be the global minimum with unit probability [21, 78, 89, 94].[1]

The random variable, $\dfrac{\|r(x_i)\| - \|r(x_i^*)\|}{\|r(x_i^*)\|}$, in (6.14) is a continuous function of $\{x_i\}$ defined on a compact set (the 'prior feasible set'), so it achieves a finite maximum on that set. This maximum, if it could be determined with probability one, is precisely ϵ in (6.14), and when this is substituted into the transfer function, (6.13), we would have determined the confidence level, σ_v, with unit probability. Later we will relax any claims of unit probability in determining ϵ, but we are permitted to

[1]The Multi-Level Single Linkage method guarantees that the global minimum will be found within a finite number of iterations with probability one, given a sufficiently large sample size of trial points. Numerical experiments with model and laboratory data for a variety of inverse problems over many years [111] suggest that 500 trial points yield a reliable estimate of the global minimum for problems with the number of variables that we are considering.

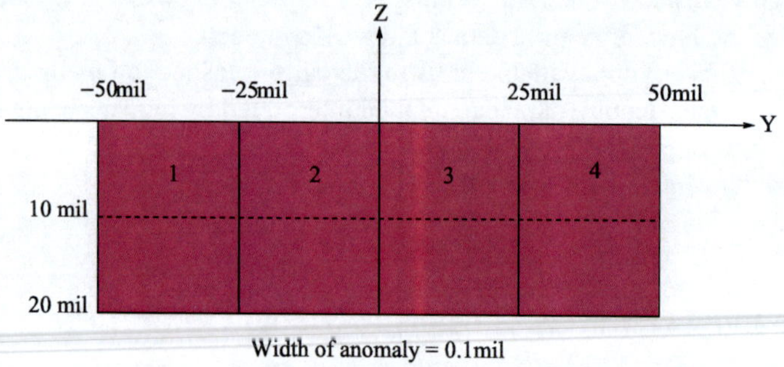

Fig. 6.3 Showing the configuration of the one-dimensional pulse basis functions for parametrizing a complex flaw. The nodes are located at depths of 0, 10, and 20 mils

make a strong statement about the confidence level, because in this formulation of a stochastic inverse problem, we are assuming prior statistical constraints of the unknown parameters, $\{x_n\}$. This approach is quite 'Bayesian', in the sense that we are combining prior information on the random variables with a likelihood estimation (which follows from the least-squares inversion process) to get posterior information on the variables.

Example: A Complex 'Flaw' The configuration of the problem is shown in Fig. 6.3. The expansion of the flaw in the $(Y, Z)-$plane is given by

$$f(y,z) = \sum_{i=1}^{4} \alpha_i \pi_i^{(1)}(y)\pi^{(1)}(z) , \tag{6.15}$$

where $\pi^{(1)}$ is a unit pulse function, and the expansion coefficients, $\{\alpha_i\}_{i=1}^4$, determine the magnitude of $\pi^{(1)}(z)$. These coefficients are the unknown degrees of freedom of the problem, and will be modeled as independent random variables with a uniform distribution over the range $[0, 20]$. They will be determined by inversion of the data, which are impedances measured by a probe that is scanned over $-100 \leq Y \leq 100$, $X = 0$. It should be understood that this formalism fixes the resolution of the flaw in the $Y-$direction to be 25 mils, as well as the width of the flaw in the $X-$direction to be 0.1 mil. These numbers are arbitrary, of course, and can be changed to suit the problem. Furthermore, with the four blocks arranged as shown, this configuration will be best suited for modeling and reconstructing midbore, throughwall, and corner bolt-hole cracks.

Figure 6.4 illustrates a complex flaw extending over the entire range in Y. We will use the output of a **VIC-3D®** model of this flaw to serve as the input data for inversion. To illustrate the inversion process and the importance of the 'surrogate' interpolation table for the $\{\alpha_i\}$, we will perform a numerical experiment in which the table has successively two, three and four nodes per dimension. In the first case, the

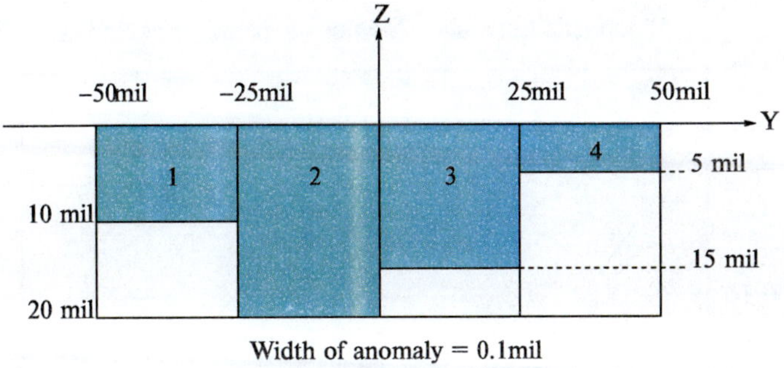

Fig. 6.4 Showing the configuration of a complex flaw extending over the entire range of Fig. 6.3

Table 6.1 Results for the example problem vs. number of nodes per dimension

# Nodes	$\|r(\alpha^*)\|$	α_1/sensit	α_2/sensit	α_3/sensit	α_4/sensit	# Points
2	0.319(−1)	11.47/4.12	20.0/3.4	16.80/3.58	4.64/4.46	272
3	0.405(−2)	10.18/0.53	19.86/0.698	15.87/0.558	3.9/0.444	226
4	0.159(−2)	11.19/0.125	20.11/0.216	15.56/0.212	6.06/0.1918	255

nodes are at [0, 20] , in the second, they are at [0, 10, 20], as in Fig. 6.3, and in the third, [0, 7, 14, 21] (in this case, we assume a uniform distribution of the variables over the range [0, 21]). Thus, the first table comprises $2^4 = 16$ nodes, the second $3^4 = 81$ nodes, and the last $4^4 = 256$ nodes. A blending function for each node is computed by **VIC-3D®**. We quickly see the 'curse of dimensionality' occurring. This curse will be obviated through the use of sparse-grid interpolation techniques to reduce the computational burden of building the new table.

The results of the experiment are shown in Table 6.1. The column labeled '# Points' lists the number of the original 500 global starting points that are attracted to the global minimum. These results show that increasing the number of nodes per dimension yields improvements in reducing the norm of the residuals, Φ, and the sensitivity coefficients of each variable. Figure 6.5 illustrates the results of Table 6.1, and clearly indicates that increasing the number of nodes beyond 4 will have little effect on the norm of the residuals, r, and only a slight reduction in the various sensitivity coefficients, sensit_i.

We ran NLSE four times, effectively sampling the $\{\alpha_i\}$ space 2000 times, yielding values of $\|r(\alpha)\|_{\max} = 0.2545,\ 0.2689,\ 0.2351,$ and 0.265. The inverted results of each of these runs were identical to those tabulated in Table 6.1, as we expected, since the algorithm in NLSE ensures convergence to the global minimum with probability one. Hence, using the data of the bottom row of Table 6.1 we have

$$\frac{\|r(\alpha)\|_{\max} - \|r(\alpha^*)\|}{\|r(\alpha^*)\|} = \frac{0.2689 - 0.00159}{0.00159} = 168.12 = \epsilon\ , \tag{6.16}$$

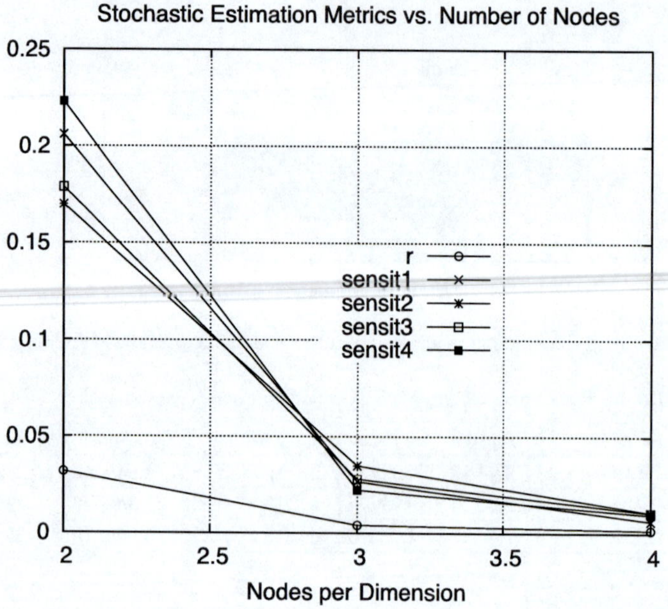

Fig. 6.5 The results of Table 6.1 indicate that increasing the number of nodes beyond 4 will have little effect on the norm of the residuals, r, and only a slight reduction in the various sensitivity coefficients, sensit(i). The sensitivity coefficients have been scaled downward by a factor of 20 to make them commensurate with the norm of the residuals for clarity of visualization

and when this is substituted into (6.13), along with the sensitivity coefficients tabulated in the bottom row of Table 6.1, we get the parameters of the confidence intervals to be $\sigma_1 = 1.62,\ \sigma_2 = 2.8,\ \sigma_3 = 2.75,\ \sigma_4 = 2.49$. These effectively define the posterior distribution of the $\{\alpha_i\}$, which is certainly much different than the prior distribution.

We summarize the results for α_i by claiming that we are 'certain' that $\alpha_i - \sigma_i \leq \alpha_i \leq \alpha_i + \sigma_i$, with the most likely value being α_i^*. In the case where one of the posterior limits on α_i exceeds the prior limit, we reject it in favor of the prior limit, because if the crack actually exceeded the prior limit, the inversion process would have been constrained at the prior limit of the interpolation table. For example, $17.31 \leq \alpha_2 \leq 21$, rather than $17.31 \leq \alpha_2 \leq 22.91$.

The Chebyshev Inequality We can improve the calculation of the confidence level, and even make its definition more precise in our example, by resorting to the Chebyshev inequality [63], which states that, if Z is a random variable, then, for every $\xi > 0$,

$$P[|Z| \geq \xi] \leq \frac{\mathrm{VAR}(Z)}{\xi^2} = \text{MAX UNCERTAINTY}(\xi)$$

Table 6.2 Results at 100 Hz–1 kHz for conductivity and permeability

Trial	$\|r(x)\|_{max}$	$\|r(x^*)\|$	σ/sensit	μ/sensit
1	0.2503	0.188(−2)	1.372(6)/2.32(4)	68.18/0.1504
2	0.2509	0.188(−2)	1.372(6)/2.32(4)	68.18/0.1504
3	0.2521	0.188(−2)	1.372(6)/2.32(4)	68.18/0.1504
4	0.2552	0.188(−2)	1.372(6)/2.32(4)	68.18/0.1504
5	0.2525	0.188(−2)	1.372(6)/2.32(4)	68.18/0.1504

$$\text{MINIMUM CERTAINTY}(\xi) = 1 - \frac{\text{VAR(Z)}}{\xi^2}\,, \tag{6.17}$$

where ξ is the threshold or decision boundary for determining the confidence interval. For example, if we want to be at least 95% confident in our assertion of the probability of the first equality in (6.17), then $1 - \frac{\text{VAR(Z)}}{\xi^2} = 0.95$, which implies that $\xi = \left(\frac{\text{VAR(Z)}}{0.05}\right)^{1/2}$.

To apply this theorem to our problem, we define $Z = \|r(\alpha)\|_{max} - \overline{\|r(\alpha)\|_{max}}$, where $\|r(\alpha)\|_{max}$ is a random variable whose sample value is the output of the following 'experiment': run a 500-sample trial, as in the Multi-Level Single Linkage algorithm, and choose the largest result for $\|r(\alpha)\|_{max}$. Repeat the experiment for the second sample, and so on. We have already given an example of this, with the result after four trials that $\{\|r(\alpha)\|_{max}\} = \{0.2545,\ 0.2689,\ 0.2351,\ 0.265\}$, from which follow $\overline{\|r(\alpha)\|_{max}} = 0.2559$, $\text{VAR(Z)} = 0.0001716$, and $\xi = (0.0001716/0.05)^{1/2} = 0.0586$ for 95% confidence level.

From the Chebyshev inequality we have, therefore, $\|r(\alpha)\|_{max} = 0.2559 + 0.0586 = 0.3145$. This replaces $\|r(\alpha)\|_{max} = 0.2689$ in (6.16), so that the 95% upper bound is given by

$$\frac{\|r(\alpha)\|_{max} - \|r(\alpha^*)\|}{\|r(\alpha^*)\|} = \frac{0.3145 - 0.00159}{0.00159} = 196.8 = \epsilon\,. \tag{6.18}$$

The new values for the parameters corresponding to the 95% confidence interval are $\{\sigma_1 = 1.75,\ \sigma_2 = 3.03,\ \sigma_3 = 2.97,\ \sigma_4 = 2.69\}$. The confidence intervals for the four variables are, therefore: $\alpha_1 : [9.44, 12.94]$, $\alpha_2 : [17.08, 21]$, $\alpha_3 : [12.59, 18.53]$, $\alpha_4 : [3.37, 8.75]$.

Joint Measurement of Conductivity and Magnetic Permeability We have taken impedance measurements over the frequency range of 100Hz–1 kHz of a ferritic heat-exchanger tube, with the intention of jointly determining the conductivity and relative magnetic permeability of the tube. The interpolation table had the following nodal values: σ : 1.0×10^6, 1.2×10^6, 1.4×10^6, 1.6×10^6, 1.8×10^6; μ : 50, 60, 70, 80, 90. We ran five trials of NLSE with the following results (Table 6.2):

Table 6.3 Inverted results for the width of a crack

Trial	$\|r(x)\|_{max}$	$\|r(x^*)\|$	W/sensit
1	0.7979(−1)	0.2495(−1)	0.0793/0.0296
2	0.7990(−1)	0.2495(−1)	0.0793/0.0296
3	0.8024(−1)	0.2495(−1)	0.0793/0.0296
4	0.7941(−1)	0.2495(−1)	0.0793/0.0296
5	0.7999(−1)	0.2495(−1)	0.0793/0.0296

Following the procedure described above with respect to the Chebyshev inequality, we calculate a value of $\epsilon = 137$, which yields $\sigma_{\text{cond}} = 0.272 \times 10^6$, and $\sigma_\mu = 1.76$. Hence, we can say that the most likely value of the conductivity is 1.372×10^6, with a 95% confidence interval of $[1.1 \times 10^6,\ 1.644 \times 10^6]$. For the permeability we get even tighter results; the most likely value is 68.18, with a 95% confidence interval of [66.42, 69.94].

The fact that the permeability is well defined at these low frequencies has been validated by use of the Cramer-Rao Lower Bound (CRLB), [111, pp. 407–410], where it is also shown that the optimum frequency for estimating conductivity is 6.0 kHz.

Estimation of Width of a Long, Thin Crack We are given data at 200 kHz for a crack in a bolt-hole. The data were obtained by a splitD probe with ferrite cores, and the crack was 100 mils long and 18 mils deep. The objective was to determine the width of the crack. The problem is described in greater detail in [111, Section 6.6].

The interpolation table for the width has nodes at 0, 0.125 mils, and 0.25 mils. The inverted results after 5 trials are shown in Table 6.3.

These results yield a value of $\epsilon = 2.31$ and $\sigma_W = 0.045$. The most likely value of W is 0.0793 mils, and the 95% confidence interval is [0.0343, 0.1243]. We should note that in these two examples, the confidence interval calculation becomes more precise with an increase in the number of nodes in the interpolation table, as indicated earlier.

6.4 Summary

We summarize the algorithm and process here.

1. $\dfrac{\|r(x)\|_{max} - \|r(x^*)\|}{\|r(x^*)\|} = \text{OBJ}(x)$ is a random variable.
2. $\|r(x^*)\|$ and the Jacobian, $J(x^*)$, are determined with prob $\to 1$ (Stochastic Global Optimization via MLSL).
3. The set $\{x(\epsilon)\} \ni \text{OBJ}(x) \leq \epsilon$ is the 'posterior feasible set at level ϵ'.
4. If $\text{OBJ}(x)$ is parabolic (ellipsoidal in N-space), then the set $\{x(\epsilon)\}$ is called the 'first-order posterior feasible set at level ϵ'.

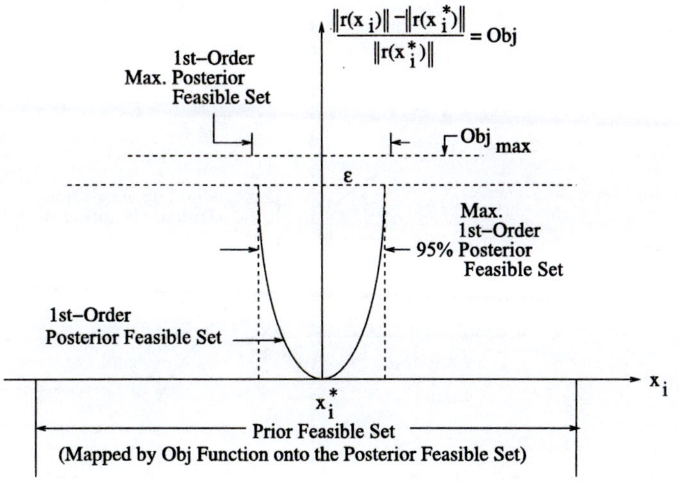

Fig. 6.6 Illustrating the interpretation and calculation of confidence intervals

5. $\sigma_v = \epsilon^{1/2}\left(\frac{\|r(x^*)\|}{\|J(x^*)\cdot v\|}\right)$ is a mapping from the 'prior feasible set' to the 'first-order posterior feasible set at level ϵ'.
6. If we choose ϵ to be at the 95% confidence level, as with the Chebyshev Inequality, then the measure of $\{x^* - \sigma \leq x \leq x^* + \sigma\}$ is at least 95% that of the maximum first-order posterior feasible set, and x^* is the most likely value of x.

Figure 6.6 illustrates the algorithm.

A 2D Example The results just given are for the situation in which each parameter is tested separately, while the others are fixed at the solution point. Now, we must consider the general case in which the totality of variables are considered jointly. This means operating in four-dimensional space. The tools that we have already set up allow us to do that with no additional expense, except for a minor enhancement to the NLSE code in **VIC-3D®**. Equation (6.13) is valid for arbitrary orientations of the unit vector, v, and ϵ has already been computed using the entire four-dimensional random parameter space in the MLSL stochastic global optimization algorithm.

Consider the 2D example shown in Fig. 6.7, which is the projection onto the (x_1, x_2)-plane of the four-dimensional hyperellipsoid associated with the complex flaw example described earlier. Using NLSE, we compute the joint sensitivity associated with the unit vector, $v = [0.5, 0.5, 0.5, 0.5]$ to be 0.129. Then, using $\epsilon = 196.8$, as before, we compute $\sigma_{0.5,0.5,0.5,0.5} = 1.81$ from (6.13) for the 95%-confidence region for this combination of variables. It should be understood

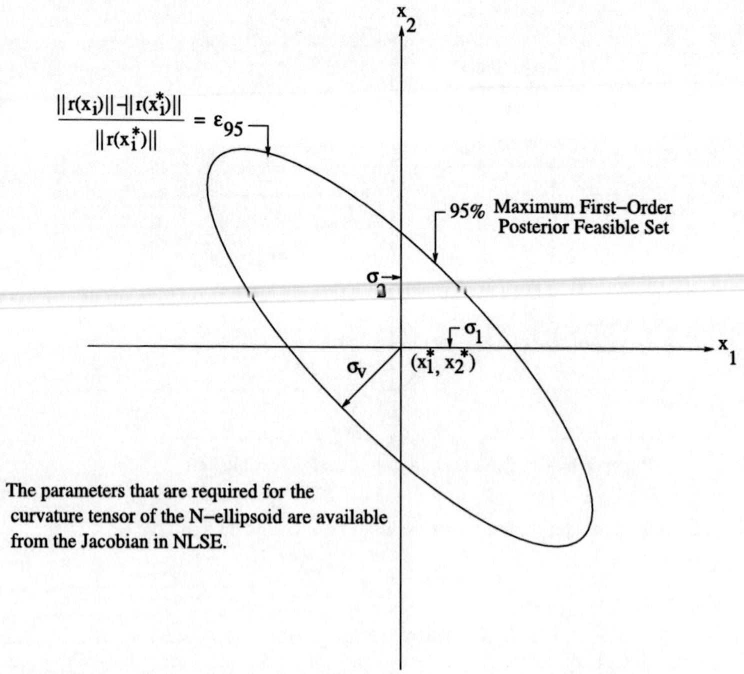

Fig. 6.7 A 2D ellipsoidal confidence region. The origin, (x_1^*, x_2^*), is the solution of the optimization problem

that NLSE already gives us the information to generate the entire N-dimensional hyperellipsoid for a given problem. This would allow us to analytically calculate such things as the volume of the ellipsoid, or cross-sectional areas, etc.

Chapter 7
Integration of Functionals, PCM and Stochastic Integral Equations

7.1 Theoretical Background

The mathematical theory of functional integration was developed in the fifties to be applied to problems in probability and quantum field theory. We're not interested in the formal theory [38], but in its more practical form which can actually be used to generate numbers. In particular, it leads to the probabilistic collocation method (PCM) and other techniques for high-dimensional model representation (HDMR) [37, 42, 65, 69, 142], which we intend to apply to eddy-current nondestructive evaluation (NDE), using the volume-integral code, **VIC-3D®**, as our vehicle.

Let $\sigma(\mathbf{r})$ be a random conductivity field, which means that at each point, $\mathbf{r}$, there exists a random variable, $\sigma(\omega, \mathbf{r})$, with a probability density. Here, ω, denotes the 'random-set' parameter that defines the 'outcome' of the experiment to determine $\sigma(\mathbf{r})$.

Given $\sigma(\mathbf{r})$, we use **VIC-3D®** to compute an impedance, $Z(l, \sigma(\mathbf{r}))$, where l is the position of the probe coil. Clearly, before we can compute Z, we must know $\sigma(\mathbf{r})$ at all of its field points, which are infinite in number (a continuous infinity!). Hence, Z is a 'functional' of $\sigma(\mathbf{r})$, and because $\sigma(\mathbf{r}) = \sigma(\omega, \mathbf{r})$ is a random field, Z becomes a random variable at each of the probe coil points, l. We should write this as $Z(l, [\sigma])$ to denote that Z is a *function* of l, and a *functional* of $\sigma(\mathbf{r})$. We could include the parameter, ω, in Z to remind us that Z is a random functional of $\sigma(\mathbf{r})$, but that would be gilding the lilly with symbols.

If we want to find the average value of $Z(l, [\sigma])$, we form

$$\left[\int_{-\infty}^{\infty}\right] Z(l, [\sigma])\, p[\sigma] d[\sigma]\,, \tag{7.1}$$

H. A. Sabbagh et al., *Advanced Electromagnetic Models for Materials Characterization and Nondestructive Evaluation*, Scientific Computation,
https://doi.org/10.1007/978-3-030-67956-9_7

where the brackets around the integral sign denote a continuous infinity of integrals, one for each **r**, and $p[\sigma]$ is the probability density of $\sigma(\mathbf{r})$ at each **r**. Equation (7.1) is what mathematicians call an 'integral of a functional' or 'functional integration.'

Clearly, we cannot compute infinitely many integrals, so we discretize the problem by discretizing the anomalous region into N cells (voxels), and then replacing $\sigma(\mathbf{r})$ by σ_n, $n = 1, \ldots, N$, where σ_n is a random variable associated with the nth cell. If there are N voxels, then (7.1) becomes

$$\int_{-\infty}^{\infty} \cdots \int_{-\infty}^{\infty} Z(l, \sigma_1, \cdots, \sigma_N) p(\sigma_1, \cdots, \sigma_N) d\sigma_1 \cdots d\sigma_N \,. \tag{7.2}$$

N will generally be large, so we must consider high-dimensional numerical quadrature routines to compute (7.2). This is where 'Gaussian quadratures' and PCM come in [37, 42, 65, 69, 142].

The 'quadrature rule' determines a few points that are optimum for evaluating (7.2), and these are the PCM points at which **VIC-3D®** must compute Z. That is, the PCM points are points in $(\sigma_1, \sigma_2, \cdots, \sigma_N)$-space that go into the **VIC-3D®**-model. We hope that the number of points is not huge, and will yield a good approximation with fewer runs than a Monte Carlo study. The PCM points will depend upon the orthogonal polynomials that are used in the approximation of the integrand, and these polynomials depend upon the nature of the density function $p(\sigma_1, \cdots, \sigma_N)$.

7.2 Probability Densities and Numerical Procedures

We will work with certain probability densities of interest to us in the context of **VIC-3D®**. Anomalous regions in **VIC-3D®**-models are defined by volume-fractions, VF, from which conductivities of the voxels are computed (see (7.62) of Appendix 1). The standard model that we will use is one in which each of the VFs is independent of the other, but each has the same density function. Thus, we have a problem with independent, identically distributed (iid) random variables. Only the density function will change from problem to problem.

Uniform Density The first statistical model is one in which each cell VF has a uniform density. This is clearly the simplest one to analyze from a statistical point of view, and it easily falls within the numerical-quadrature model described above, because orthogonal polynomials and their zeros ('Gauss points') that are compatible with the uniform density are known and tabulated. The 'quadrature weights' for the uniform density are tabulated, as well.

Quadratic Density In Appendix 1 we describe a model for random anisotropic grain noise that arises from the random orientation of crystals with, say, 6 mm symmetry, as in Ti64. The important result is (7.63), and the conclusion that $VF = \cos^2\phi$. In the model, however, it is $\cos\phi$ that is uniformly distributed, so that VF is derived via a quadratic transformation. It can be shown, therefore [141, pp. 61-62], that

$$p_{\mathrm{VF}}(\alpha) = \begin{cases} \dfrac{1}{2\sqrt{\alpha}}\left[p_{\cos\phi}(\sqrt{\alpha}) + p_{\cos\phi}(-\sqrt{\alpha})\right]; & \alpha \geq 0 \\ 0; & \alpha < 0. \end{cases}$$

$$= \begin{cases} \dfrac{1}{2\sqrt{\alpha}}; \ 0 < \alpha \leq 1 \\ 0; \quad \text{otherwise.} \end{cases}, \tag{7.3}$$

where the final result follows because $p_{\cos\phi}(\alpha) = 1$ for $0 \leq \alpha \leq 1$ and vanishes elsewhere. p_{VF} has a weak (integrable) singularity, but it does satisfy the requirements for a probability density, namely that it is positive, and that its integral is unity:

$$\int_0^1 p_{\mathrm{VF}}(\alpha)d\alpha = \frac{1}{2}\int_0^1 \frac{d\alpha}{\sqrt{\alpha}} = 1\,. \tag{7.4}$$

7.3 Second-Order Random Functions

If we don't know the a priori probability density for the volume-fractions or conductivities, then we resort to the next best thing, namely certain second-order properties of these random functions. The usual property that is either given, or can be measured, is the covariance function, and this is the point of departure in calculating properties of random surfaces in [2, 19, 53, 145].

We still need to transform knowledge of the covariance function into probability densities for the volume fractions in a **VIC-3D®** model, and we use the Karhunen-Loève expansion to do this. Using Loève's notation [63, pp. 478–479] for a one-dimensional problem, the expansion is given by the following

Theorem 7.1 (Proper Orthogonal Decomposition) *A random function, $X(t)$, continuous in quadratic mean (q.m.) on a closed interval, I, has on I an orthogonal decomposition*

$$X(t) = \sum \lambda_n \xi_n \psi_n(t) \tag{7.5}$$

with

$$E\xi_m\xi_n^* = \delta_{mn}, \quad \int \psi_m(t)\psi_n^*(t)dt = \delta_{mn}\,, \tag{7.6}$$

if, and only if, the λ_n are the proper (eigen-) values and the $\psi_n(t)$ are the orthonormalized proper (eigen-) functions of its covariance. Then the series converges in q.m. uniformly on I.

The significance of this expansion is that it is a method for reducing the infinite-dimensional random process, $X(t)$, to a finite-dimensional one, $\{\xi_n\}$, $n = 1, \cdots, M$. This is another way of transforming expressions such as (7.1) into (7.2), except that the appropriate joint probability density, $p(\sigma_1, \cdots, \sigma_N)$, becomes $\pi(\xi_1, \cdots, \xi_M)$, where we expect that $M << N$. The joint density, π, yields the PCM quadrature points.

7.4 A One-Dimensional Random Surface

The text, [143, pp. 47–50], contains a discussion of the Karhunen-Loève expansion as it applies to the double-exponential covariance function. We will apply this theory to the problem posed in [145], in which the authors postulate a pseudorandom realization of a surface by using the spectral method of S. O. Rice, and then attempting to fit it to the correlation function shown in Fig. 7.1. In this problem, we are given the correlation function, and will use the K-L expansion to derive the realization of the surface. We will be in a better position to look at numerical questions of convergence of the approximations, as well as stability of computations.

The double-exponential correlation function shown in Fig. 7.1 is given by

$$C(x, x') = \exp(-|x - x'|/L)\,, \tag{7.7}$$

and the appropriate equations out of [143] for the eigensolutions of the K-L expansion for this covariance function are

$$\lambda_i = \begin{cases} \dfrac{2L}{1 + L^2 w_i^2}, & \text{if } i \text{ is even} \\ \dfrac{2L}{1 + L^2 v_i^2}, & \text{if } i \text{ is odd} \end{cases}$$

$$\psi_i(x) = \begin{cases} \sin(w_i x)\Big/\sqrt{b - \dfrac{\sin(2w_i b)}{2w_i}}, & \text{if } i \text{ is even} \\ \cos(v_i x)\Big/\sqrt{b + \dfrac{\sin(2v_i b)}{2v_i}}, & \text{if } i \text{ is odd} \end{cases}$$

$$\begin{array}{l} Lw + \tan(wb) = 0, \ \text{if } i \text{ is even} \\ 1 - Lv\tan(vb) = 0, \ \text{if } i \text{ is odd} \end{array}, \tag{7.8}$$

where $x \in [-b, b]$, and L is the correlation length of the double-exponential function. We are changing notation from t in (7.5) and (7.6) so that x is our independent variable, and the closed interval, I, in (7.5) and (7.6) is $[-b, b]$.

With the expansion using (7.8) in hand, we can project the continuous eigenfunction solution onto the space spanned by unit pulse functions, and derive the appropriate expressions for the volume-fractions that are assigned to the random

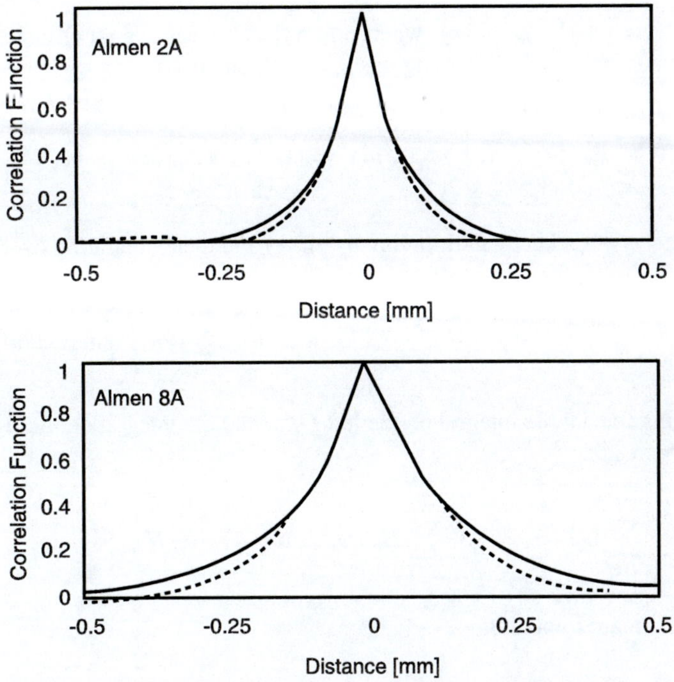

Fig. 7.1 Double-exponential measured correlation function (dashed lines) in shot-peened copper and the best-fitting exponential curves (solid lines) for two different peening intensities: top Almen 2A; bottom Almen 8A (from [145])

conductivity voxels. That's relatively straightforward, but now we want to consider a numerical version of the Karhunen-Loève expansion that may be easier to use in the context of assigning volume-fractions for **VIC-3D®**.

For an arbitrary covariance, $C(x, x')$, the eigenvalue problem

$$\int_{-b}^{b} C(x, x')\psi(x')dx' = |\lambda|^2\psi(x), \quad x \in [-b, b] \,, \tag{7.9}$$

can be transformed into the vector-matrix generalized eigenvalue problem

$$\mathbf{G} \cdot \mathbf{v} = |\lambda|^2 \mathbf{H} \cdot \mathbf{v} \,, \tag{7.10}$$

where $\mathbf{G}$ and $\mathbf{H}$ are matrices and $\mathbf{v}$ is a column-vector. We'll derive this result now, but first we note that any covariance function is non-negative definite and Hermitian, i.e., $C(x, x') = C^*(x', x)$, where the asterisk, $*$, denotes the complex conjugate. Because of its non-negative definiteness, it follows that the eigenvalues in (7.9) are real and non-negative, which allows us to write them as shown.

Let $\psi(x) = \sum_{n=1}^{N} \psi_n f_n(x)$, where $\{f_n(x)\}$ is a basis for $\psi(x)$, and $\{\psi_n\}$ are expansion coefficients. Substituting this into (7.9) yields

$$\sum_{n=1}^{N} \psi_n \int_{-b}^{b} C(x, x') f_n(x') dx' = |\lambda|^2 \sum_{n=1}^{N} \psi_n f_n(x) . \tag{7.11}$$

Take moments of (7.11) by multiplying by $f_m(x)$ and then integrating over $[-b, b]$:

$$\sum_{n=1}^{N} \psi_n \int_{-b}^{b} \int_{-b}^{b} C(x, x') f_m(x) f_n(x') dx dx' = |\lambda|^2 \sum_{n=1}^{N} \psi_n \int_{-b}^{b} f_m(x) f_n(x) dx . \tag{7.12}$$

Upon calling the double integral on the left G_{mn}, and the integral on the right H_{mn}, we have the result

$$\sum_{n=1}^{N} G_{mn} \psi_n = |\lambda|^2 \sum_{n=1}^{N} H_{mn} \psi_n, \quad m = 1, \cdots, N , \tag{7.13}$$

or, in vector-matrix notation

$$\mathbf{G} \cdot \mathbf{v} = |\lambda|^2 \mathbf{H} \cdot \mathbf{v} , \tag{7.14}$$

where $\mathbf{v} = [\psi_1, \cdots, \psi_N]^T$. This completes the derivation of (7.10). If $\{f_n(x)\}$ is orthogonal, then $\mathbf{H}$ is diagonal, and if $\{f_n(x)\}$ is normalized to unity, then $\mathbf{H}$ is the identity matrix, and the generalized eigenvalue problem, (7.10) reduces to the standard form

$$\mathbf{G} \cdot \mathbf{v} = |\lambda|^2 \mathbf{v} . \tag{7.15}$$

The discrete version of the K-L expansion then becomes

$$\mathbf{V} = \sum_{n=1}^{N} \lambda_n \xi_n \mathbf{v}_n , \tag{7.16}$$

where $\{\lambda_n\}$ and $\{\mathbf{v}_n\}$ are the eigenpair solutions of (7.15), and $\{\xi_n\}$ are uncorrelated random variables, $E\xi_m\xi_n^* = \delta_{mn}$, that define the PCM points which we'll talk about later.

In order to make this theory amenable to **VIC-3D®** we will discretize the interval $[-b, b]$ into $2N$ cells, each of length $\delta = b/N$, and let the basis functions, $\{f_n(x)\}$, be unit pulses defined over this grid

$$f_n(x) = \pi(x/\delta - n) = \begin{cases} 1, & \text{if } n\delta \leq x < (n+1)\delta, \quad -N \leq n \leq N-1 \\ 0, & \text{otherwise} \end{cases} \tag{7.17}$$

Note that this system is also orthogonal.

Applying this to the double-exponential covariance function (7.7), we have for the components of the matrix, **G**:

$$G_{mn} = \int_{-b}^{b}\int_{-b}^{b} e^{-|x-x'|/L}\pi(x/\delta - m)\pi(x'/\delta - n)dxdx'$$

$$= \int_{n\delta}^{(n+1)\delta} dx' \int_{m\delta}^{(m+1)\delta} e^{-|x-x'|/L}dx$$

$$= I_1 + I_2 + I_3 + I_4\,, \tag{7.18}$$

where $[I_1, I_2, I_3, I_4]$ are given by [99, eqn(50c)]

$$\begin{aligned}
I_1 &= -2x_1L - L^2e^{-x_1/L},\ x_1 > 0,\ x_1 = (m-n)\delta \\
&= -L^2e^{x_1/L}, \qquad\qquad x_1 < 0 \\
I_2 &= -2x_2L - L^2e^{-x_2/L},\ x_2 > 0,\ x_2 = (m-n)\delta \\
&= -L^2e^{x_2/L}, \qquad\qquad x_2 < 0 \\
I_3 &= 2x_3L + L^2e^{-x_3/L}, \quad x_3 > 0,\ x_3 = (m-n-1)\delta \\
&= L^2e^{x_3/L}, \qquad\qquad x_3 < 0 \\
I_4 &= 2x_4L + L^2e^{-x_4/L}, \quad x_4 > 0,\ x_4 = (m-n+1)\delta \\
&= L^2e^{x_4/L}, \qquad\qquad x_4 < 0\ .
\end{aligned} \tag{7.19}$$

When this is substituted into (7.18) we get the final expressions for G_{mn}:

$$\begin{aligned}
G_{mn} &= -2L^2\left(1 - e^{-\delta/L} - \delta/L\right), \qquad m = n \\
&= -2L^2e^{-|n-m|\delta/L}\left(1 - \cosh(\delta/L)\right), \text{ otherwise}\ .
\end{aligned} \tag{7.20}$$

This is a symmetric Töplitz (convolution) matrix, whose critical parameter is the ratio, δ/L. For this same system of expansion functions, the **H** matrix is diagonal with the constant value of δ.

By referring to (7.20) we can draw some conclusions about the eigenvalues for extreme cases of δ/L without solving the eigenvalue problem, (7.15). For example, if $\delta/L \approx 0$, i.e., L is very large compared to the cell length, we can see from (7.20) that the exponentials in the bottom two equations tend to unity, and the expressions within the parentheses in all three equations tend to the same approximate value of $-(1/2)(\delta/L)^2$, as can be verified by expanding each expression in a Taylor series. This means that in the limit of small δ/L, $G_{mn} = \delta^2$ for all $-N \leq m, n \leq N$. Hence, this matrix is singular, and has a constant (normalized) eigenvector, $v_n = 1/\sqrt{2N}$ for all $-N \leq n \leq N$, and a single nonzero eigenvalue, $\lambda^2 = 2N\delta = 2b$, where $2b$ is the total length of the anomalous region. Thus, this process is fully correlated, and the expansion, (7.16), has only a single term with just one random variable. From the perspective of (7.9) the operator equation on the left-hand side is a simple integrator, which means that the eigenfunction must be a constant.

On the other hand, in the limit of vanishing correlation length, for which δ/L become very large, we have the opposite situation. In this case the off-diagonal terms in (7.20) vanish to exponential order, and we are left with a diagonal matrix that is a multiple of the identity matrix. Actually, the expression for G_{mm} in (7.20) vanishes as $L \to 0$, because the double-exponential covariance function vanishes everywhere except at $x = x'$, where it is assigned the value of unity. The integral of a function that vanishes everywhere except at one point is zero, unless at that one point the function becomes 'infinite' in some manner. If we normalize the covariance by dividing by $2L$, then as can be seen from (7.20) the diagonal elements all take on the value of δ as $L \to 0$.

Thus, there are $2N$ identical eigenvalues, and $2N$ orthogonal eigenvectors that can be chosen arbitrarily. The expansion, (7.16), therefore, has the full number of terms with uncorrelated random variables; there will be no decay of the eigenvalues with n. This is an example of an uncorrelated process, that we could call 'discrete white noise'; each voxel in the **VIC-3D®** grid is uncorrelated with each other voxel. This is essentially the way that we are currently treating random problems with **VIC-3D®**.

Figure 7.2 shows the normalized eigenvalue spectrum for a 32×32 matrix (7.20) for three values of the ratio, L/δ. Clearly, our surmise is confirmed: the spectrum dies out faster for larger values of this ratio, and this will have a profound effect on the profile of conductivity, as will be discussed next.

In the preceding discussion of the eigenvalue spectrum, and in what follows, we assume a one-dimensional grid that is 6.4 mm long and contains 32 cells, each of length $\delta = 0.2$ mm. Further, we assume that the correlation function shown in Fig. 7.1 is multiplied by a variance, σ^2, to produce the covariance function associated with the random surface.

The random conductivity profiles to be shown next are computed from (7.16), with $\{\xi_n\}$ being independent, identically distributed, zero-mean, unit-variance random variables with a uniform density function. These are generated using the Fortran RANDOM_NUMBER subroutine, as described in Appendix 2. The profile shows the (uniform) conductivity of each cell, which is the basis for defining anomalous regions in **VIC-3D®**. Sample functions for the case, $L/\delta = 0.1$, are shown in Fig. 7.3, while Fig. 7.4 shows three sample functions for the conductivity profile when $L/\delta = 1$, and Fig. 7.5 shows three sample functions for the conductivity profile when $L/\delta = 3$.

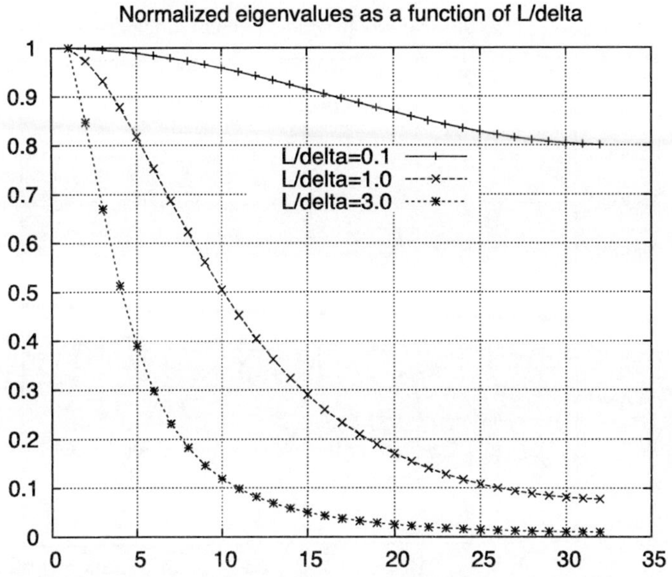

Fig. 7.2 Normalized eigenvalue spectrum for the double-exponential covariance function of Fig. 7.1 for three values of the critical ratio, L/δ

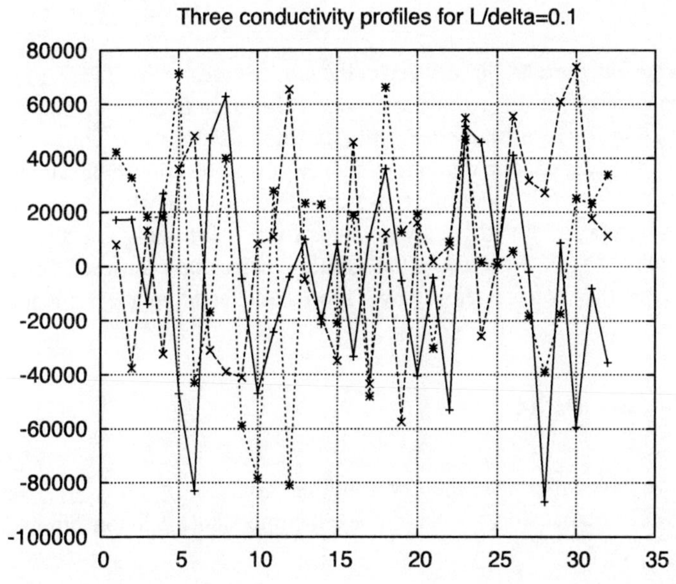

Fig. 7.3 Three sample functions for the conductivity profile when $L/\delta = 0.1$. These functions are the departure from the mean value of $\sigma_{\text{host}} = 3.02 \times 10^5$ S/m. We assume a uniform probability density function, centered at zero and with variance $= 1$, for the random variables, $\{\xi_i\}$, in the Karhunen-Loève expansion

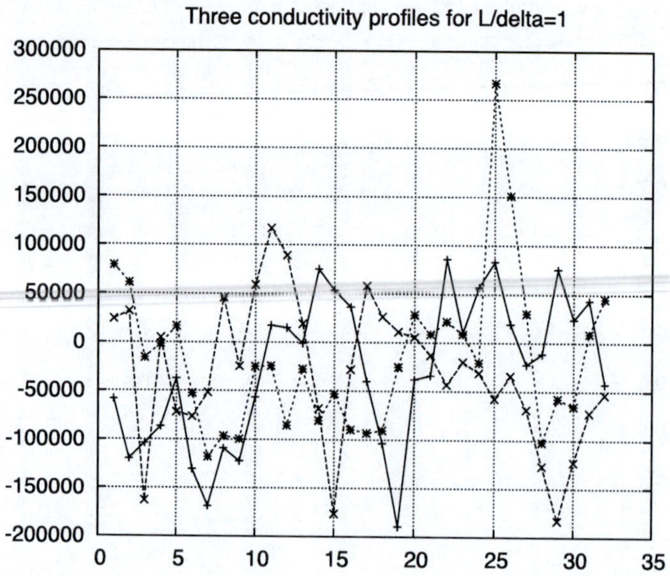

Fig. 7.4 Three sample functions for the conductivity profile when $L/\delta = 1$. These functions are the departure from the mean value of $\sigma_{\text{host}} = 3.02 \times 10^5$ S/m. We assume a uniform probability density function, centered at zero and with variance = 1, for the random variables, $\{\xi_i\}$, in the Karhunen-Loève expansion

The increasing correlation between cell conductivities with L/δ is quite apparent. This is consistent with the eigenvalue spectra shown in Fig. 7.2. Because the spectrum for $L/\delta = 0.1$ is virtually constant (sort of a discrete 'white noise') it follows that all of the independent random variables in (7.16) contribute almost equally to the profile, thereby generating the greatest 'chaos,' in which the conductivity jumps between positive and negative values from cell-to-cell. On the other hand, the condition $L/\delta = 3$ produces the greatest correlation between cells, thereby maintaining positivity or negativity over more cells, yielding a less chaotic profile.

7.5 gPC and PCM[1]

Now that we have introduced the new random variables, $\{\xi_1, \cdots, \xi_M\}$, that are the result of the K-L expansion, we can replace the impedance relationship in (7.2) with

[1]The theoretical treatment in this section largely follows [143]. A related problem of stochastic electromagnetic modeling with uncertain dielectric properties is discussed in [140].

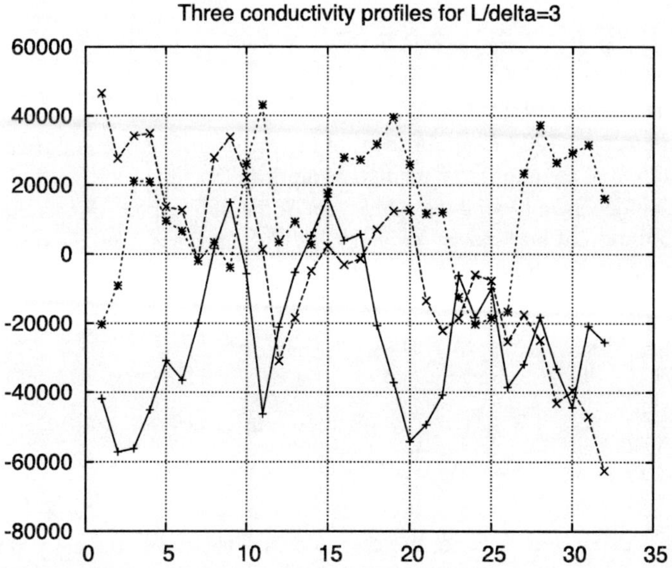

Fig. 7.5 Three sample functions for the conductivity profile when $L/\delta = 3$. These functions are the departure from the mean value of $\sigma_{\text{host}} = 3.02 \times 10^5$ S/m. We assume a uniform probability density function, centered at zero and with variance = 1, for the random variables, $\{\xi_i\}$, in the Karhunen-Loève expansion

$$\int_{-\infty}^{\infty} \cdots \int_{-\infty}^{\infty} Z(l, \xi_1, \cdots, \xi_M) \pi_M(\xi_1, \cdots, \xi_M) d\xi_1 \cdots d\xi_M$$
$$= \int_{-\infty}^{\infty} \cdots \int_{-\infty}^{\infty} Z(l, \xi_1, \cdots, \xi_M) \pi(\xi_1) \cdots \pi(\xi_M) d\xi_1 \cdots d\xi_M \,, \qquad (7.21)$$

where we are assuming that the $\{\xi_i\}$, $i = 1, \cdots, M$, are independent, identically distributed random variables (iid), with the common density function, $\pi(\xi)$. This allows us to replace the joint $M-$dimensional density function, π_M, with the product of M copies of the univariate density function, π, in order to make the following developments numerically feasible. Furthermore, we are free to decide on π to suit the purposes of our problem; the K-L expansion specifies only that the $\{\xi_i\}$ are uncorrelated.

The generalized polynomial chaos (gPC) expansion of degree N for $Z(l, \xi_1, \cdots, \xi_M)$ is given by Xiu [143]:

$$Z(l, \xi_1, \cdots, \xi_M) = \sum_{0 \le |\mathbf{i}| \le N} \hat{Z}_{i_1}(l) \cdots \hat{Z}_{i_M}(l) \phi_{i_1}(\xi_1) \cdots \phi_{i_M}(\xi_M) \,, \qquad (7.22)$$

where $|\mathbf{i}| = i_1 + \cdots + i_M$ and $\{\phi_k(\xi_i)\}_{k=0}^{N}$ is the set of univariate gPC basis functions in ξ_i of degree $0 \le k \le N$. They are orthogonal polynomials associated with the density function, $\pi(\xi)$, in the sense that

$$E[\phi_m(\xi_i)\phi_n(\xi_i)] = \int \phi_m(\xi)\phi_n(\xi)\pi(\xi)d\xi = \delta_{mn}\gamma_m, \quad 0 \le m, n \le N, \tag{7.23}$$

where γ_m is a normalizing constant.

The expansion, (7.22), is reminiscent of the method of separation of variables in partial differential equations that results in solutions that are products of functions in each of the independent variables. This form of the gPC expansion is a direct result of the assumption of iid random variables, $\{\xi_i\}_1^M$. It follows from the orthogonality property, (7.23), that the expansion coefficients of (7.22) are given by the $M-$fold integral

$$\begin{aligned} &\hat{Z}_{i_1}(l)\cdots\hat{Z}_{i_M}(l) \\ &= \frac{1}{\gamma_{i_1}\cdots\gamma_{i_M}}\int\cdots\int Z(l,\xi_1,\cdots,\xi_M)\phi_{i_1}(\xi_1)\cdots\phi_{i_M}(\xi_M)\pi(\xi_1)\cdots\pi(\xi_M)d\xi_1\cdots d\xi_M\,. \end{aligned} \tag{7.24}$$

With this expansion in hand, we can calculate various statistical properties of $Z(l, \xi_1, \cdots, \xi_M)$ [143]. For example, letting $i_1 = \cdots = i_M = 0$ in (7.24) yields $[\hat{Z}_0(l)]^M$ to be the average value of Z. This follows because the zero-order gPC polynomials are unity, and the gammas are all unity as well. Similarly, the variance and covariance are given by:

$$\begin{aligned} \text{Var}[Z(l,\xi_1,\cdots,\xi_M)] &= \sum_{0<|i|\le N} \gamma_{i_1}\cdots\gamma_{i_M}\hat{Z}_{i_1}^2(l)\cdots\hat{Z}_{i_M}^2(l) \\ C_Z(l_1,l_2) &= \sum_{0<|i|\le N} \gamma_{i_1}\cdots\gamma_{i_M}\hat{Z}_{i_1}(l_1)\cdots\hat{Z}_{i_M}(l_1)\hat{Z}_{i_1}(l_2)\cdots\hat{Z}_{i_M}(l_2)\,, \end{aligned} \tag{7.25}$$

where the sums exclude $|i| = 0$.

Now we must turn our attention to the probabilistic collocation method (PCM) in order to develop practical schemes to numerically evaluate the integral in (7.24). The most straightforward scheme is to adopt an integration rule based on the same set of orthogonal functions, $\{\phi_k(\xi_i)\}_{k=0}^N$, that are already present in (7.24). Thus, we have

$$\hat{Z}_{i_1}\cdots\hat{Z}_{i_M} \approx \frac{1}{\gamma_{i_1}\cdots\gamma_{i_M}}\sum_{j_1=1}^{q_1}\cdots\sum_{j_M=1}^{q_M} Z(l,\xi_1^{j_1},\cdots,\xi_M^{j_M})\phi_{i_1}(\xi_1^{j_1})\cdots\phi_{i_M}(\xi_M^{j_M})\alpha_1^{j_1}\cdots\alpha_M^{j_M}\,, \tag{7.26}$$

where $\{\xi_i^{j_i}\}_{j_i=1}^{q_i}$ are the nodes of the one-dimensional rule, and $\{\alpha_i^{j_i}\}_{j_i=1}^{q_i}$ are the corresponding weights.

Clearly, Q is the total number of nodal points required for the $M-$variate integration rule, and if we choose the same number of points in each dimension, $q_1 = \cdots = q_M = q$, then $Q = q^M$, which for $M >> 1$ can grow enormously. For

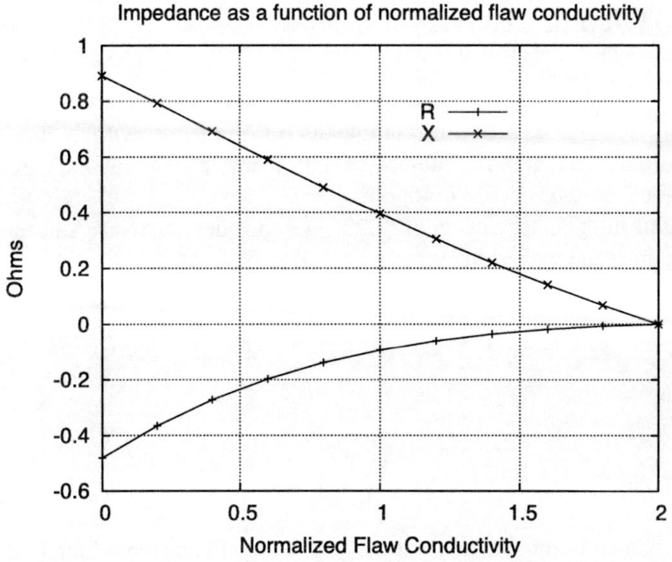

Fig. 7.6 Showing Z as a function of normalized flaw conductivity for a flaw in an infinite host with $\sigma_H = 2 \times 10^7$ S/m. The resistance component is reasonably approximated by $R = -0.12(2-\hat{\sigma}_f)^2$, and the reactance by $X = -0.44521\hat{\sigma}_f + 0.89042$, where $\hat{\sigma}_f = \sigma_f/10^7$ is the normalized flaw conductivity

example, if $q = 3$ and $M = 10$, then $Q = 3^{10} \approx 5.9 \times 10^4$. If we keep $M \leq 5$, we can reduce $Q \leq 243$, for the same q, which is quite reasonable.

The value of M depends upon the result of the K-L expansion, whereas q depends upon the degree of the approximating polynomial. We know that in one dimension the integration rule with q points is exact for any polynomial of degree $\leq 2q-1$. In order to get some insight into the degree of the approximating polynomial that we can expect, refer to Fig. 7.6, which shows a **VIC-3D®**-computed model impedance, Z, as a function of normalized flaw conductivity for a flaw in an infinite host with $\sigma_H = 2 \times 10^7$ S/m. The resistance component is reasonably approximated by $R = -0.12(2 - \hat{\sigma}_f)^2$, and the reactance by $X = -0.44521\hat{\sigma}_f + 0.89042$, where $\hat{\sigma}_f = \sigma_f/10^7$ is the normalized flaw conductivity. These results are typical of other model results.

Thus, the expansion polynomials in (7.22) need be of second order, only, and, therefore, the integrands in (7.24) are at most fourth-order polynomials. From what was just stated, the number of nodal points to exactly compute the integral must be $q \geq (p+1)/2 = 5/2 = 2.5$, or the minimum $q = 3$. We'll leave gPC and PCM at this point, but point out that the recently-developed Uncertainty Quantification Toolkit (see [33] for references to the toolkit) contains software for executing common algorithms for gPC and PCM.

7.6 HDMR and ANOVA

High-dimensional model representation (HDMR) and analysis of variance (ANOVA)[2] seek to reduce the complexity of problems with a large number of dimensions, and are the subject of considerable contemporary research in computational methods [15, 37, 46, 65].

Instead of the gPC expansion of (7.22), we consider an ANOVA decomposition as (we suppress the scan variable, l, to simplify notation):

$$Z(\xi_1, \cdots, \xi_M) = Z_0 + \sum_{j_1}^{M} Z_{j_1}(\xi_{j_1}) + \sum_{j_1<j_2}^{M} Z_{j_1,j_2}(\xi_{j_1}, \xi_{j_2})$$

$$+ \sum_{j_1<j_2<j_3} Z_{j_1,j_2,j_3}(\xi_{j_1}, \xi_{j_2}, \xi_{j_3}) + \cdots + Z_{j_1,\cdots,j_M}(\xi_{j_1}, \cdots, \xi_{j_M}), \qquad (7.27)$$

where Z_0 is a constant function, the $\{Z_{j_1}\}$ are one-dimensional functions, $\{Z_{j_1,j_2}\}$ are two-dimensional functions, and so on, yielding 2^M different terms [46]. For example, the expansion in five-dimensional space is given by

$$\begin{aligned} Z(\xi_1, \cdots, \xi_5) &= Z_0 + Z_1(\xi_1) + Z_2(\xi_2) + Z_3(\xi_3) + Z_4(\xi_4) + Z_5(\xi_5) \\ &+ Z_{12}(\xi_1, \xi_2) + Z_{13}(\xi_1, \xi_3) + Z_{14}(\xi_1, \xi_4) + Z_{15}(\xi_1, \xi_5) \\ &+ Z_{23}(\xi_2, \xi_3) + Z_{24}(\xi_2, \xi_4) + Z_{25}(\xi_2, \xi_5) + Z_{34}(\xi_3, \xi_4) \\ &+ Z_{35}(\xi_3, \xi_5) + Z_{45}(\xi_4, \xi_5) + Z_{123}(\xi_1, \xi_2, \xi_3) + Z_{124}(\xi_1, \xi_2, \xi_4) + Z_{125}(\xi_1, \xi_2, \xi_5) \\ &+ Z_{134}(\xi_1, \xi_3, \xi_4) + Z_{135}(\xi_1, \xi_3, \xi_5) + Z_{145}(\xi_1, \xi_4, \xi_5) + Z_{234}(\xi_2, \xi_3, \xi_4) \\ &+ Z_{235}(\xi_2, \xi_3, \xi_5) + Z_{245}(\xi_2, \xi_4, \xi_5) + Z_{345}(\xi_3, \xi_4, \xi_5) + Z_{1234}(\xi_1, \xi_2, \xi_3, \xi_4) \\ &+ Z_{1235}(\xi_1, \xi_2, \xi_3, \xi_5) + Z_{1245}(\xi_1, \xi_2, \xi_4, \xi_5) + Z_{1345}(\xi_1, \xi_3, \xi_4, \xi_5) \\ &+ Z_{2345}(\xi_2, \xi_3, \xi_4, \xi_5) + Z_{12345}(\xi_1, \xi_2, \xi_3, \xi_4, \xi_5). \end{aligned} \qquad (7.28)$$

The higher-order terms in (7.27) and (7.28) express the effects of correlations between the random variables, and if only a few of them are non-negligible, then we have a good shot at breaking the 'curse of dimensionality' that haunts approximation theory in high-dimensional spaces, but first we will give an algorithm for generating the terms in the ANOVA expansion [46], [42].

Define a one-dimensional projection operator

$$Pf(x) = \int f(x) d\mu(x), \qquad (7.29)$$

[2] ANOVA is often referred to in the mathematics literature as Kolmogorov's superposition theorem.

where, if the 'density measure' $d\mu(x) = dx$, results in the usual expression for an average

$$Pf(x) = \int f(x)dx\ . \tag{7.30}$$

If $d\mu(x) = \delta(x-a)dx$, where $\delta(x-a)$ is the one-dimensional Dirac delta function centered at $x = a$, then the average is simply given by the evaluation of $f(x)$ at $x = a$:

$$Pf(x) = \int f(x)\delta(x-a)dx = f(a)\ . \tag{7.31}$$

From here on we will work in the $M-$dimensional Dirac measure, $\delta(x_1-a_1)\delta(x_2-a_2)\cdots\delta(x_M-a_M)$ where $\mathbf{a}$ is the $M-$dimensional 'anchor point' of the expansion.

Decompose the $M-$dimensional identity operator into the tensor product of M one-dimensional operators:

$$\begin{aligned} I^{(M)} &= \otimes_{j=1}^{M}\left(P_j + (I_j - P_j)\right) \\ &= \prod_i P_i + \sum_i (I_i - P_i)\prod_{i\neq j} P_j \\ &+ \sum_{i<j}(I_i - P_i)(I_j - P_j)\prod_{k\neq i,j} P_k + \cdots + \prod_i (I_i - P_i)\ , \end{aligned} \tag{7.32}$$

then each term of (7.27) is given by applying this operator to $Z(\xi_1, \cdots, \xi_M)$:

$$\begin{aligned} Z_0 &= \textstyle\prod_i P_i Z \\ Z_{j_1} &= (I_{j_1} - P_{j_1})\textstyle\prod_{j_1\neq k} P_k Z \\ \vdots\ &\ \vdots\ \vdots \\ Z_{j_1,\cdots,j_M} &= \textstyle\prod_i (I_i - P_i) Z\ , \end{aligned} \tag{7.33}$$

where $P_i Z(\xi_1, \cdots, \xi_M) = Z(\xi_1, \cdots, \xi_{i-1}, a_i, \xi_{i+1}, \cdots, \xi_M)$.

This can be expressed in the following recursive form after expanding the projection operators [46]:

$$\begin{aligned} Z_0 &= Z(\boldsymbol{\xi})|_{\boldsymbol{\xi}=\mathbf{a}} \\ Z_{j_1}(\xi_{j_1}) &= Z(\boldsymbol{\xi})|_{\boldsymbol{\xi}=\mathbf{a}\backslash\xi_{j_1}} - Z_0 \\ Z_{j_1,j_2}(\xi_{j_1}, \xi_{j_2}) &= Z(\boldsymbol{\xi})|_{\boldsymbol{\xi}=\mathbf{a}\backslash\{\xi_{j_1},\xi_{j_2}\}} - Z_{j_1}(\xi_{j_1}) - Z_{j_2}(\xi_{j_2}) - Z_0 \\ \cdots &= \cdots \\ Z_{j_1,\cdots,j_k}(\xi_{j_1}, \cdots, \xi_{j_k}) &= Z(\boldsymbol{\xi})|_{\boldsymbol{\xi}=\mathbf{a}\backslash\{\xi_{j_1},\cdots,\xi_{j_k}\}} \end{aligned}$$

$$
\begin{aligned}
&- \sum_{\{i_1,\cdots,i_{k-1}\}\subset\{j_1,\cdots,j_k\}} Z_{i_1,\cdots,i_{k-1}}(\xi_{i_1},\cdots,\xi_{i_{k-1}}) \\
&- \sum_{\{i_1,\cdots,i_{k-2}\}\subset\{j_1,\cdots,j_k\}} Z_{i_1,\cdots,i_{k-2}}(\xi_{i_1},\cdots,\xi_{i_{k-2}}) \\
&\vdots \\
&- \sum_{j_1} Z_{j_1}(\xi_{j_1}) - Z_0 \\
&\cdots = \cdots
\end{aligned}
\tag{7.34}
$$

where we have introduced the notation

$$
Z(\boldsymbol{\xi})|_{\boldsymbol{\xi}=\mathbf{a}\backslash\{\xi_{j_1},\cdots,\xi_{j_k}\}} = Z(a_1,\cdots,a_{j_1-1},\xi_{j_1},\cdots,\xi_{j_k},a_{j_k+1},\cdots,a_M)\,. \tag{7.35}
$$

As an example, we'll expand $Z(\xi_1,\xi_2) = Z_0 + Z_1(\xi_1) + Z_2(\xi_2) + Z_{12}(\xi_1,\xi_2)$, where:

$$
\begin{array}{lll}
Z_0 & = P_1P_2Z(\xi_1,\xi_2) & = Z(a_1,a_2) \\
Z_1(\xi_1) & = (I-P_1)P_2Z(\xi_1,\xi_2) & = Z(\xi_1,a_2) - Z(a_1,a_2) \\
Z_2(\xi_2) & = P_1(I-P_2)Z(\xi_1,\xi_2) & = Z(a_1,\xi_2) - Z(a_1,a_2) \\
Z_{12}(\xi_1,\xi_2) & = (I-P_1)(I-P_2)Z(\xi_1,\xi_2) & = Z(\xi_1,\xi_2) - Z(\xi_1,a_2) - Z(a_1,\xi_2) + Z(a_1,a_2) \\
 & & = Z(\xi_1,\xi_2) - Z_1(\xi_1) - Z_2(\xi_2) - Z_0\,.
\end{array}
\tag{7.36}
$$

Thus, we see that (7.36), and by extension, (7.27), is a multi-scale expansion of an $M-$dimensional function along points, lines, faces, hyperplanes, etc., which pass through the anchor point, $\mathbf{a}$.

We can draw some general conclusions from (7.36). First, we note that the mean value of the higher-order terms (beyond Z_0) is zero, which means that these terms are orthogonal to Z_0, a constant. It also means that Z_0 is the mean value of Z. Recall that to compute the mean, we simply replace the 'free variable,' ξ_i, by the fixed anchor point, a_i. Secondly, the correlation between each term vanishes; i.e., (7.27) is an expansion in orthogonal functions. Finally, it is easy to show that

$$
\begin{aligned}
\overline{Z_1^2(\xi_1)} &= \overline{Z^2(\xi_1,a_2)} - Z_0^2 \\
\overline{Z_2^2(\xi_2)} &= \overline{Z^2(a_1,\xi_2)} - Z_0^2 \\
\overline{Z_{12}^2(\xi_1,\xi_2)} &= \overline{Z^2(\xi_1,\xi_2)} - \overline{Z_1^2(\xi_1)} - \overline{Z_2^2(\xi_2)} - Z_0^2\,,
\end{aligned}
\tag{7.37}
$$

from which we draw the important conclusion that

$$
\overline{Z_1^2} + \overline{Z_2^2} + \overline{Z_{12}^2} = \overline{Z^2} - Z_0^2 = \mathrm{VAR}[Z]\,. \tag{7.38}
$$

Equation (7.38) allows us to identify the terms beyond Z_0 as contributing partial variances to the overall variance. With this we can define 'global sensitivity indices' as the ratio of each of these partial variances to the overall variance. These indices describe the contribution of the corresponding inputs, $\{\xi_{j_1}, \cdots, \xi_{j_k}\}$, to the variance of the output [46].

It is pointed out in [46] and [65], and the references therein, that there is a close relationship between the multi-dimensional Taylor expansion and the ANOVA expansion, (7.27). The infinite number of terms in the Taylor expansion

$$\begin{aligned} Z(\boldsymbol{\xi}) &= Z(\mathbf{a}) + \sum_{j=1}^{\infty} \frac{1}{j!} \sum_{i=1}^{M} \frac{\partial^j Z}{\partial \xi_i^j}(\mathbf{a})(\xi_i - a_i)^j \\ &\quad + \sum_{j_1, j_2 > 0}^{\infty} \frac{1}{j_1! j_2!} \sum_{i_1 < i_2} \frac{\partial^{j_1 + j_2} Z}{\partial \xi_{i_1}^{j_1} \partial \xi_{i_2}^{j_2}}(\mathbf{a})(\xi_{i_1} - a_{i_1})^{j_1} (\xi_{i_2} - a_{i_2})^{j_2} \\ &\quad + \cdots \end{aligned} \quad (7.39)$$

are partitioned into a finite number of groups, with each group corresponding to one of the component functions of (7.27). For example, the first-order function, $Z_{j_1}(\xi_{j_1})$, is the sum of all of the Taylor series terms that contain only the variable, ξ_{j_1}, and so on. This suggests that a truncated ANOVA expansion should be more accurate than a truncated Taylor series of the same order [65].

7.7 Determining the ANOVA Anchor Point

There are a number of options for choosing the anchor point [37, 42, 46, 65]. We will use the point in $\boldsymbol{\xi}$-space that corresponds to the mean of a ten-sample Monte Carlo run. This will require a straight-forward inverse problem of the type that we have done before in other random-characterization problems. The idea is to replace the random surface with an equivalent homogeneous (non-random) surface that produces the mean output of the ten **VIC-3D®** runs. This procedure is called 'homogenization,' and is an active area of mathematics research [13].

We use the ten sample functions for $L/\delta = 1$, shown in Fig. 7.7, as the input to **VIC-3D®**, which then produces the ten-sample set of impedances responses, shown in Fig. 7.8. The mean of these ten impedances is shown in Fig. 7.9, and this will be the input to NLSE to finish the homogenization process.

The best fit to the data of Fig. 7.9 is given when the homogeneous conductivity of the surface is $\sigma_H = 2.8 \times 10^5$ S/m, and the resulting impedance response when this conductivity is used is compared with the original data in Fig. 7.10. Clearly, there is a good fit. Furthermore, we note that the equivalent homogeneous conductivity

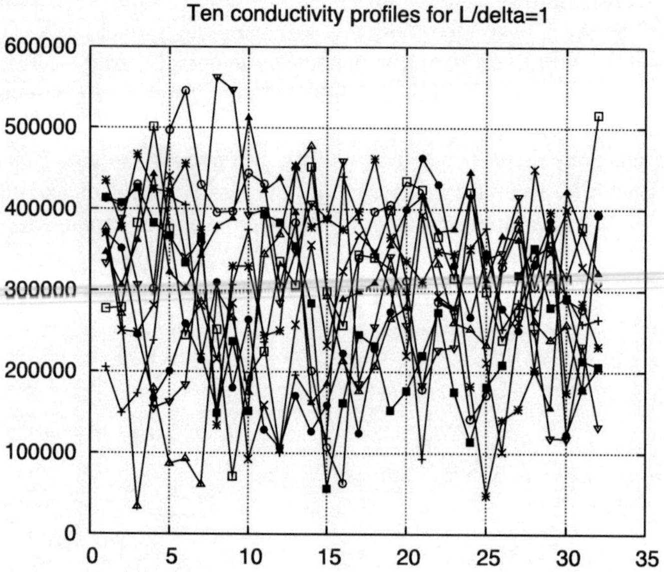

Fig. 7.7 Ten sample functions for $L/\delta = 1$ that extend the set shown in Fig. 7.4, except that we have restored the mean host conductivity, $\sigma_{\text{host}} = 3.02 \times 10^5$. These ten will be input to **VIC-3D®** in order to generate ten samples for NLSE that will be used to determine the homogeneous (nonrandom) conductivity of the surface

differs from the statistical mean conductivity of the surface, $\sigma_{\text{host}} = 3.02 \times 10^5$ S/m, as is typical of nonlinear random problems.

In the language of probability theory [127], σ_{host} is the 'prior mean' of the model and σ_{H} is the 'posterior mean' of the model. The significance of the terms is that the prior mean is a property of the known material, and is known at the outset of the problem ('a priori'), whereas the posterior mean is that equivalent homogeneous conductivity that produces the mean of the *impedances*, or measured data. The posterior mean is the more important concept when it comes to applying inverse methods to characterize flaws in the random patch. In fact, the random patch will be replaced by a homogeneous, nonrandom patch whose conductivity is σ_{H}.

The left-hand vector in (7.16) is the difference between two constant vectors, $[2.8 \times 10^5]^{32} - [3.02 \times 10^5]^{32} = [-0.22 \times 10^5]^{32}$, where the notation implies a 32-row vector, each entry of which is the constant shown within the brackets. The eigenvectors, $\{\mathbf{v}_n\}$, consitute a complete orthonormal system (cons) of basis vectors for the 32-dimensional space, which means that the expansion coefficients are given by the simple inner-product, $\lambda_n \xi_n =< [-0.22 \times 10^5]^{32}, \mathbf{v}_n >$, from which we get the anchor point in $\boldsymbol{\xi}$-space

$$\boldsymbol{\xi}_{\text{anchor}} = \left\{ \frac{< [-0.22 \times 10^5]^{32}, \mathbf{v}_n >}{\lambda_n} \right\}_{n=1}^{32} . \tag{7.40}$$

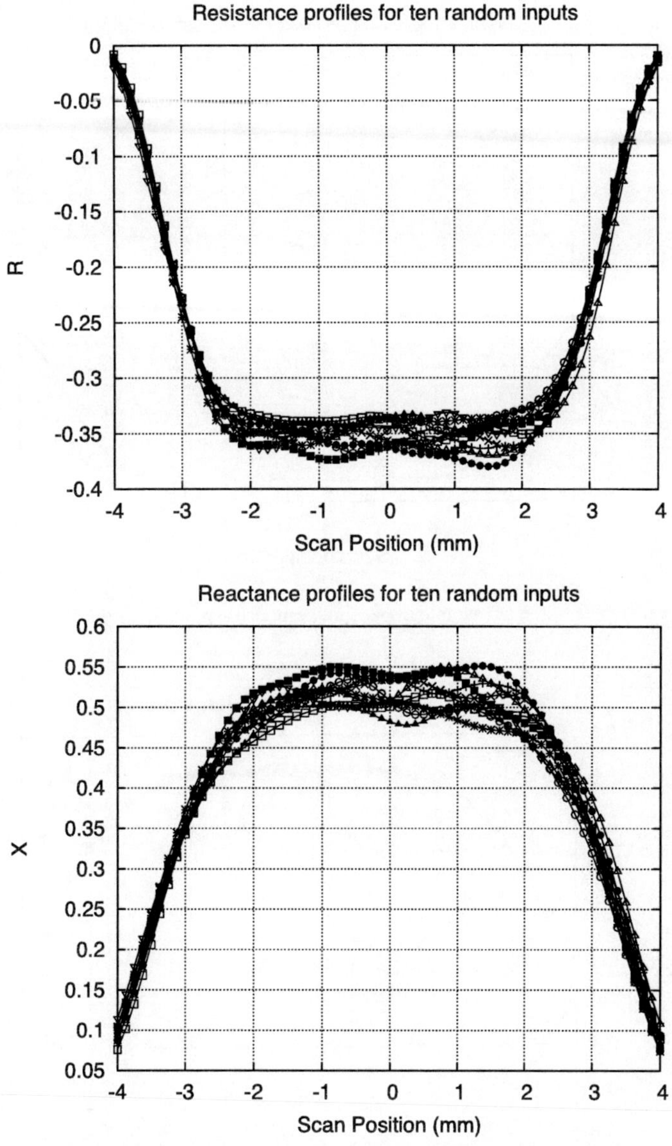

Fig. 7.8 Ten sample impedance responses corresponding to the ten input sample functions of Fig. 7.7

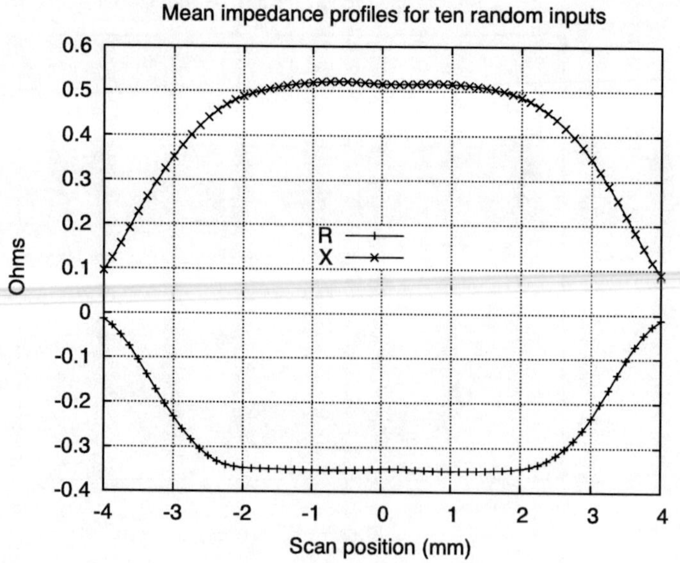

Fig. 7.9 Mean values of the ten sample impedance responses shown in Fig. 7.8

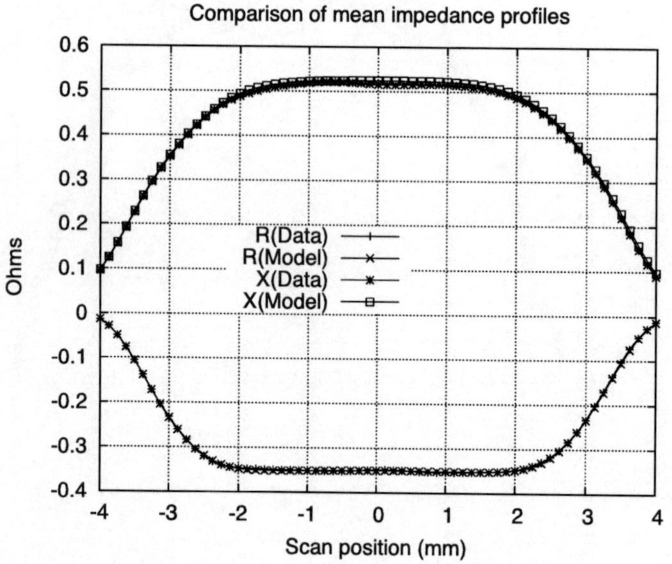

Fig. 7.10 Comparing the mean values of the ten sample impedance responses shown in Fig. 7.9 with the model response for the homogeneous conductivity, $\sigma_H = 2.8 \times 10^5$ S/m

Table 7.1 Coordinates for the ANOVA anchor point, **a**, in $\boldsymbol{\xi}$-space. Coordinates for even numbered indices vanish due to symmetry, so coordinates for odd-numbered indices only are shown

Index	a_n
1	−0.1917E+00
3	−0.6392E−01
5	0.3836E−01
7	−0.2738E−01
9	−0.2125E−01
11	−0.1730E−01
13	−0.1452E−01
15	−0.1241E−01
17	−0.1072E−01
19	−0.9277E−02
21	−0.7978E−02
23	−0.6742E−02
25	−0.5506E−02
27	−0.4227E−02
29	−0.2879E−02
31	−0.1461E−02

The eigenvectors with an even index are antisymmetric, so their inner-products in (7.40) vanish, which leaves only inner-products with odd indices that contribute to the anchor point. The resulting coordinates in $\boldsymbol{\xi}$-space are shown in Table 7.1. Note that they cluster near the origin, with the exception of a_1, which means that the response at $\sigma_H = 2.8 \times 10^5$ S/m is not too different from the response at the mean conductivity of 3.02×10^5 S/m. Calculations (not shown here) confirm this conclusion.

7.8 Interpolation Theory Using Splines Based Upon Higher-Order Convolutions of the Unit Pulse[3]

Consider a regular one-dimensional grid, whose spacing is $h = 1$. Relative to this grid we define $\pi(x)$ to be the unit pulse

$$\pi(x) = \begin{cases} 1, & \text{if } 0 \leq x < 1 \\ 0, & \text{otherwise,} \end{cases} \tag{7.41}$$

and $\pi_{m+1}(x)$ to be the mth-order convolution of $\pi(x)$ (we define $\pi_1(x) = \pi(x)$). The $\pi_{m+1}(x)$ are shown in Fig. 7.11 for $m = 0, 1, 2, 3$.

[3]See [12, 101] for additional examples.

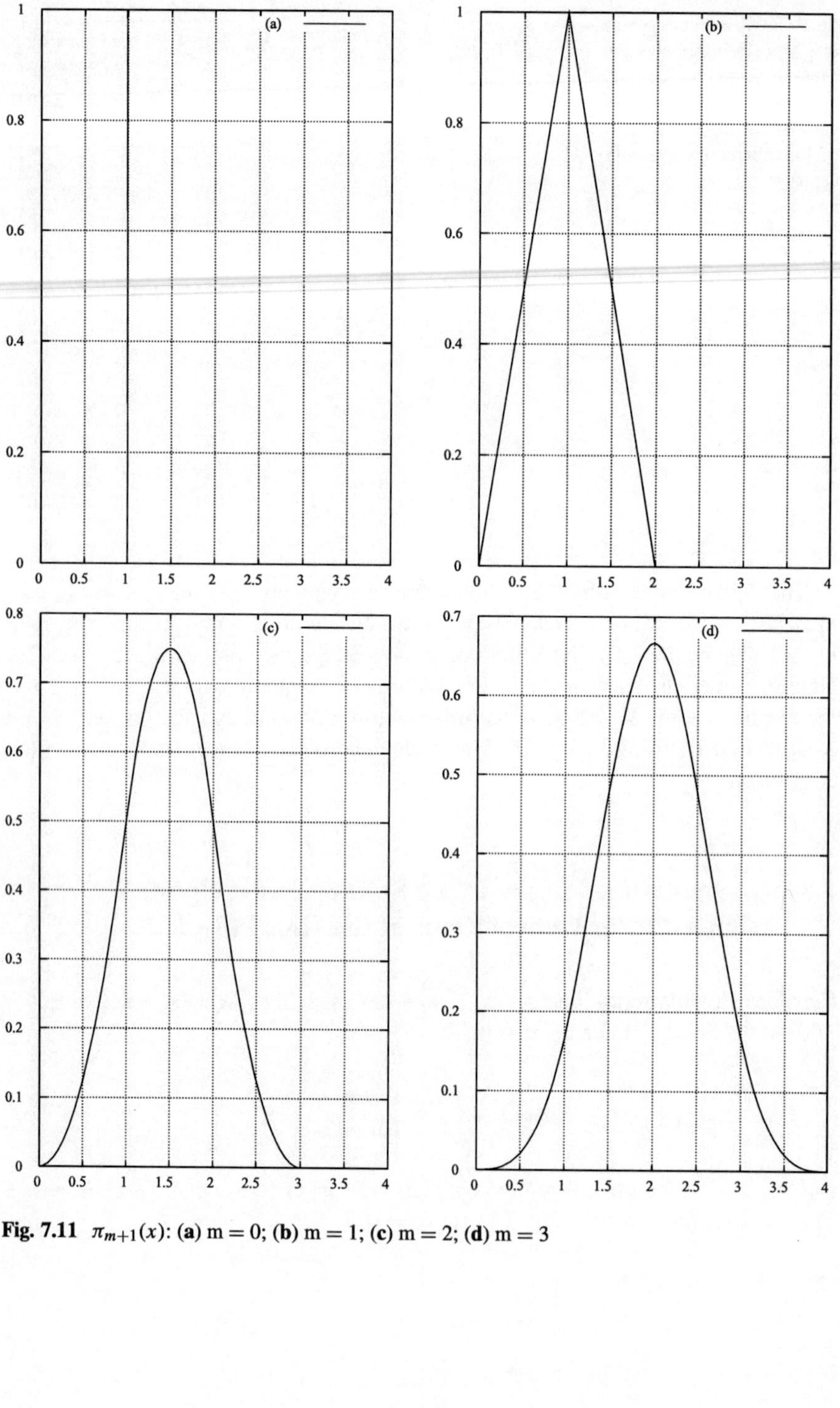

Fig. 7.11 $\pi_{m+1}(x)$: **(a)** m = 0; **(b)** m = 1; **(c)** m = 2; **(d)** m = 3

Table 7.2 Piecewise polynomials, $\alpha_m^k(x)$, in $\pi_{m+1}(x) = \sum_{k=0}^{m} \alpha_m^k(x-k)\theta(x-k)$

Order (m)	$\alpha_m^k(x) = \sum_{j=0}^{m} a_m(k,j)x^j/j!$
m = 0	$\alpha_0^0(x) = 1$
m = 1	$\alpha_1^0(x) = x$
	$\alpha_1^1(x) = 1 - x$
m = 2	$\alpha_2^0(x) = x^2/2$
	$\alpha_2^1(x) = 1/2 + x - x^2$
	$\alpha_2^2(x) = 1/2 - x + x^2/2$
m = 3	$\alpha_3^0(x) = x^3/6$
	$\alpha_3^1(x) = 1/6 + x/2 + x^2/2 - x^3/2$
	$\alpha_3^2(x) = 4/6 - x^2 + x^3/2$
	$\alpha_3^3(x) = 1/6 - x/2 + x^2/2 - x^3/6$

The higher-order convolutions of the unit pulse can be written in terms of the piecewise polynomials, $\alpha_m^k(x)$, as

$$\pi_{m+1}(x) = \sum_{k=0}^{m} \alpha_m^k(x-k)\theta(x-k),$$

where $\theta(z)$ is the characteristic function of the unit interval, and the $\alpha_m^k(x) = \sum_{j=0}^{m} a_m(k,j)x^j/j!$ are tabulated in Table 7.2. The coefficients $a_m(k,j)$ satisfy the recursion relation

$$a_m(k,j) = a_{m-1}(k,j-1) - a_{m-1}(k-1,j-1), \;\; 1 \le k \le m-1, \;\; 1 \le j \le m. \tag{7.42}$$

Numerical values for $a_m(k,j)$, for $1 \le m \le 3$, can be easily inferred from Table 7.2. For example, $a_3(k,j)$ is given by the following matrix,

$$\begin{matrix} 0 & 0 & 0 & 1 \\ 1/6 & 1/2 & 1 & -3 \\ 4/6 & 0 & -2 & 3 \\ 1/6 & -1/2 & 1 & -1. \end{matrix}$$

Keeping in mind the polynomial relationship shown in Fig. 7.6, we will use piecewise polynomials of the second order, which correspond to $m = 2$ in Table 7.2, for interpolating within the ANOVA expansion. Hence, we write the general expression

$$Z(\xi) = z_0\pi_3(\xi) + z_1\pi_3(\xi-1) + z_2\pi_3(\xi-2)\,, \tag{7.43}$$

which is just a superspoition of $\pi_3(\xi)$ and its translates of unit amounts. The expansion coefficients, $z_0,\ z_1,\ z_2$, will be given in terms of computed values of $Z(\xi)$ at the nodes, $\xi = 2,\ \xi = 2.5,\ \xi = 3$.

Over the interval, $2 \le \xi \le 3$, the expansion, (7.43), is given by

$$Z(\xi) = z_0\left[1/2 - (\xi - 2) + (\xi - 2)^2/2\right] + z_1\left[1/2 + (\xi - 2) - (\xi - 2)^2\right]$$
$$+z_2(\xi - 2)^2/2$$
$$= \frac{z_0+z_1}{2} + (z_1 - z_0)(\xi - 2) + (z_0 - 2z_1 + z_2)\frac{(\xi - 2)^2}{2}, \quad 2 \le \xi \le 3 . \quad (7.44)$$

Calling Z_2 the computed nodal impedance at $\xi = 2$ and similarly $Z_{2.5}$ and Z_3 at the other two nodes, we see immediately from (7.44) that

$$Z_2 = \frac{z_0 + z_1}{2}$$
$$Z_{2.5} = \frac{z_0}{8} + \frac{3z_1}{4} + \frac{z_2}{8}$$
$$Z_3 = \frac{z_1 + z_2}{2}, \quad (7.45)$$

which yields

$$z_0 = 2.5Z_2 - 2Z_{2.5} + 0.5Z_3$$
$$z_1 = -0.5Z_2 + 2Z_{2.5} - 0.5Z_3$$
$$z_2 = 0.5Z_2 - 2Z_{2.5} + 2.5Z_3 . \quad (7.46)$$

Substituting these results into (7.44) gives the result in terms of the nodal values of the impedances:

$$Z(\xi) = Z_2 - (3Z_2 - 4Z_{2.5} + Z_3)(\xi - 2) + 2(Z_2 - 2Z_{2.5} + Z_3)(\xi - 2)^2, \quad 2 \le \xi \le 3 . \quad (7.47)$$

We will be interested in applying this expansion to the case in which ξ has a uniform density between $[-0.5, +0.5]$, so we will translate it to[4]

$$Z(\xi) = Z_{-0.5} - (3Z_{-0.5} - 4Z_0 + Z_{0.5})(\xi + 0.5) + 2(Z_{-0.5} - 2Z_0 + Z_{0.5})(\xi + 0.5)^2,$$
$$-0.5 \le \xi \le 0.5 . \quad (7.48)$$

We can compute the quantities in (7.37) and (7.38) directly by integrating the square of (7.48):

[4]Do not confuse Z_0 in this expression with the impedance computed at the anchor point in (7.37) and (7.38).

$$\int_{-0.5}^{0.5} Z^2(\xi)d\xi = a^2 + ab + \frac{2ac + b^2}{3} + \frac{bc}{2} + \frac{c^2}{5}, \tag{7.49}$$

where

$$\begin{aligned} a &= Z_{-0.5} \\ b &= -3Z_{-0.5} + 4Z_0 - Z_{0.5} \\ c &= 2(Z_{-0.5} - 2Z_0 + Z_{0.5}) . \end{aligned} \tag{7.50}$$

With this result, we can calculate the partial variances (see (7.37) and (7.38)) due to the one-dimensional functions defined in (7.27) and (7.34) associated with the six random variables, $\xi_1, \cdots, \xi_5, \xi_{32}$, and plot the results in Fig. 7.12.

The total variances associated with these one-dimensional functions are plotted in Fig. 7.13. These results are obtained by adding each of the variances shown in Fig. 7.12, because, as we have already stated, these variances are independent of each other.

Beyond the question of variances, these results are important because they indicate the importance of the higher-order terms in (7.27). It seems clear that the dominant variables are ξ_1, ξ_2, ξ_3, and ξ_5, at least in the middle of the scan, so that helps in keeping the problem tractable. Indeed, from the pattern displayed in (7.28), we see that there is a total of 16 terms in the expansion with these four variables.

7.9 Two-Dimensional Functions

In terms of these four variables, the two-dimensional functions to be computed in (7.27) and (7.34) are $Z_{12}(\xi_1, \xi_2)$, $Z_{13}(\xi_1, \xi_3)$, $Z_{15}(\xi_1, \xi_5)$, $Z_{23}(\xi_2, \xi_3)$, $Z_{25}(\xi_2, \xi_5)$, $Z_{35}(\xi_3, \xi_5)$. We will expand each of these functions over the unit-square centered at the origin of the appropriate two-dimensional space. Using (7.44) as a template, the expansion is given by the tensor product

$$\begin{aligned} Z(\xi_1, \xi_2) = {} & z_{00} P_0(\xi_1) P_0(\xi_2) + z_{01} P_0(\xi_1) P_1(\xi_2) + z_{02} P_0(\xi_1) P_2(\xi_2) \\ & + z_{10} P_1(\xi_1) P_0(\xi_2) + z_{11} P_1(\xi_1) P_1(\xi_2) + z_{12} P_1(\xi_1) P_2(\xi_2) \\ & + z_{20} P_2(\xi_1) P_0(\xi_2) + z_{21} P_2(\xi_1) P_1(\xi_2) + z_{22} P_2(\xi_1) P_2(\xi_2) , \end{aligned} \tag{7.51}$$

where

$$\begin{aligned} P_0(\xi) &= 1/2 - (\xi + 0.5) + (\xi + 0.5)^2/2 \\ P_1(\xi) &= 1/2 + (\xi + 0.5) - (\xi + 0.5)^2 \\ P_2(\xi) &= (\xi + 0.5)^2/2 . \end{aligned} \tag{7.52}$$

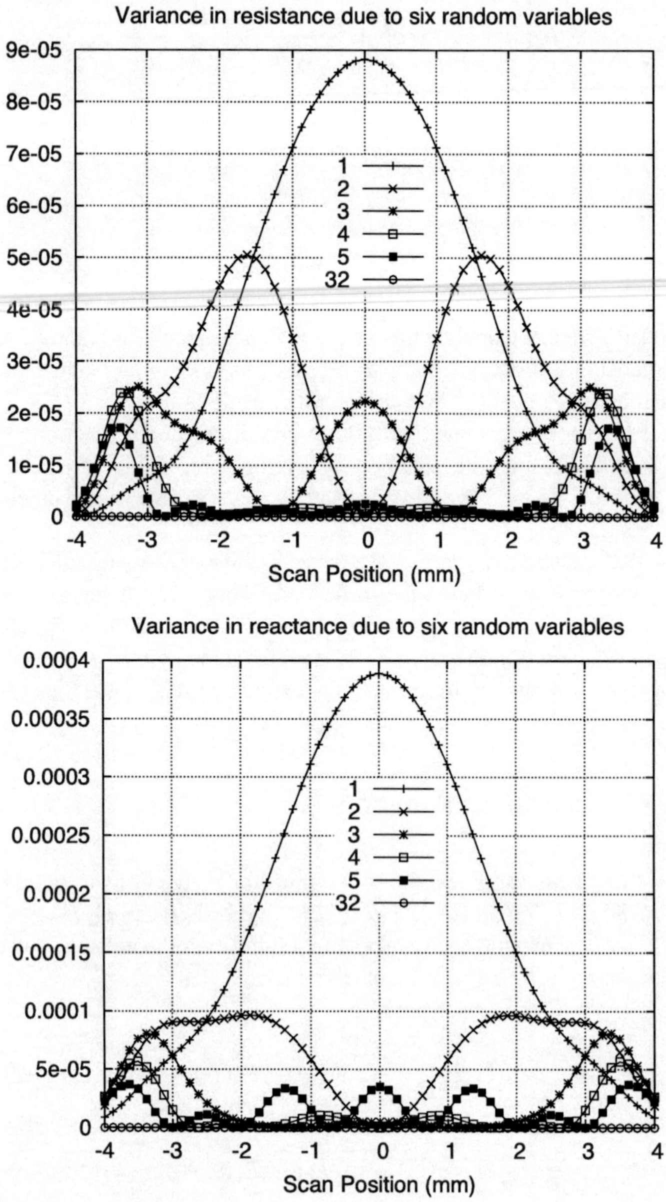

Fig. 7.12 Partial variances in resistance and reactance due to the one-dimensional functions defined in (7.27) and (7.34) associated with the six random variables, $\xi_1, \cdots, \xi_5, \xi_{32}$

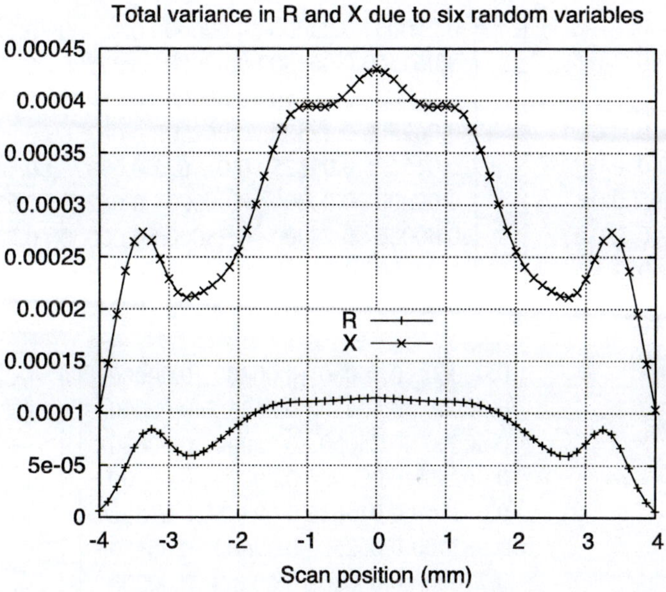

Fig. 7.13 Total variances in resistance and reactance due to the one-dimensional functions defined in (7.27) and (7.34) associated with the six random variables, $\xi_1, \cdots, \xi_5, \xi_{32}$

Fig. 7.14 Two-dimensional interpolation nodes

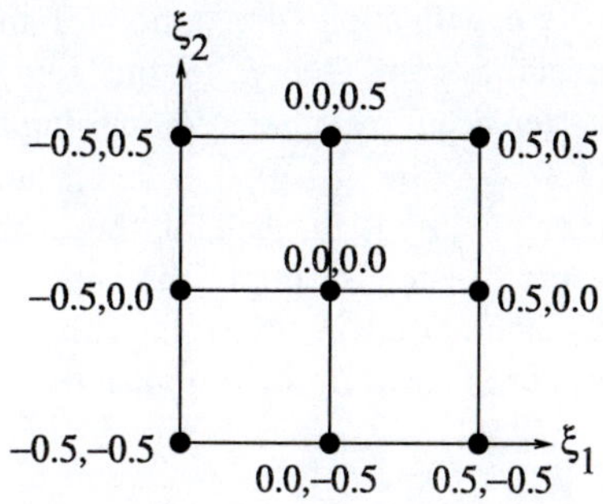

The interpolation nodes for the two-dimensional functions are the tensor product of the one-dimensional nodes used above (see Fig. 7.14). With this in mind, we can determine the expansion coefficients in (7.51) as the solution of the vector-matrix equation:

$$
\begin{bmatrix} Z(-0.5,-0.5) \\ Z(0.0,-0.5) \\ Z(0.5,-0.5) \\ Z(-0.5,0.0) \\ Z(0.0,0.0) \\ Z(0.5,0.0) \\ Z(-0.5,0.5) \\ Z(0.0,0.5) \\ Z(0.5,0.5) \end{bmatrix} = \begin{bmatrix} 0.25000 & 0.25000 & 0.00000 & 0.25000 & 0.25000 \\ 0.062500 & 0.062500 & 0.00000 & 0.37500 & 0.37500 \\ 0.00000 & 0.00000 & 0.00000 & 0.25000 & 0.25000 \\ 0.062500 & 0.37500 & 0.062500 & 0.062500 & 0.37500 \\ 0.015625 & 0.093750 & 0.015625 & 0.093750 & 0.56250 \\ 0.00000 & 0.00000 & 0.00000 & 0.062500 & 0.37500 \\ 0.00000 & 0.25000 & 0.25000 & 0.00000 & 0.25000 \\ 0.00000 & 0.062500 & 0.062500 & 0.00000 & 0.37500 \\ 0.00000 & 0.00000 & 0.00000 & 0.00000 & 0.25000 \end{bmatrix}
$$

$$
\begin{matrix} 0.00000 & 0.00000 & 0.00000 & 0.00000 \\ 0.00000 & 0.062500 & 0.062500 & 0.00000 \\ 0.00000 & 0.25000 & 0.25000 & 0.00000 \\ 0.062500 & 0.00000 & 0.00000 & 0.00000 \\ 0.093750 & 0.015625 & 0.093750 & 0.015625 \\ 0.062500 & 0.062500 & 0.37500 & 0.062500 \\ 0.25000 & 0.00000 & 0.00000 & 0.00000 \\ 0.37500 & 0.00000 & 0.062500 & 0.062500 \\ 0.25000 & 0.00000 & 0.25000 & 0.25000 \end{matrix} \Bigg] \begin{bmatrix} z_{00} \\ z_{01} \\ z_{02} \\ z_{10} \\ z_{11} \\ z_{12} \\ z_{20} \\ z_{21} \\ z_{22} \end{bmatrix} . \tag{7.53}
$$

We will apply the notion of the 'impulse response,' that is well-known in linear system theory, to the solution of (7.53). The left-hand side of (7.53) is replaced by nine 'source impulses,' which are simply the basis vectors, $[1, 0, \cdots, 0], \cdots, [0, 0, \cdots, 1]$, and one then computes the nine responses, $[z_{00}, \cdots, z_{22}]$, to each of these basis vectors. Then the response to a general source vector is simply the superposition of the impulse responses weighted by the appropriate coefficients. This follows because of the expansion of an arbitrary vector in terms of the 'impulse basis':

$$
\begin{bmatrix} Z(-0.5,-0.5) \\ Z(0.0,-0.5) \\ Z(0.5,-0.5) \\ Z(-0.5,0.0) \\ Z(0.0,0.0) \\ Z(0.5,0.0) \\ Z(-0.5,0.5) \\ Z(0.0,0.5) \\ Z(0.5,0.5) \end{bmatrix} = Z(-0.5,-0.5) \begin{bmatrix} 1 \\ 0 \\ \vdots \\ 0 \end{bmatrix} + \cdots + Z(0.5,0.5) \begin{bmatrix} 0 \\ 0 \\ \vdots \\ 1 \end{bmatrix} . \tag{7.54}
$$

Hence, the solution of (7.53) is given by

Table 7.3 Impulse response of the linear system (7.53)

1	2	3	4	5	6	7	8	9
6.25	−5.00	1.25	−5.00	4.00	−1.00	1.25	−1.00	0.25
−1.25	1.00	−0.25	5.00	−4.00	1.00	−1.25	1.00	−0.25
1.25	−1.00	0.25	−5.00	4.00	−1.00	6.25	−5.00	1.25
−1.25	5.00	−1.25	1.00	−4.00	1.00	−0.25	1.00	−0.25
0.25	−1.00	0.25	−1.00	4.00	−1.00	0.25	−1.00	0.25
−0.25	1.00	−0.25	1.00	−4.00	1.00	−1.25	5.00	−1.25
1.25	−5.00	6.25	−1.00	4.00	−5.00	0.25	−1.00	1.25
−0.25	1.00	−1.25	1.00	−4.00	5.00	−0.25	1.00	−1.25
0.25	−1.00	1.25	−1.00	4.00	−5.00	1.25	−5.00	6.25

$$\begin{bmatrix} z_{00} \\ z_{01} \\ z_{02} \\ z_{10} \\ z_{11} \\ z_{12} \\ z_{20} \\ z_{21} \\ z_{22} \end{bmatrix} = Z(-0.5,-0.5)\mathscr{M}^{-1}\begin{bmatrix} 1 \\ 0 \\ \vdots \\ 0 \end{bmatrix} + \cdots + Z(0.5,0.5)\mathscr{M}^{-1}\begin{bmatrix} 0 \\ 0 \\ \vdots \\ 1 \end{bmatrix}, \tag{7.55}$$

where $\mathscr{M}$ is the coefficient matrix in (7.53). The operation of $\mathscr{M}^{-1}$ on the basis vectors is the 'impulse response' of the system, which is tabulated in Table 7.3.

The final expression for the expansion coefficients of (7.51) is obtained when the results of Table 7.3 are substituted into (7.55):

$$\begin{aligned} z_{00} = {} & 6.25Z(-0.5,-0.5) - 5.00[Z(0.0,-0.5) + Z(-0.5,0.0)] \\ & +1.25[Z(0.5,-0.5) + Z(-0.5,0.5)] \\ & -[Z(0.5,0.0) + Z(0.0,0.5)] + 0.25Z(0.5,0.5) + 4.00Z(0.0,0.0) \\ z_{01} = {} & -1.25[Z(-0.5,-0.5) + Z(-0.5,0.5)] \\ & +[Z(0.0,-0.5) + Z(0.5,0.0) + Z(0.0,0.5)] \\ & -0.25[Z(0.5.-0.5) + Z(0.5,0.5)] + 5.00Z(-0.5,0.0) - 4.00Z(0.0,0.0) \\ z_{02} = {} & 1.25[Z(-0.5,-0.5) + Z(0.5,0.5)] - [Z(0.0,-0.5) + Z(0.5,0.0)] \\ & +0.25Z(0.5,-0.5) \\ & -5.00[Z(-0.5,0.0) + Z(0.0,0.5)] + 4.00Z(0.0,0.0) + 6.25Z(-0.5,0.5) \\ z_{10} = {} & -1.25[Z(-0.5,-0.5) + Z(0.5,-0.5)] + 5.00Z(0.0,-0.5) \\ & +[Z(-0.5,0.0) + Z(0.5,0.0) + Z(0.0,0.5)] - 4.00Z(0.0,0.0) \\ & -0.25[Z(-0.5,0.5) + Z(0.5,0.5)] \end{aligned}$$

$$
\begin{aligned}
z_{11} &= 0.25[Z(-0.5, -0.5) + Z(0.5, -0.5) + Z(-0.5, 0.5) + Z(0.5, 0.5)] \\
&\quad -[Z(0.0, -0.5) + Z(-0.5, 0.0) + Z(0.5, 0.0) + Z(0.0, 0.5)] \\
&\quad +4.00Z(0.0, 0.0) \\
z_{12} &= -0.25[Z(-0.5, -0.5) + Z(0.5, -0.5)] \\
&\quad +[Z(0.0, -0.5) + Z(-0.5, 0.0) + Z(0.5, 0.0)] \\
&\quad -4.00Z(0.0, 0.0) - 1.25[Z(-0.5, 0.5) + Z(0.5, 0.5)] + 5.00Z(0.0, 0.5) \\
z_{20} &= 1.25[Z(-0.5, -0.5) + Z(0.5, 0.5)] - 5.00[Z(0.0, -0.5) + Z(0.5, 0.0)] \\
&\quad +6.25Z(0.5, -0.5) \\
&\quad -[Z(-0.5, 0.0) + Z(0.0, 0.5)] + 4.00Z(0.0, 0.0) + 0.25Z(-0.5, 0.5) \\
z_{21} &= -0.25[Z(-0.5, -0.5) + Z(-0.5, 0.5)] + [Z(0.0, -0.5) + Z(-0.5, 0.0) \\
&\quad +Z(0.0, 0.5)] \\
&\quad -1.25[Z(0.5, -0.5) + Z(0.5, 0.5)] - 4.00Z(0.0, 0.0) + 5.00Z(0.5, 0.0) \\
z_{22} &= 0.25Z(-0.5, -0.5) - [Z(0.0, -0.5) + Z(-0.5, 0.0)] + 1.25[Z(0.5, -0.5) \\
&\quad +Z(-0.5, 0.5)] \\
&\quad +4.00Z(0.0, 0.0) - 5.00[Z(0.5, 0.0) + Z(0.0, 0.5)] + 6.25Z(0.5, 0.5) \qquad (7.56)
\end{aligned}
$$

We compute the nine nodal responses for each of the six two-dimensional functions listed above (7.51) using **VIC-3D®**, and in Fig. 7.15 we show ten sample functions of R and X for $Z_{12}(\xi_1, \xi_2)$ computed following this procedure. Figure 7.16 shows the variance computed from these ten sample functions.

In a similar manner, we compute the variances of $Z_{13}(\xi_1, \xi_3)$, $Z_{15}(\xi_1, \xi_5)$, $Z_{23}(\xi_2, \xi_3)$, $Z_{25}(\xi_2, \xi_5)$, and $Z_{35}(\xi_3, \xi_5)$ using ten sample functions of each, and plot the results in Fig. 7.17. It is interesting to note that only those pairs of random variables that involve ξ_2 contribute a non-null response, and this indicates why this algorithm is referred to as 'analysis of variance.' It clarifies which variables, whether singly or jointly, contribute significantly to the variance of the process.

The total variance associated with the one- and two-dimensional functions is obtained by adding the results of Figs. 7.13 and 7.17, and is shown in Fig. 7.18.

7.10 Probability of Detection and the Chebychev Inequality

If we assume, as is typical, that a flaw is 'detected' if its response exceeds the uncertainty in the background, as in Fig. 7.19, then it is clear from Fig. 7.18 that a flaw whose peak signal is away from the center will be obscured by the random clutter to a greater extent than a flaw whose peak signal is centered on the random surface. We assume that the variance of a random process defines its uncertainty or 'noise level.'

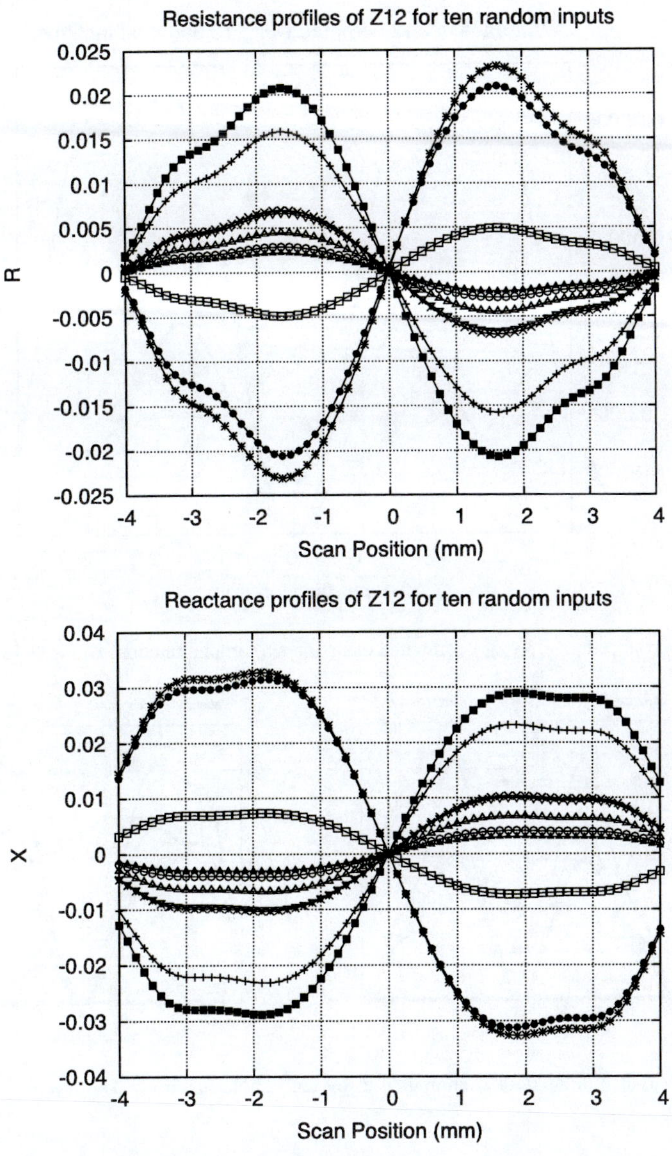

Fig. 7.15 Ten sample impedance responses of $Z_{12}(\xi_1, \xi_2)$, computed according (7.34). Each sample is computed analytically using the interpolation algorithm of (7.51)

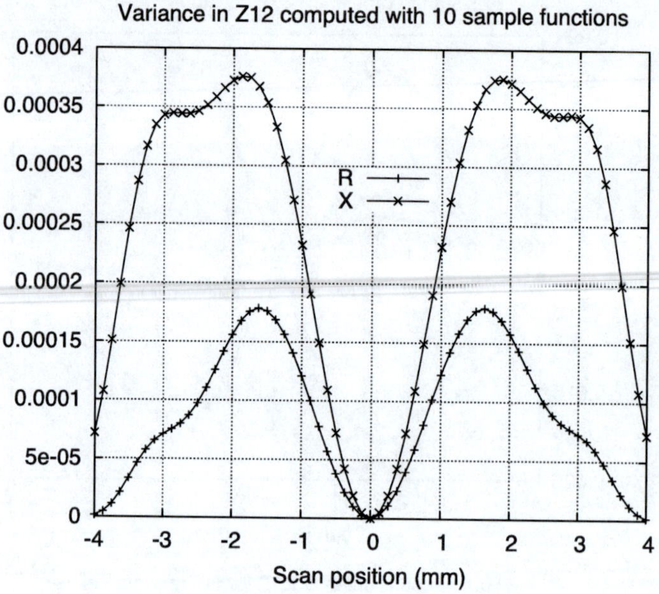

Fig. 7.16 Variance in $Z_{12}(\xi_1, \xi_2)$ computed using the ten sample functions of Fig. 7.15

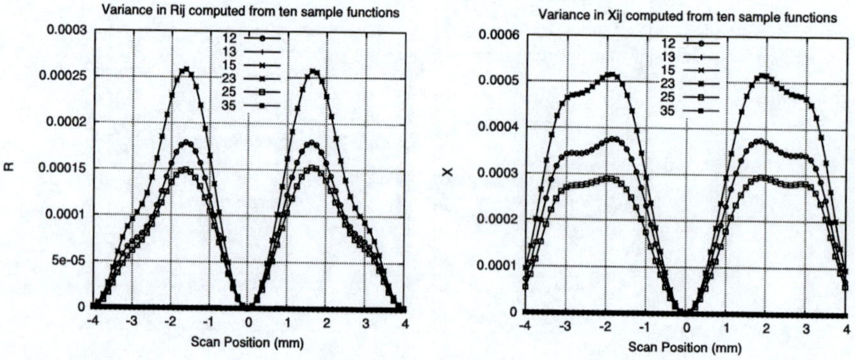

Fig. 7.17 Variance in $Z_{ij}(\xi_i, \xi_j)$ computed using ten sample functions. The legend denotes the index-pair, ij

The problem that we are going to address in this section is that of determining metrics for estimating the 'detectability' of a flaw located at, say, 1.625 mm in Fig. 7.18, compared with one located at the center of the scan. This is a problem of 'signal detection,' as described, for example, in [48]. Following [48], we introduce the notion of 'hypothesis testing,' in which we designate by H_0 the 'null hypothesis', that we have detected only random noise in our measurement, and the alternate hypothesis that we have detected a flaw by H_1. There is a probability associated with a true response for each hypothesis, and that is what we will determine next.

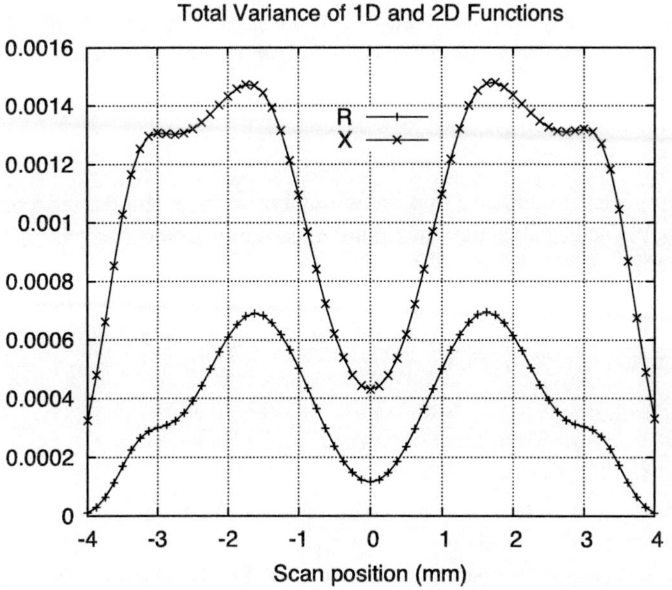

Fig. 7.18 Total variance associated with the one- and two-dimensional functions

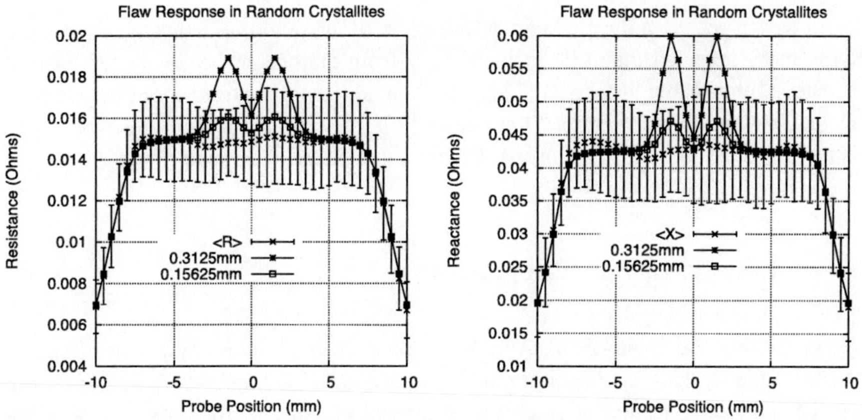

Fig. 7.19 Illustrating a situation in which one flaw, labeled 0.3125 mm, is 'detectable' against background random noise, while the other, labeled 0.15625 mm, is not. The first flaw has a volume four times greater than the second, and its peak extends significantly above the variance of the background noise due to a random patch in an otherwise pristine workpiece. The smaller flaw's response is buried within the noise level. The figure is illustrative only, and the data shown are completely unrelated to the models in this report

We use the Chebychev Inequality [63] to quantify the probability of H_0. This inequality states that for a random variable, Z, and a real number, ϵ, we have

$$P[|Z| \geq \epsilon] \leq \frac{\mathrm{VAR}(Z)}{\epsilon^2} . \tag{7.57}$$

Z is the impedance measured due to the random surface, and the term on the right side of (7.57) can be called the 'maximum uncertainty' associated with Z for a given ϵ. This is the metric to be associated with H_0.

Because H_0 and H_1 are complementary processes (their sample spaces are disjoint), it follows that their probabilities must sum to unity. Hence, we can assign a 'minimum certainty' metric to H_1 by

$$\text{Minimum Certainty}(\epsilon) = 1 - \frac{\mathrm{VAR}(Z)}{\epsilon^2} . \tag{7.58}$$

By referring to this as the 'minimum certainty,' we are claiming that we are at least this certain of a correct decision.

Figure 7.20 shows 'probability of detection' (POD) curves for the detectability of a flaw located at two different points within the random surface. These curves are plots of (7.58), in which ϵ plays the role of a 'threshold variable' or 'decision boundary,' that will allow us to determine whether to choose H_0 or H_1 (see Fig. 7.21).

As we suspected, a flaw located at the center of the random surface is much more likely to be detected than one located away from the center, because its threshold of detectability is much smaller. This does not address the question of how the size of the flaw enters the picture. That can only be determined by solving a series of forward problems for a given probe, frequency and size and shape of the flaw.

7.11 Consistency of Calculations

We know that the mean of the ANOVA expansion is obtained by substituting the anchor point, **a**, for the generic point in $\boldsymbol{\xi}$-space in the general ANOVA expansion. We get the same result by analytically integrating the expressions for the first-order and second-order functions given by the quadratic polynomial interpolation expansions. This yields

$$\begin{aligned}
\overline{Z}(\xi) &= (Z_{-0.5} + 4Z_0 + Z_{0.5})/6 - Z(\mathbf{a}) \\
\overline{Z}(\xi_1, \xi_2) &= (z_{00} + 4z_{01} + z_{02} + 4z_{10} + 16z_{11} + 4z_{12} \\
&\quad + z_{20} + 4z_{21} + z_{22})/36 - \overline{Z}(\xi_1) - \overline{Z}(\xi_2) - Z(\mathbf{a}) .
\end{aligned} \tag{7.59}$$

The mean values vanish when numerical values for $Z_{-0.5}$, Z_0, and $Z_{0.5}$, as well as $z_{00}, \cdots, z_{22}$, as presented in (7.56), are used.

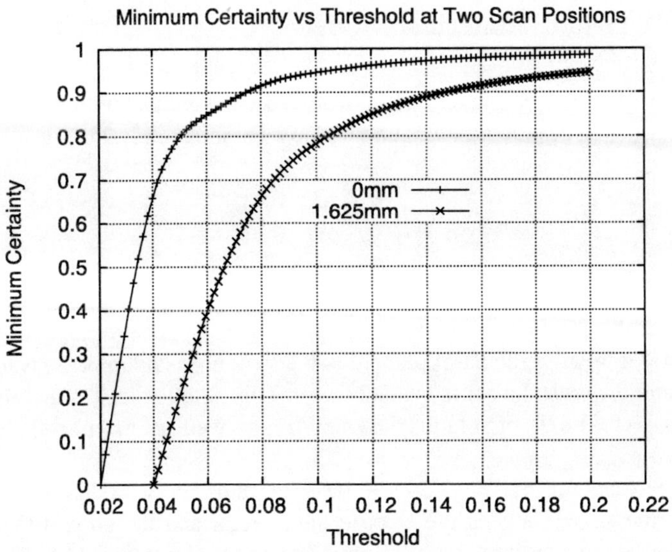

Fig. 7.20 Illustrating 'POD' curves for detectability of a flaw located at two different points within the random surface. The first point, 1.625 mm from the center, is associated with the maximum variance of the random signal, as shown in Fig. 7.18, and the second point is at the center of the surface

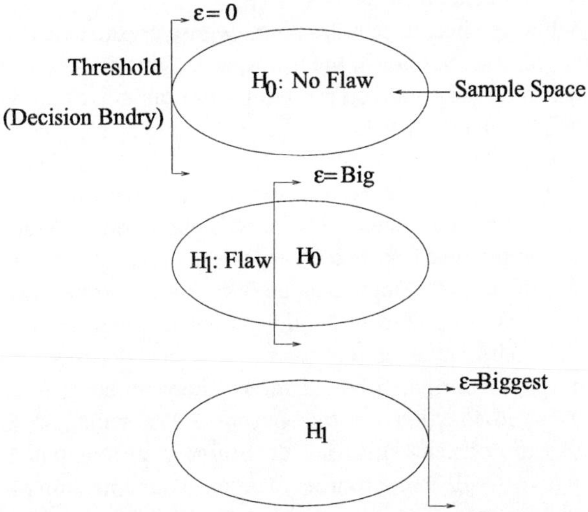

Fig. 7.21 Illustrating the separation of the two regions of the sample space corresponding to hypotheses H_0 and H_1. The decision boundary, or threshold, is determined by the magnitude of ϵ in (7.58)

Appendix 1: The Numerical Model

In its principal-axis coordinate system, the electrical conductivity tensor for hexagonal crystals, such as pure titanium and its most common alloys, is given by

$$\sigma = \begin{bmatrix} \sigma_{11} & 0 & 0 \\ 0 & \sigma_{11} & 0 \\ 0 & 0 & \sigma_{33} \end{bmatrix} , \tag{7.60}$$

where σ_{11} is the conductivity in the basal plane (plane of isotropy), and σ_{33} is the conductivity normal to the basal plane. Such a system is said to possess transverse isotropy, and the crystal class is labeled 6 mm. This notation means that the crystal contains a sixfold axis of rotational symmetry, as well as six mirror planes that contain that axis.

For pure titanium, $\sigma_{11} = 2.205 \times 10^6$ S/m, and $\sigma_{33} = 2.083 \times 10^6$ S/m. If ϕ denotes the angle between the electric field vector and the normal to the basal plane, then the conductivity in the $\phi-$direction can be represented by the ellipsoid of revolution

$$\sigma(\phi) = \sigma_{33} \cos^2 \phi + \sigma_{11} \sin^2 \phi . \tag{7.61}$$

We will use (7.61) in establishing our numerical model. See the next section for a further discussion and generalization of (7.61).

Consider a half-space host of pure titanium, whose crystal axis is oriented in the $z-$direction, normal to the surface of the half-space. Lying at the surface of this host is a rough patch of randomly oriented crystallites of titanium, as shown in Fig. 7.22. The use of a single conductivity value for the host would be rigorously correct if the coil were the only current source in the problem. The electric field induced into the host by this source lies within the basal plane, so only the basal-plane conductivity, 2.205×10^6 S/m, enters the picture. The anomalous currents within the random crystallites, however, produce a scattered field within the host that is not confined to the basal plane, so the use of a single conductivity is an approximation.

We can use any number of cells in the model in order to get a statistically reasonable answer. The value of the conductivity to be assigned each cell (or crystallite or grain) is determined by randomly choosing $\cos\phi$ in (7.61). We use a uniform distribution function for this purpose. The volume-fractions for the cells, from which the conductivities are determined, are computed off-line, and then imported into **VIC-3D®** in a routine manner. After simulating the impedance response of the bad patch, we can run several small slot responses to gain insight into how deleterious the grain noise is to detecting a crack in its presence.

The relationship between volume-fractions, VF, and conductivity of a cell is given by

$$\sigma = \sigma_{\max} + VF(\sigma_{\min} - \sigma_{\max}) , \tag{7.62}$$

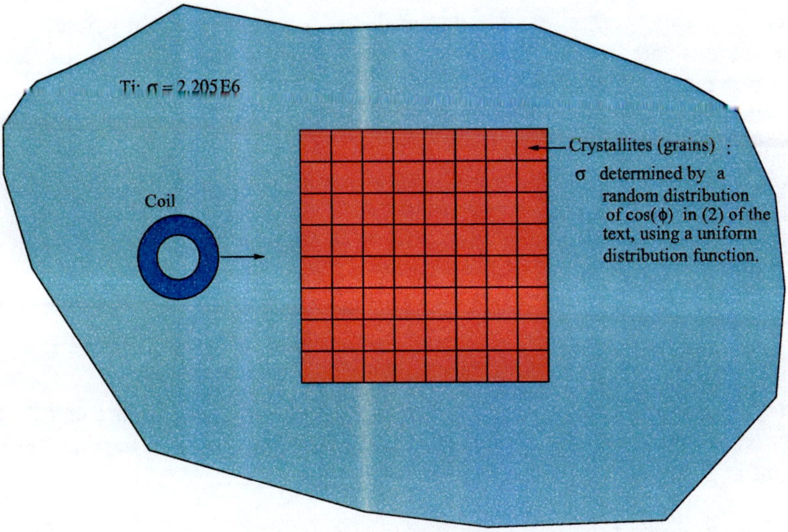

Fig. 7.22 Illustrating the model setup for **VIC-3D®**. There is a second layer identical to the one shown immediately below the one shown. This is due to the fact that **VIC-3D®** requires a minimum of two cells in any direction of the grid

where $\sigma_{\max}$ is the maximum conductivity in the region of interest (host plus anomaly), $\sigma_{\min}$ is the minimum conductivity, and $0 \leq VF \leq 1$. As Fig. 7.22 indicates, $\sigma_{\max} = \sigma_{\text{host}} = 2.205 \times 10^6$ in this setup.

Returning to (7.61), we can easily determine a relationship for volume-fractions:

$$\begin{aligned} \sigma &= \sigma_{\max} \sin^2 \phi + \sigma_{\min} \cos^2 \phi \\ &= \sigma_{\max}(1 - \cos^2 \phi) + \sigma_{\min} \cos^2 \phi \\ &= \sigma_{\max} + \cos^2 \phi (\sigma_{\min} - \sigma_{\max}) , \end{aligned} \tag{7.63}$$

from which we conclude that $VF = \cos^2 \phi$.

More Physics Let $\boldsymbol{\sigma}$ be the conductivity tensor of a material in an arbitrary coordinate system, so that it will have (in general) nonzero off-diagonal entries. In this coordinate system we still have $\mathbf{J} = \boldsymbol{\sigma} \cdot \mathbf{E}$, where $\mathbf{J}$ is the electric current density, and $\mathbf{E}$ is the electric field. The electric power density dissipated within the material is $P = \mathbf{E} \cdot \mathbf{J} = \mathbf{E} \cdot \boldsymbol{\sigma} \cdot \mathbf{E}$, which defines an ellipsoid in $\mathbf{E}$-space.

By definition, the conductivity in the direction of the electric field, $\mathbf{E}$, is given by

$$\begin{aligned} \sigma_E &= \frac{P}{\mathbf{E} \cdot \mathbf{E}} \\ &= \frac{P}{|\mathbf{E}|^2} , \end{aligned} \tag{7.64}$$

which, because it comprises only dot-products, is a scalar under rotations. Hence, it can be evaluated in any coordinate system, such as the principal-axis system described in the Introduction. This means that (7.61) is rigorously correct when applied to the randomly-oriented crystallites of the numerical model that we have described earlier in this section.

We'll apply this result to a crystal of lower symmetry, such that in its principal axis system its conductivity tensor becomes:

$$\sigma = \begin{bmatrix} \sigma_{11} & 0 & 0 \\ 0 & \sigma_{22} & 0 \\ 0 & 0 & \sigma_{33} \end{bmatrix} . \tag{7.65}$$

Therefore, $\mathbf{J} = \sigma_{11} E_x \mathbf{a_x} + \sigma_{22} E_y \mathbf{a_y} + \sigma_{33} E_z \mathbf{a_z}$, where $\mathbf{a_x},\ \mathbf{a_y},\ \mathbf{a_z}$ are unit vectors in the $x,\ y,$ and z directions, respectively. The electric power dissipated per unit volume is given by

$$\begin{aligned} P &= \mathbf{E} \cdot \mathbf{J} \\ &= \sigma_{11} E_x^2 + \sigma_{22} E_y^2 + \sigma_{33} E_z^2 \\ &= E^2 \left(\sigma_{11} \sin^2 \phi \cos^2 \theta + \sigma_{22} \sin^2 \phi \sin^2 \theta + \sigma_{33} \cos^2 \phi \right) , \end{aligned} \tag{7.66}$$

where E is the magnitude of the electric-field vector, and $\theta,\ \phi$ are the azimuthal and polar angles in spherical coordinates, respectively.

Hence,

$$\sigma_E = \sigma_{11} \sin^2 \phi \cos^2 \theta + \sigma_{22} \sin^2 \phi \sin^2 \theta + \sigma_{33} \cos^2 \phi . \tag{7.67}$$

In this case, we require two variables, θ and ϕ, to define σ_E, and the calculation of the volume-fractions requires a separate step. If $\sigma_{11} < \sigma_{22} < \sigma_{33}$, then, from (7.62)

$$VF = \frac{\sigma_E - \sigma_{33}}{\sigma_{11} - \sigma_{33}} . \tag{7.68}$$

If $\sigma_{11} = \sigma_{22}$ in (7.67), then we recover the case of transverse isotropy in (7.61).

Appendix 2: The Fortran RANDOM_NUMBER Subroutine

The Fortran 90 RANDOM_NUMBER Subroutine [34] returns uniformly distributed pseudorandom number(s) over the range $0 \le x < 1$. This is the subroutine that we use to generate the random variables that are required in the Karhunen-Loève expansion. As such, it is necessary that we demonstrate that its output is consistent with the requirements of the expansion, namely that the random variables are uncorrelated and have a unit variance.

First, we note that the variance of a random variable that is uniformly distributed over [0, 1) is 1/12, so that we must multiply the output of RANDOM_NUMBER by $\sqrt{12}$ in order to generate a unit-variance random variable. Secondly, we transform the range of the output to $[-0.5, 0.5)$ in order to generate a zero-mean random variable, which will be useful in our later work.

In order to demonstrate that the output of RANDOM_NUMBER is uncorrelated, we perform the following experiment. We generate a 32-element random vector which is the output of RANDOM_NUMBER, and identify the 17th and 32nd elements as two typical random variables. We repeat this experiment 10, 100, 1000, 10,000 and 100,000 times to generate five sample spaces. We then compute the means and variances of each of the two random variables, as well as their covariance. The results, obtained using the usual equations of statistics [18],

$$\begin{aligned}
\text{MEAN}(17) &= \frac{1}{N}\sum_{i=1}^{N} RV_{17}(i) \\
\text{MEAN}(32) &= \frac{1}{N}\sum_{i=1}^{N} RV_{32}(i) \\
\text{COV}(17,32) &= \frac{1}{N}\sum_{i=1}^{N}(RV_{17}(i) - \text{MEAN}(17)) \times (RV_{32}(i) - \text{MEAN}(32)) \\
\text{VAR}(17) &= \frac{1}{B}\sum_{i=1}^{N}(RV_{17}(i) - \text{MEAN}(17))^2 \\
\text{VAR}(32) &= \frac{1}{B}\sum_{i=1}^{N}(RV_{32}(i) - \text{MEAN}(32))^2 ,
\end{aligned} \tag{7.69}$$

are shown in Table 7.4. It is clear that the required conditions are met, especially the very small covariance and unit variances, with increasing sample size. Thus, we can confidently use the Fortran RANDOM_NUMBER Subroutine to generate numbers that are consistent with the statement of the Karhunen-Loève expansion.

Table 7.4 Convergence of the Fortran RANDOM_NUMBER Subroutine

Trials	MEAN(17)	MEAN(32)	COV(17,32)	VAR(17)	VAR(32)
10	0.2664	0.3056	−0.2649	1.117	0.8505
100	−0.6011(−1)	−0.2929	0.3290(−1)	1.056	1.031
1000	0.7366(−2)	0.3469(−1)	−0.1442(−1)	1.004	1.027
10,000	−0.1173(−2)	0.5354(−2)	0.1377(−1)	0.9972	0.9926
100,000	0.7556(−2)	0.1951(−2)	−0.4245(−2)	0.9962	1.001

Chapter 8
A Model for Microstructure Characterization

8.1 Introduction

Titanium is noisy. Contrary to the situation discussed in Chap. 7 in which the surface of an isotropic body, copper, is made artificially noisy through shot-peening, titanium is intrinsically noisy because of its microstructure consisting of a random distribution of anisotropic crystallites. Figure 8.1 illustrates this condition by comparing the response of the same flaw in the alloy, Ti-6Al-4V, to aluminum, which is isotropic.

Figure 8.2 contrasts noise in a titanium alloy with a large crack response. The origin of the noise lies in the random, anisotropic microstructure of the alloy, and the objective of the present chapter is to develop a model for computing this noise, starting with a realistic model of the microstructure.

Figures 8.3 and 8.4 further illustrate the microstructure of a titanium alloy, in this case, Ti-7Al. The clustering of grains of similar orientation in Fig. 8.4 produces continuum effects that are different than those of the individual grains. We expect to see conductivity variations that are smoothed and elongated in the direction of the clusters, rather than chaotic from grain to grain. Thus, the grain size used for a particular property model is no longer actually the grain size measured from optical or backscattered electron (BSE) images.

8.2 Stochastic Euler Space

In Appendix 1 of Chap. 7 we pointed out that the anisotropic model was an approximate one. In this chapter we intend to rectify that, and we will extend the work of Chap. 7 to two dimensions, with the intention of developing an approach to characterizing the microstructure of Ti and its alloys. We'll start with **Euler Angles**.

H. A. Sabbagh et al., *Advanced Electromagnetic Models for Materials Characterization and Nondestructive Evaluation*, Scientific Computation,
https://doi.org/10.1007/978-3-030-67956-9_8

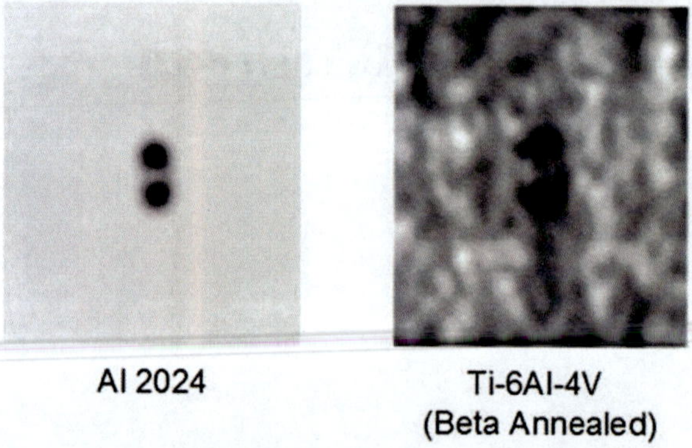

Fig. 8.1 Illustrating the response of the same flaw in 'noisy' Ti-6Al-4V (right) and 'noiseless' Al. (Image courtesy of M. Blodgett, Air Force Research Laboratory)

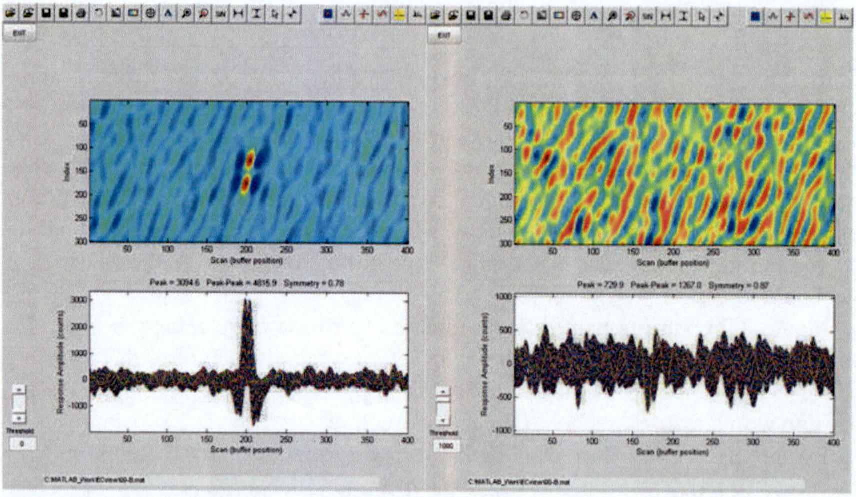

Fig. 8.2 Contrasting noise in a titanium alloy (right) with a large (~28 × 15 mils) crack response (left). The sample has a duplex microstructure (~50% spherical alpha, ~50% lamellar secondary alpha platelets) (Image courtesy of E. Shell, Wyle Labs)

Let a rotation about O carry the orthogonal triad $(\mathbf{I}, \mathbf{J}, \mathbf{K})$ into $(\mathbf{i}, \mathbf{j}, \mathbf{k})$. We break this rotation into three rotations. First, rotate about $\mathbf{K}$ so as to make the new position of the plane $(\mathbf{I}, \mathbf{K})$ contain $\mathbf{k}$, say through an angle ϕ; this gives a transformation

$$(\mathbf{I}, \mathbf{J}, \mathbf{K}) \to (\mathbf{I}_1, \mathbf{J}_1, \mathbf{K}_1) \left\{ \begin{array}{l} \mathbf{I}_1 = \mathbf{I}\cos\phi + \mathbf{J}\sin\phi \\ \mathbf{J}_1 = -\mathbf{I}\sin\phi + \mathbf{J}\cos\phi \\ \mathbf{K}_1 = \mathbf{K} \end{array} \right\}. \tag{8.1}$$

Secondly, rotate about $\mathbf{J}_1$ to bring $\mathbf{K}_1$ to $\mathbf{k}$, say through an angle θ; this gives a transformation

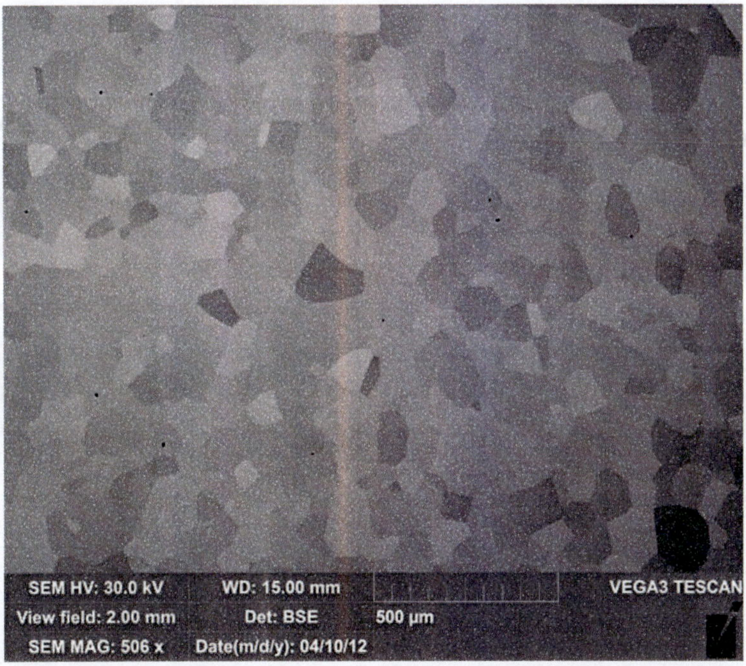

Fig. 8.3 Microstructure of Ti-7Al: backscattered electron image (Image courtesy of M. Cherry, Air Force Research Laboratory)

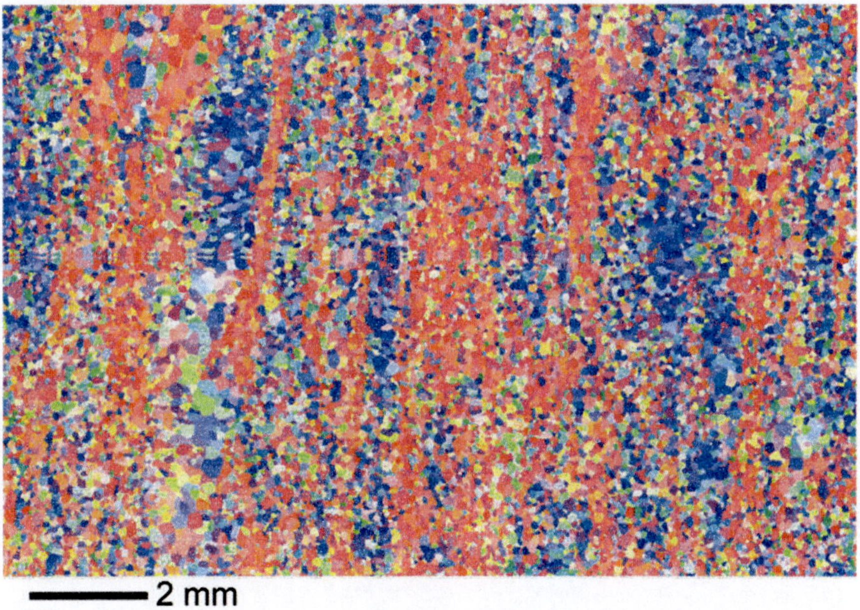

Fig. 8.4 Microstructure of Ti-7Al: crystal orientation map (electron backscatter diffraction) (Image courtesy of M. Cherry, Air Force Research Laboratory)

$$(\mathbf{I}_1, \mathbf{J}_1, \mathbf{K}_1) \rightarrow (\mathbf{I}_2, \mathbf{J}_2, \mathbf{k}) \left\{ \begin{aligned} \mathbf{I}_2 &= \mathbf{I}_1 \cos\theta - \mathbf{K}_1 \sin\theta \\ \mathbf{J}_2 &= \mathbf{J}_1 \\ \mathbf{k} &= \mathbf{I}_1 \sin\theta + \mathbf{K}_1 \cos\theta \end{aligned} \right\} . \tag{8.2}$$

Finally, rotate about $\mathbf{k}$ to bring $\mathbf{I}_2$ to $\mathbf{i}$ and $\mathbf{J}_2$ to $\mathbf{j}$, say through an angle ψ; this gives the transformation

$$(\mathbf{I}_2, \mathbf{J}_2, \mathbf{k}) \rightarrow (\mathbf{i}, \mathbf{j}, \mathbf{k}) \left\{ \begin{aligned} \mathbf{i} &= \mathbf{I}_2 \cos\psi + \mathbf{J}_2 \sin\psi \\ \mathbf{j} &= -\mathbf{I}_2 \sin\psi + \mathbf{J}_2 \cos\psi \\ \mathbf{k} &= \mathbf{k} \end{aligned} \right\} . \tag{8.3}$$

The angles (θ, ϕ, ψ) are the Euler angles. Their values determine the position of the triad $(\mathbf{i}, \mathbf{j}, \mathbf{k})$ relative to $(\mathbf{I}, \mathbf{J}, \mathbf{K})$. The angles range over the following values:

$$\begin{aligned} 0 &\le \theta \le \pi \\ 0 &\le \phi < 2\pi \\ 0 &\le \psi < 2\pi \end{aligned} . \tag{8.4}$$

From the above equations of transformation, we can obtain the matrix, $\mathbf{M}$, that defines orthogonal rotations. We use the notation, $c = \cos$, $s = \sin$ and let the subscripts, 1, 2, 3, refer to θ, ϕ, ψ, respectively:

	$\mathbf{i}$	$\mathbf{j}$	$\mathbf{k}$
$\mathbf{I}$	$c_1c_2c_3 - s_2s_3$	$-c_1c_2s_3 - s_2c_3$	s_1c_2
$\mathbf{J}$	$c_1s_2c_3 + c_2s_3$	$-c_1s_2s_3 + c_2c_3$	s_1s_2
$\mathbf{K}$	$-s_1c_3$	s_1s_3	c_1

(8.5)

Now, we'll apply this to the problem at hand. Let the host conductivity have the transverse isotropy associated with Ti-6Al-4V in its principal coordinate system:

$$\boldsymbol{\sigma}(\mathbf{r}) = \begin{bmatrix} \sigma_1 & 0 & 0 \\ 0 & \sigma_1 & 0 \\ 0 & 0 & \sigma_2 \end{bmatrix} . \tag{8.6}$$

Data supplied by P. B. Nagy[1] give values of $\sigma_1 = 6.04 \times 10^5$ S/m (1.04% IACS) and $\sigma_2 = 5.90 \times 10^5$ S/m (1.02%IACS) for the basal plane and normal-to-the-basal-plane conductivities, respectively.

In its rotated coordinate system, this conductivity tensor becomes

$$\sigma(\mathbf{r}) = \mathbf{M}\begin{bmatrix} \sigma_1 & 0 & 0 \\ 0 & \sigma_1 & 0 \\ 0 & 0 & \sigma_2 \end{bmatrix}\mathbf{M}^T$$

$$= \begin{bmatrix} \sigma_1 + m_{13}^2(\sigma_2 - \sigma_1) & (\sigma_2 - \sigma_1)m_{13}m_{23} & (\sigma_2 - \sigma_1)m_{13}m_{33} \\ (\sigma_2 - \sigma_1)m_{23}m_{13} & \sigma_1 + (\sigma_2 - \sigma_1)m_{23}^2 & (\sigma_2 - \sigma_1)m_{23}m_{33} \\ (\sigma_2 - \sigma_1)m_{13}m_{33} & (\sigma_2 - \sigma_1)m_{23}m_{33} & \sigma_1 + (\sigma_2 - \sigma_1)m_{33}^2 \end{bmatrix}. \quad (8.7)$$

Note that this is a symmetric tensor, with components that are independent of ψ, in the rotated coordinate system. We identify the Euler angles, θ and ϕ, as the random variables that define the orientation of each grain (or voxel) of the anomalous region. Once these two variables are given for each realization, the random conductivity field follows from (8.7).

Preliminary Stochastic Calculations We will assume that θ and ϕ are independent random variables whose first order probability densities are uniform over the ranges shown in (8.4). The mean and variance of these two variables are easily computed to be

$$\begin{aligned} \overline{\theta} &= \frac{1}{\pi}\int_0^{\pi} \theta d\theta \\ &= \frac{\pi}{2} \\ \mathrm{VAR}(\theta) &= \frac{1}{\pi}\int_0^{\pi} \theta^2 d\theta - \frac{\pi^2}{4} \\ &= \frac{\pi^2}{12} \\ \overline{\phi} &= \frac{1}{2\pi}\int_0^{2\pi} \phi d\phi \\ &= \pi \\ \mathrm{VAR}(\phi) &= \frac{1}{2\pi}\int_0^{2\pi} \phi^2 d\phi - \pi^2 \\ &= \frac{\pi^2}{3} \end{aligned} \quad (8.8)$$

[1]Private communication.

Furthermore, we have the following results for the mean values of the random coefficients of the conductivity tensor in (8.7):

$$\begin{aligned}
\overline{m_{13}m_{23}} &= \frac{1}{2\pi^2}\int_0^{\pi}\sin^2\theta d\theta\int_0^{2\pi}\sin\phi\cos\phi d\phi \;;\; \overline{m_{13}^2} = \frac{1}{2\pi^2}\int_0^{\pi}\sin^2\theta d\theta\int_0^{2\pi}\cos^2\phi d\phi\\
&= 0 \qquad ; \qquad = 1/4\\
\overline{m_{13}m_{33}} &= \frac{1}{2\pi^2}\int_0^{\pi}\sin\theta\cos\theta d\theta\int_0^{2\pi}\cos\phi d\phi \;;\; \overline{m_{23}^2} = \frac{1}{2\pi^2}\int_0^{\pi}\sin^2\theta d\theta\int_0^{2\pi}\sin^2\phi d\phi\\
&= 0 \qquad ; \qquad = 1/4\\
\overline{m_{23}m_{33}} &= \frac{1}{2\pi^2}\int_0^{\pi}\sin\theta\cos\theta d\theta\int_0^{2\pi}\sin\phi d\phi \;;\; \overline{m_{33}^2} = \frac{1}{\pi}\int_0^{\pi}\cos^2\theta d\theta\\
&= 0 \qquad ; \qquad = 1/2
\end{aligned} \tag{8.9}$$

Thus, the mean value of the conductivity tensor is

$$\begin{aligned}
\overline{\boldsymbol{\sigma}(\mathbf{r})} &= \begin{bmatrix} \sigma_1 + 1/4(\sigma_2-\sigma_1) & 0 & 0\\ 0 & \sigma_1 + 1/4(\sigma_2-\sigma_1) & 0\\ 0 & 0 & \sigma_1 + 1/2(\sigma_2-\sigma_1)\end{bmatrix}\\
&= \begin{bmatrix} 3/4\sigma_1 + 1/4\sigma_2 & 0 & 0\\ 0 & 3/4\sigma_1 + 1/4\sigma_2 & 0\\ 0 & 0 & 1/2(\sigma_1+\sigma_2)\end{bmatrix}.
\end{aligned} \tag{8.10}$$

This result indicates that stochastic mixing of the eigenvalues produces a slightly less anisotropic host in the mean, with the difference in the eigenvalues given by $1/4(\sigma_2 - \sigma_1)$.

Expand the random variables, θ and ϕ, in the usual manner:

$$\begin{aligned}
\theta &= \overline{\theta} + \hat{\theta} = \pi/2 + \hat{\theta}\\
\phi &= \overline{\phi} + \hat{\phi} = \pi + \hat{\phi}\,,
\end{aligned} \tag{8.11}$$

where $\hat{}$ denotes the random, zero-mean, residual that will be computed using the Karhunen-Loève expansion to be discussed next. Substituting these into the expressions for the random coefficients of the conductivity tensor yields:

$$\begin{aligned}
m_{13}^2 &= \sin^2(\pi/2+\hat{\theta})\cos^2(\pi+\hat{\theta}) = \cos^2\hat{\theta}\cos^2\hat{\phi}\\
m_{23}^2 &= \sin^2(\pi/2+\hat{\theta})\sin^2(\pi+\hat{\phi}) = \cos^2\hat{\theta}\sin^2\hat{\phi}\\
m_{33}^2 &= \cos^2(\pi/2+\hat{\theta}) = \sin^2\hat{\theta}\\
m_{13}m_{23} &= \sin^2(\pi/2+\hat{\theta})\sin(\pi+\hat{\phi})\cos(\pi+\hat{\phi}) = 1/2\cos^2\hat{\theta}\sin 2\hat{\phi}\\
m_{13}m_{33} &= \sin(\pi/2+\hat{\theta})\cos(\pi/2+\hat{\theta})\cos(\pi+\hat{\phi}) = 1/2\sin 2\hat{\theta}\cos\hat{\phi}\\
m_{23}m_{33} &= \sin(\pi/2+\hat{\theta})\cos(\pi/2+\hat{\theta})\sin(\pi+\hat{\phi}) = 1/2\sin 2\hat{\theta}\sin\hat{\phi}\,.
\end{aligned} \tag{8.12}$$

As an aside, we note that if $\hat{\theta} = \hat{\phi} = 0$, then (8.12) gives us the values of the coefficients at the mean of their random arguments, which, as can be seen by comparison with (8.9), is not equal to the mean of the coefficients. This is due to the fact that the random coefficients are nonlinear functions of their random arguments.

8.3 The Karhunen-Loève Model

The two-dimensional K-L expansion that corresponds to (7.9) for either variable is

$$\int_{-b}^{b}\int_{-b}^{b} C(x, x'; y, y')\psi(x', y')dx'dy' = \lambda^2\psi(x, y)\,. \tag{8.13}$$

We will deal only with stationary covariances that are separable in (x, x') and (y, y'): $C(x, x'; y, y') = C_x(x, x')C_y(y, y')$. Examples are:

$$\begin{aligned}
\text{Double-Exponential}: C(x, x'; y, y') &= s^2 \exp[-\frac{|x-x'|+|y-y'|}{L}] \\
&= s^2 \exp[-\frac{|x-x'|}{L}]\exp[-\frac{|y-y'|}{L}] \\
\text{Gaussian}: \qquad C(x, x'; y, y') &= s^2 \exp[-\frac{(x-x')^2+(y-y')^2}{L^2}] \\
&= s^2 \exp[-\frac{(x-x')^2}{L^2}]\exp[-\frac{(y-y')^2}{L^2}]\,,
\end{aligned} \tag{8.14}$$

where L is the correlation length of the process, and s^2 is the variance of the random variable, taking on the value $\pi^2/12$ for θ, and $\pi^2/3$ for ϕ (recall (8.8)). Obviously, in the special cases of (8.14), we have $C_x(.,.) = C_y(.,.)$.

We will follow our earlier one-dimensional development, as shown in (7.11)–(7.17), to transform (8.13) into a discrete two-dimensional model. Let $\psi(x, y) = \sum_{m=1}^{N}\sum_{n=1}^{N}\psi_{mn}f_m(x)f_n(y)$, where $\{f_m(x)f_n(y)\}$ is a basis for $\psi(x, y)$, and $\{\psi_{mn}\}$ are expansion coefficients. Substituting this into (8.13) yields

$$\sum_{m=1}^{N}\sum_{n=1}^{N}\psi_{mn}\int_{-b}^{b} C_x(x, x')f_m(x')dx'\int_{-b}^{b} C_y(y, y')f_n(y')dy' = |\lambda|^2\sum_{m=1}^{N}\sum_{n=1}^{N}\psi_{mn}f_m(x)f_n(y)\,. \tag{8.15}$$

Take moments of (8.15) by multiplying by $f_{m'}(x)f_{n'}(y)$ and then integrating over $[-b, b]\otimes[-b, b]$ ($\otimes$ denotes the direct (or tensor) product of two entities):

$$\begin{aligned}
&\sum_{m=1}^{N}\sum_{n=1}^{N}\psi_{mn}\int_{-b}^{b}\int_{-b}^{b} C_x(x, x')f_m(x)f_{m'}(x')dxdx'\int_{-b}^{b}\int_{-b}^{b} C_y(y, y')f_n(y)f_{n'}(y')dydy' \\
&= |\lambda|^2\sum_{m=1}^{N}\sum_{n=1}^{N}\psi_{mn}\int_{-b}^{b} f_m(x)f_{m'}(x)dx\int_{-b}^{b} f_n(y)f_{n'}(y)dy
\end{aligned} \tag{8.16}$$

Upon calling the double integrals on the left $G^{(x)}_{m'm}$, $G^{(y)}_{n'n}$, respectively, and the integrals on the right $H^{(x)}_{m'm}$ and $H^{(y)}_{n'n}$, respectively, we have the result

$$\sum_{m=1}^{N}\sum_{n=1}^{N} G^{(x)}_{m'm} G^{(y)}_{n'n} \psi_{mn} = |\lambda|^2 \sum_{m=1}^{N}\sum_{n=1}^{N} H^{(x)}_{m'm} H^{(y)}_{n'n} \psi_{mn}, \quad m', n' = 1, \cdots, N, \tag{8.17}$$

or, in vector-matrix notation

$$\mathbf{G} \cdot \mathbf{v} = |\lambda|^2 \mathbf{H} \cdot \mathbf{v}, \tag{8.18}$$

where $\mathbf{G} = \mathbf{G}^{(x)} \otimes \mathbf{G}^{(y)}$, $\mathbf{H} = \mathbf{H}^{(x)} \otimes \mathbf{H}^{(y)}$, and $\mathbf{v} = [\psi_{11}, \cdots, \psi_{NN}]^T$. For the separable covariances shown in (8.14), we have $\mathbf{G}^{(x)} = \mathbf{G}^{(y)}$. This completes the derivation of the generalized eigenvalue problem. If $\{f_n(x)\}$ is orthogonal, then $\mathbf{H}$ is diagonal, and if $\{f_n(x)\}$ is normalized to unity, then $\mathbf{H}$ is the identity matrix, and the generalized eigenvalue problem reduces to the standard form

$$\mathbf{G} \cdot \mathbf{v} = |\lambda|^2 \mathbf{v}. \tag{8.19}$$

We have already calculated the matrix elements for the double-exponential function of (8.14) in (7.20), so we will proceed with the computation of the matrix elements of the Gaussian function. Using the one-dimensional basis functions that were defined in (7.17), we have

$$\begin{aligned}
G^{(x)}_{mm'} &= s \int_{-b}^{b} \int_{-b}^{b} e^{-(x-x')^2/L^2} \pi(x/\delta - m)\pi(x'/\delta - m')dxdx' \\
&= s \int_{m'\delta}^{(m'+1)\delta} dx' \int_{m\delta}^{(m+1)\delta} e^{-(x-x')^2/L^2} dx \\
&= sL \int_{m'\delta}^{(m'+1)\delta} dx' \int_{m\delta/L - x'/L}^{(m+1)\delta/L - x'/L} e^{-u^2} du \\
&= s\frac{\sqrt{\pi}}{2} L \int_{m'\delta}^{(m'+1)\delta} \left\{\operatorname{erf}\left[x'/L - m\delta/L\right] - \operatorname{erf}\left[x'/L - (m+1)\delta/L\right]\right\} dx' \\
&= s\frac{\sqrt{\pi}}{2} L^2 \int_{(m'-m)\delta/L}^{(m'-m+1)\delta/L} \left[\operatorname{erf} v - \operatorname{erf}(v - \delta/L)\right] dv,
\end{aligned} \tag{8.20}$$

where the error function, erf, and its properties are defined in [81, Chapter 7].

We can get an explicit analytical expression for $G^{(x)}_{mm'}$ by substituting the power series representation for erf z,

$$\operatorname{erf} z = \frac{2}{\sqrt{\pi}} \sum_{n=0}^{\infty} \frac{(-1)^n z^{2n+1}}{n!(2n+1)}, \tag{8.21}$$

into (8.20):

$$\begin{aligned}
G^{(x)}_{mm'} &= sL^2 \sum_{n=0}^{\infty} \frac{(-1)^n}{n!(2n+1)} \int_{(m'-m)\delta/L}^{(m'-m+1)\delta/L} \left[v^{2n+1} - (v-\delta/L)^{2n+1}\right] dv \\
&= sL^2 \sum_{n=0}^{\infty} \frac{(-1)^n}{n!(2n+1)} \left[\frac{v^{2n+2}}{2n+2} - \frac{(v-\delta/L)^{2n+2}}{2n+2}\right]_{(m'-m)\delta/L}^{(m'-m+1)\delta/L} \\
&= sL^2 \sum_{n=0}^{\infty} \frac{(-1)^n}{n!(2n+1)(2n+2)} \Big\{[(m'-m+1)(\delta/L)]^{2n+2} - 2[(m'-m)(\delta/L)]^{2n+2} \\
&\quad + [(m'-m-1)(\delta/L)]^{2n+2}\Big\}
\end{aligned} \tag{8.22}$$

Note that the final term within the wiggly braces is the discrete second derivative operator, which means that $G^{(x)}_{mm'}$ can be computed by evaluating, at the point $(m'-m)(\delta/L)$, the second derivative of the integral of the error function. Following this logic, we see from (8.20) that this yields the derivative of the error function, which is precisely the discrete version of the Gaussian kernel in (8.14). Thus,

$$G^{(x)}_{mm'} = sL^2 \exp[-(m-m')^2(\delta^2/L^2)]\,. \tag{8.23}$$

This result, as with that for the double-exponential covariance function, is a symmetric Töplitz matrix. For the same pulse basis functions, $\mathbf{H}^{(x)} = \mathbf{H}^{(y)}$ is diagonal with the constant value of δ. Therefore, in order to transform (8.18) into (8.19), we must divide $\mathbf{G}$ by δ^2; equivalently, we divide $\mathbf{G}^{(x)}$ and $\mathbf{G}^{(y)}$ each by δ. Hence, (8.23) becomes

$$G^{(x)}_{mm'} = \frac{sL^2}{\delta} \exp[-(m-m')^2(\delta^2/L^2)]\,. \tag{8.24}$$

Eigenfunction Results For the problem that we are considering here, with a 32×32 grid of cells for **VIC-3D®**, the eigenvalue problem becomes one with a 1024×1024 matrix. This yields 1024 eigenvalues, and 1024 normalized eigenvectors that form a complete orthonormal system in a 1024-dimension vector space.

Figure 8.5 shows the first 100 normalized eigenvalues for the two-dimensional Gaussian and double-exponential covariance functions with $L/\delta = 5$. We note that the spectrum of the Gaussian covariance converges to zero beyond 50 faster than that of the double-exponential. This is due to the fact that the Gaussian covariance is an analytic function whose derivatives of all orders exist, whereas the double-exponential covariance is singular at the origin, in the sense that its first derivative

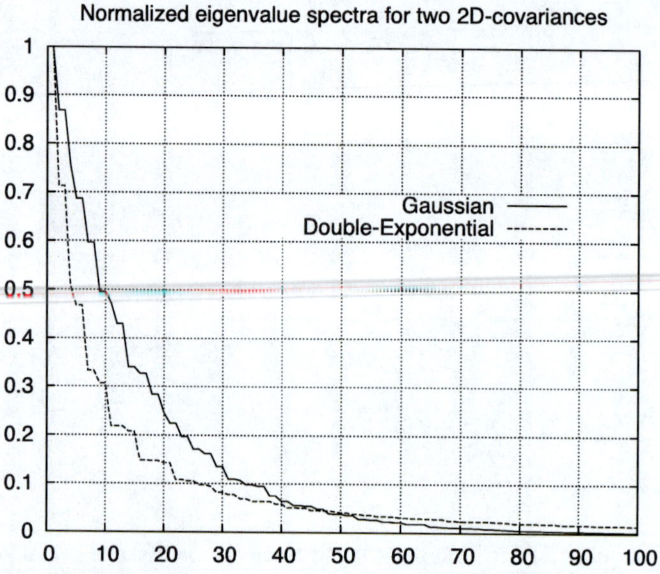

Fig. 8.5 The normalized eigenvalue spectra for the two-dimensional Gaussian and double-exponential covariance functions with $L/\delta = 5$

is discontinuous there. To be sure, however, it would be reasonable to use either function with the spectrum truncated at 50 for this particular value of L/δ.

Once we have the eigenvectors, we must assign their components to the spatial grid in a manner consistent with the direct-product decomposition of (8.16) and (8.17). Figure 8.6 shows the correct ordering. This is especially important when we consider anisotropic covariances, in which the correlation lengths of each component matrix will differ.

8.4 Anisotropic Covariances

If the correlation lengths in (8.14) are different in the orthogonal directions, x and y, then we will call the covariances 'anisotropic.' We will need this level of generality in our work, so we'll consider it to be the default in what follows. The form for the anisotropic covariances of (8.14) becomes:

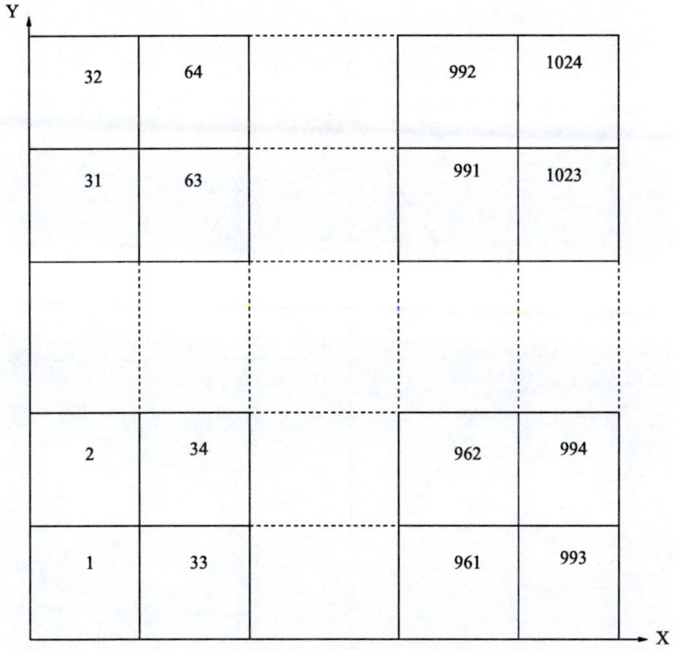

Fig. 8.6 Showing the ordering of the 1024-dimension eigenvector on the 32 × 32 spatial grid

$$\begin{aligned}
\text{Double-Exponential}: C(x, x'; y, y') &= s^2 \exp[-\left\{\frac{|x-x'|}{L_x} + \frac{|y-y'|}{L_y}\right\}] \\
&= s^2 \exp[-\frac{|x-x'|}{L_x}] \exp[-\frac{|y-y'|}{L_y}] \\
\text{Gaussian}: \quad C(x, x'; y, y') &= s^2 \exp[-\left\{\frac{(x-x')^2}{L_x^2} + \frac{(y-y')^2}{L_y^2}\right\}] \\
&= s^2 \exp[-\frac{(x-x')^2}{L_x^2}] \exp[-\frac{(y-y')^2}{L_y^2}],
\end{aligned} \tag{8.25}$$

where L_x and L_y are the correlation lengths in the x and y directions, respectively. Note that these anisotropic covariances remain separable.

The form of the exponential for both covariances is

$$\frac{|x-x'|^p}{L_x^p} + \frac{|y-y'|^p}{L_y^p}, \text{ where } p = \begin{cases} 1 \text{ for the double-exponential} \\ 2 \text{ for the Gaussian}, \end{cases} \tag{8.26}$$

and level curves for these functions are shown in Fig. 8.7.

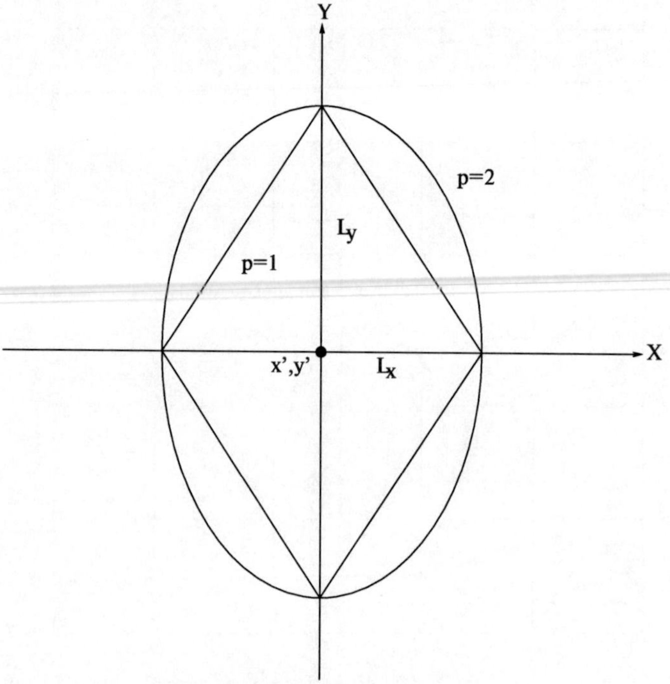

Fig. 8.7 Showing the level curves, $\frac{|x-x'|^p}{L_x^p} + \frac{|y-y'|^p}{L_y^p} = 1$, for the covariances of (8.25); $p = 1$ for the double-exponential function, and $p = 2$ for the Gaussian (See [127, p. 83] for similar curves for other values of p)

8.5 The Geometric Autocorrelation Function

Adam Pilchak and Matt Cherry of the Air Force Research Laboratory[2] have used orientation imaging microscopy (OIM) to acquire data for determining the 'geometric autocorrelation function' for Ti-7Al(wt%) , with the aim of determining the orientation and (average) size of the crystallites that make up Ti-7Al. This is important in our modeling of the random crystallite noise, because Ti-7Al has an anisotropic (6 mm) crystal structure, which we believe is the origin of this noise. That is to say, we believe that the random orientation of these crystallites is the source of the noise. Furthermore, we observe that there is a preferred 'clumping' of these crystallites in a certain direction due to the rolling process which produces the final workpiece.

There is a formal mathematical theory for generating the geometric autocorrelation function of polycrystalline materials [68], which we will refer to shortly in order to draw some conclusions from the data. The aim is to measure the probability that

[2] Private communication.

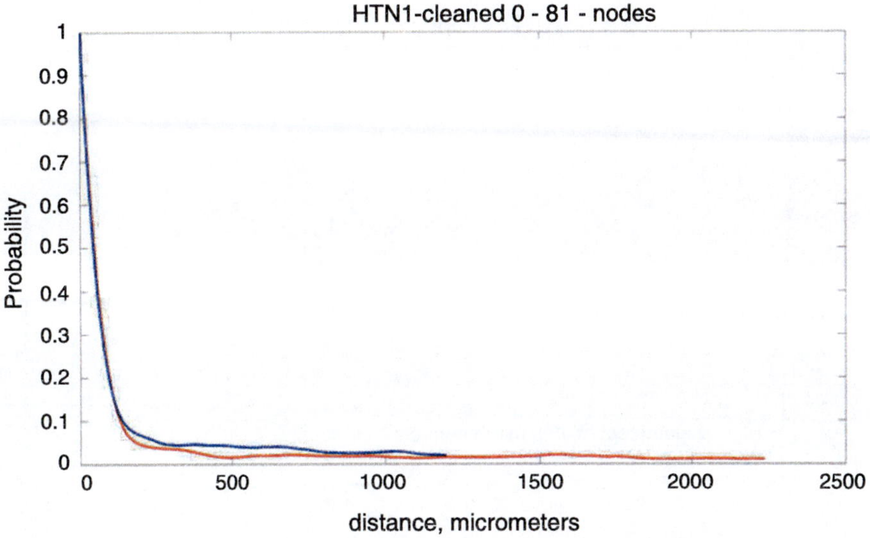

Fig. 8.8 The probability that any two points at the head and tail of an arbitrary vector have similarly oriented c-axes, within $\pm 10.9°$. The x-axis labels the length of the vector

any two points at the head and tail of an arbitrary vector have similarly oriented c-axes, within $\pm 10.9°$, which is by definition the geometric autocorrelation function of Ti-7Al. Figure 8.8 gives results for two orientations; the x-axis labels the length of the vector. The result is closely approximated by an exponential function, at least near the origin.

According to the theory developed in [68], the directional derivative of the autocorrelation function in the direction, n, of the vector evaluated at the $r = 0$, where r is the distance from the head of the vector, is equal to the negative of the reciprocal of the mean linear intercept in the direction n. By 'mean linear intercept' is meant the length of all intercepts with the boundary of the crystallite; in short, it gives us the mean size of the crystallite. Figure 8.9 depicts an application of this theorem to Fig. 8.8, with the result that the nominal size of the crystallites in Ti-7Al is $\approx 43\,\mu$m. We will use this value in our model. By the way, this theorem is an application of the well-known mathematical fact that the time-constant of a simple RC circuit is given by the intersection of the slope at $t = 0$ with the time axis. Of course, this holds because the response of a simple RC circuit is an exponential.

Figure 8.10 illustrates the geometric autocorrelation of Ti-7Al. The bright circular center is essentially the size of the crystallite at the origin, which we are taking to be $43\,\mu$m (the scale of the axes is in micrometers). Of more interest, however, is the very faint outline of a structure surrounding the center that shows the approximate diamond shape of the level curve for the double-exponential function in Fig. 8.7. Assuming a unit value for the level curve, we conclude that $L_x = 430\,\mu$m,

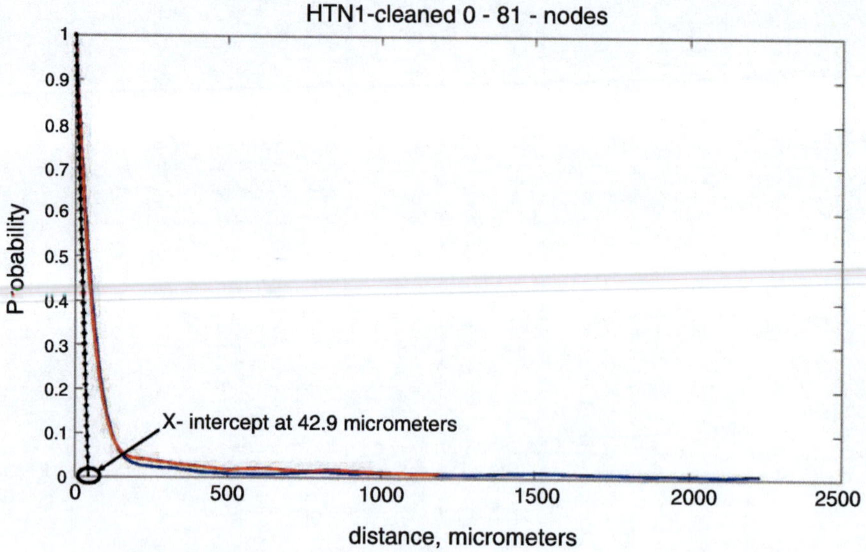

Fig. 8.9 The dashed line is tangent to the probability curve at $X = 0$, and its intercept with the X-axis is at a distance $\approx 43\,\mu\text{m}$. This is what we will call the nominal size of the crystallite (see [68])

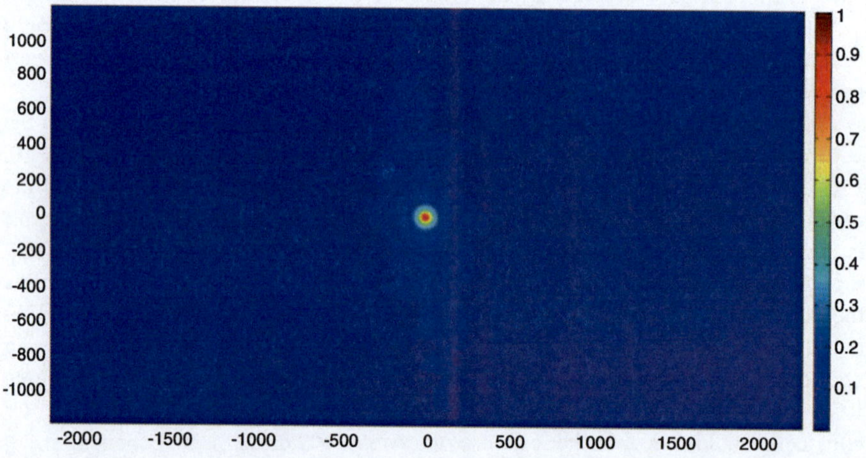

Fig. 8.10 Color graph of the geometric autocorrelation function of Ti-7Al. The circular center gives the mean size of a crystallite. The faint outline surrounding the center appears to have the diamond shape of the level curve of the double-exponential function of Fig. 8.7 with $L_x = 430\,\mu\text{m}$ and $L_y = 1290\,\mu\text{m}$ (Image courtesy of M. Cherry, Air Force Research Laboratory)

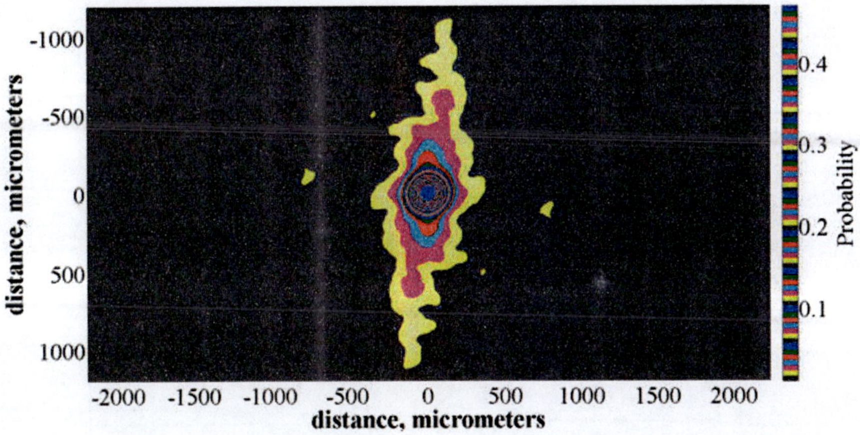

Fig. 8.11 Expanded view of the geometric autocorrelation function of Fig. 8.10 (Image courtesy of M. Cherry, Air Force Research Laboratory)

$L_y = 1290\,\mu\text{m}$. An expanded view of Fig. 8.10 is given in Fig. 8.11, which confirms our estimates of L_x and L_y.

8.6 Results for the Anisotropic Double-Exponential Model

Assuming that we are working with a double-exponential covariance function with parameters $L_x = 430\,\mu\text{m}$, $L_y = 1290\,\mu\text{m}$, and $\delta = 43\,\mu\text{m}$, we compute the normalized eigenvalue spectrum shown in Fig. 8.12. It is clear that we can get good results by using only 25 eigenvalues in our computations. Thus, we have already achieved a reduction in the order of our problem from 1024 variables (32×32 cells) to 25.

Using a 25-term eigenfunction expansion with the parameters of Fig. 8.12, we have computed the random coefficients of the conductivity tensor of (8.7). The results are that m_{13}^2 is virtually unity for all cells, $m_{13}m_{23}$ and $m_{13}m_{33}$ are both of the order of 10^{-4} to 10^{-5}, and the remaining coefficients are four to five orders of magnitude smaller. Therefore, the conductivity tensor becomes

$$\sigma(\mathbf{r}) = \begin{bmatrix} \sigma_2 & (\sigma_2-\sigma_1)m_{13}m_{23} & (\sigma_2-\sigma_1)m_{13}m_{33} \\ (\sigma_2-\sigma_1)m_{23}m_{13} & \sigma_1 & 0 \\ (\sigma_2-\sigma_1)m_{13}m_{33} & 0 & \sigma_1 \end{bmatrix}. \tag{8.27}$$

Note that the diagonal elements differ from those of (8.7) in that they are the result of rotations through the mean values of the Euler angles, ϕ and θ: $\overline{\phi} = \pi$ and $\overline{\theta} = \pi/2$ (see Fig. 8.13). It is clear from (8.27) that the stochastic properties of the conductivity tensor are wrapped up in $m_{13}m_{23}$ and $m_{13}m_{33}$.

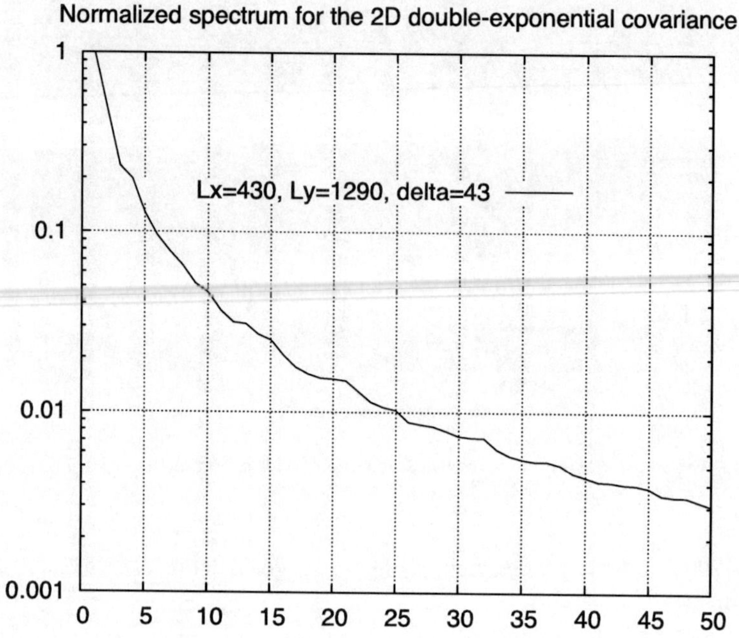

Fig. 8.12 The normalized eigenvalue spectrum for the 2D double-exponential covariance with $L_x = 430\,\mu\text{m}$, $L_y = 1290\,\mu\text{m}$, and $\delta = 43\,\mu\text{m}$

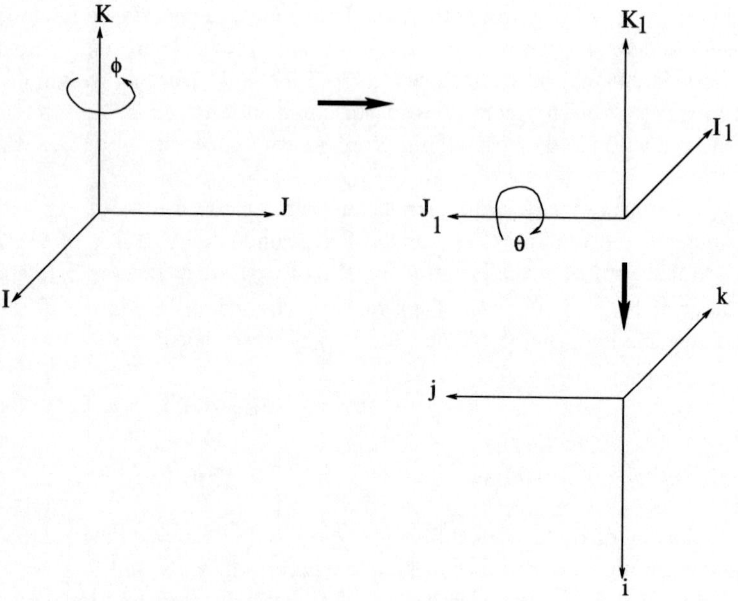

Fig. 8.13 Illustrating the Euler angle progression that leads to (8.27)

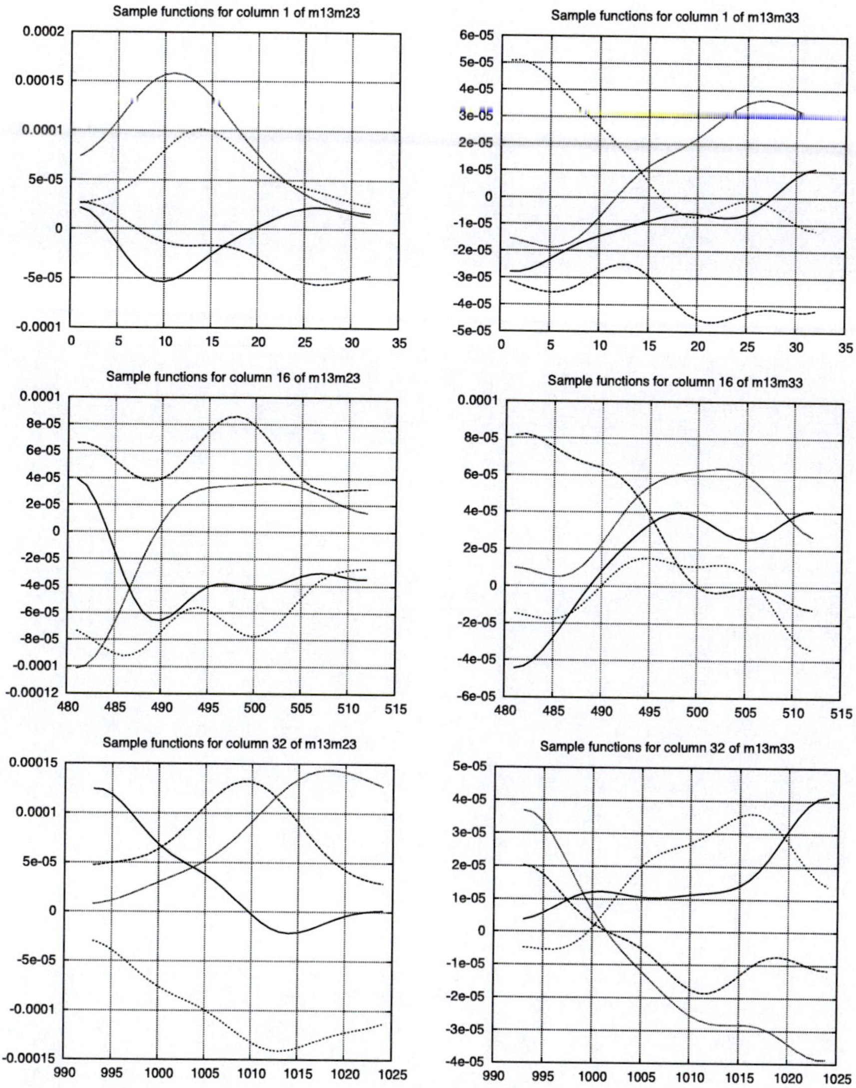

Fig. 8.14 Sample functions of $m_{13}m_{23}$ and $m_{13}m_{33}$, over the 1st, 16th, and 32nd columns of Fig. 8.6

Sample functions of the coefficients, $m_{13}m_{23}$ and $m_{13}m_{33}$, over the 1st, 16th, and 32nd columns of Fig. 8.6 are given in Fig. 8.14. The smooth behavior, which is due to the large ratio of correlation length to cell size, stands in contrast to the behavior shown in Figs. 7.3, 7.4, and 7.5.

Chapter 9
High-Dimension Model Representation via Sparse Grid Techniques

9.1 Introduction

The question of high-dimension model representation (HDMR) is of increasing importance in computational mathematics and science. We introduced this subject in Chaps. 7 and 8, where we invoked the Karhunen-Loève expansion to reduce the number of random parameters that are required to define the stochastic model. We continue the discussion of HDMR in this chapter by turning our attention to the question of determining a suitable surrogate model for computing the response of the forward problem via **VIC-3D®**. This surrogate takes the form of an interpolation table, which is then transformed into the conventional table used in NLSE for solving inverse problems. The surrogate model that we seek falls under the rubric *sparse grids*, and has been the subject of intensive research in a number of areas in recent years [15, 22, 43, 44, 46, 57, 58, 73, 121, 123, 146]. We will apply it to solving problems of model-based inversion as was developed in [111]. Sparse grids can also be used to effectively calculate high-dimensional integrals of the form (7.2).

9.2 Mathematical Structure of the Problem

The problems in this set are based on Fig. 6.3, and required 81 **VIC-3D®** runs to establish the interpolating grid. Thus, we say that the grid has 81 nodes in four-dimensional space, speaking abstractly. Each variable (the slab depth in Fig. 6.3) defines a dimension of the grid. The fifth problem introduced another variable, the width, with three possible values, making the overall grid a hypercube of 243 nodes in five-dimensional space. As we add dimensions (variables), we will soon encounter the 'curse of dimensionality,' because each node requires a **VIC-3D®** run to produce the corresponding blending function. The question arises as to

H. A. Sabbagh et al., *Advanced Electromagnetic Models for Materials Characterization and Nondestructive Evaluation*, Scientific Computation,
https://doi.org/10.1007/978-3-030-67956-9_9

whether we can reduce the number of runs by reducing the number of nodes in the interpolating grid. The answer lies in the notion of 'sparse grids.' This is an important area of study in numerical methods, and we refer the reader to [43] for a brief tutorial. We will follow the presentation and nomenclature of [43], using the complex-flaw model of Fig. 6.3 as an example. Further interesting applications of the sparse grid approach can be found in [44] and [146].

Referring to Fig. 6.3, we can define the mathematical structure of the problem by the abstract formula $(0,\ 10,\ 20) \otimes (0,\ 10,\ 20) \otimes (0,\ 10,\ 20) \otimes (0,\ 10,\ 20)$, where $\otimes$ represents the Cartesian or 'direct' or 'tensor' product. Thus, we have a problem that is defined on a four-dimensional hypercube with 16 corner nodes, given by $(0,\ 20) \otimes (0,\ 20) \otimes (0,\ 20) \otimes (0,\ 20)$ and 65 interior nodes given by the 'direct-difference' between the total nodes given above and the corner nodes. These involve the intermediate 10mil levels in Fig. 6.3. The question then becomes, are all of the interior nodes required for an accurate representation of the function, and if not, how do we choose which ones to keep? The answer to this is given by the sparse grid algorithm.

The sparse grid algorithm of [43] relies on a refinement of the interval of interest through successive halving of the previous interval, and then using the 'hierarchical' basis system of Fig. 9.1 as the interpolants. This system comprises, of course, our famous one-dimensional, first-order spline tent functions:

$$\phi_{l,j} = \begin{cases} 1 - |x/h_l - j_l|, & x \in [(j_l - 1)h_l,\ (j_l + 1)h_l] \cap |0, 1|; \\ 0, & \text{otherwise}, \end{cases} \quad , \tag{9.1}$$

where h_l is the length of an interval in the lth level, and $j_l = 0, \ldots, 2^l$ determines the position of a node. It is assumed in [43] that the grid is defined on the unit cube, which is the reason for the appearance of the interval, [0, 1]. For our problem, we are only interested in levels 0 and 1 of the hierarchy, because we only use at most two intervals for each dimension (variable) of the problem, as shown in Fig. 6.3.

Multidimensional functions are obtained by taking products of the one-dimensional splines:

$$\phi_{\mathbf{l},\mathbf{j}} = \Pi_{t=1}^{d} \phi_{l_t, j_t}(x_t) \ , \tag{9.2}$$

where $\mathbf{l} = (l_1, \ldots, l_d)$ denotes the number of intervals at the lth level in each dimension. Each entry is an integer. Similarly, $\mathbf{j} = (j_1, \ldots, j_d)$, with $j_t = 0, \ldots, 2^{l_t}$, denotes the nodal ordering at the lth level in each dimension. Associated with the multidimensional functions is the index set

$$\mathbf{B}_\mathbf{l} = \begin{cases} j_t = 1, \ldots, 2^{l_t} - 1,\ j_t \text{ odd}, & t = 1, \ldots, d, \text{ if } l_t > 0, \\ j_t = 0, 1, & t = 1, \ldots, d, \text{ if } l_t = 0\ . \end{cases} \tag{9.3}$$

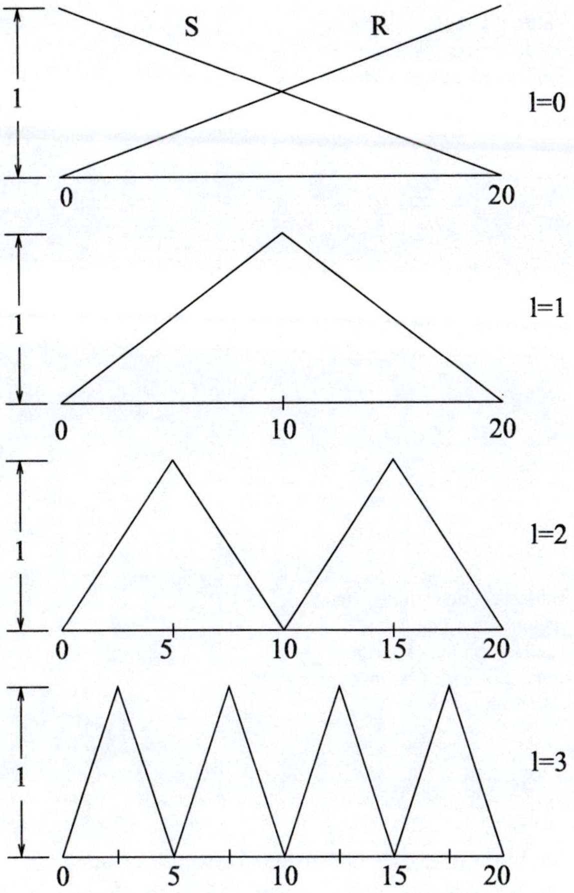

Fig. 9.1 Illustrating the one-dimensional hierarchical basis system for the function space, V_3. Each level has 2^l intervals and $2^l + 1$ nodes. Note that in our current problem, we are working in V_1, so that we are only interested in $l = 0,\ 1$. These are the usual tent functions with which we are well familiar from **VIC-3D**®. 'R' and 'S' denote 'ramp' and 'slide', respectively. The numbers along the abscissa refer to the values of the test depths of Fig. 6.3. Multidimensional functions are obtained by taking products of these functions

We define two norms for discrete multi-index vectors: $|\mathbf{l}|_\infty = \max_{1 \le t \le d} l_t$, and $|\mathbf{l}|_1 = \sum_{t=1}^{d} l_t$. Using these norms, together with (9.3), we can define a full-grid function as an expansion in terms of the basis system (9.2),

$$f(\mathbf{x}) = \sum_{\mathbf{l}_\infty \le n} \sum_{\mathbf{j} \in \mathbf{B_l}} \alpha_{\mathbf{l},\mathbf{j}} \phi_{\mathbf{l},\mathbf{j}}(\mathbf{x}) \,, \tag{9.4}$$

and a sparse-grid function similarly,

$$f(\mathbf{x}) = \sum_{\mathbf{l}_1 \le n} \sum_{\mathbf{j} \in \mathbf{B_l}} \alpha_{\mathbf{l},\mathbf{j}} \phi_{\mathbf{l},\mathbf{j}}(\mathbf{x}) \,. \tag{9.5}$$

Before we discuss either expansion in detail, let's take a look at the distinction between them, which is tied up with the definition of the norms above. In the complex-flaw case that we are considering, we have $d = 4$ and $n = 1$. Hence, the

Table 9.1 Four-dimensional vectors satisfying $|\mathbf{l}|_\infty \leq 1$. The last column gives the number of nodes associated with each vector

0	0	0	0	16
1	0	0	0	8
0	1	0	0	8
0	0	1	0	8
0	0	0	1	8
1	1	0	0	4
1	0	1	0	4
1	0	0	1	4
0	1	0	1	4
0	0	1	1	4
0	1	1	0	4
1	1	0	1	2
1	0	1	1	2
0	1	1	1	2
1	1	1	0	2
1	1	1	1	1
Total				81

Table 9.2 Four-dimensional vectors satisfying $|\mathbf{l}|_1 \leq 1$. The last column gives the number of nodes associated with each vector

0	0	0	0	16
1	0	0	0	8
0	1	0	0	8
0	0	1	0	8
0	0	0	1	8
Total				48

condition $|\mathbf{l}|_\infty \leq 1$ is satisfied 16 ways, as shown in Table 9.1. The resulting total number of nodes agrees with what we knew before for the Cartesian product of three nodes in each of four variables, $3^4 = 81$. Contrast this with the vector condition for sparse grids shown in Table 9.2. It is clear that the sparse grid algorithm gives a significant reduction in the number of nodes in the interpolation grid. The difference is even more striking in the case of the five-dimensional complex flaw of Test Problem No. 5. In that case the full grid had 243 nodes, whereas the sparse grid has only $2^5 + 5 \times 16 = 112$, which is less than half the full-grid complement.

We'll interpret Table 9.2 geometrically, using Fig. 6.3 to motivate the development. The entries in Tables 9.1 and 9.2 are the exponents, l, that yield the number of intervals, 2^l, in each slab of Fig. 6.3. Hence, the first entry in Table 9.2 corresponds to the situation in which the midpoint, 10 mil, is missing in each of the slabs (yielding a single interval in each slab) and we are working with the Cartesian product $(0, 20) \otimes (0, 20) \otimes (0, 20) \otimes (0, 20)$, which are the 16 corner points of a four-dimensional hypercube of length 20 on a side.

The next entry in Table 9.2 indicates that we have introduced the 10-mil midpoint in the first slab, yielding a slab with two intervals. Geometrically, this corresponds to the intersection of the hyperplane, $x_1 = 10$, with the four-dimensional hypercube, resulting in eight nodes at the boundaries of the hypercube. Similarly, the third through fifth entries correspond to the intersection of hyperplanes with the hypercube, resulting in eight nodes for each entry, yielding a total of 48 nodes. Each node carries a blending function that **VIC-3D®** must compute using the appropriate parameters of Fig. 6.3. For example, the blending function corresponding to the second entry in the table would have the first slab of Fig. 6.3 fixed at 10 mils depth, and the other three cycling through 0 and 20 mils, each, giving a **VIC-3D®** problem with 8 range values.

We turn our attention, now, to the hierarchical structure of the algorithm, which lies at the heart of (9.5). Using the format of Table 9.2, we expand (9.5) as follows:

$$\begin{aligned} f(\mathbf{x}) &= \sum_{|\mathbf{l}|_1 \le 1} \sum_{\mathbf{j} \in \mathbf{B_l}} \alpha_{\mathbf{l},\mathbf{j}} \phi_{\mathbf{l},\mathbf{j}}(\mathbf{x}) \\ &= \sum_{\mathbf{j} \in B_{0000}} \alpha_{0000,\mathbf{j}} \phi_{0000,\mathbf{j}}(\mathbf{x}) + \sum_{\mathbf{j} \in B_{1000}} \alpha_{1000,\mathbf{j}} \phi_{1000,\mathbf{j}}(\mathbf{x}) + \sum_{\mathbf{j} \in B_{0100}} \alpha_{0100,\mathbf{j}} \phi_{0100,\mathbf{j}}(\mathbf{x}) \\ &+ \sum_{\mathbf{j} \in B_{0010}} \alpha_{0010,\mathbf{j}} \phi_{0010,\mathbf{j}}(\mathbf{x}) + \sum_{\mathbf{j} \in B_{0001}} \alpha_{0001,\mathbf{j}} \phi_{0001,\mathbf{j}}(\mathbf{x}) \,. \end{aligned} \tag{9.6}$$

Because the expansion functions, $\{\phi_{\mathbf{l},\mathbf{j}}(\mathbf{x})\}$, are nonoverlapping for a given level, l, and have a unit amplitude, the expansion coefficients, $\{\alpha_{\mathbf{l},\mathbf{j}}\}$, are simply equal to the blending function associated with the node of the appropriate function at level l.

Figure 9.2 illustrates the situation in one dimension at levels 0 and 1. In this example, we have $\alpha_{0,0} = BF(0)$, $\alpha_{0,20} = BF(20)$, $\alpha_{1,10} =$ SURPLUS, where SURPLUS $= BF(10) - 1/2(BF(20) + BF(0))$. Hence, the expansion shown in Fig. 9.2 is given by

$$f(x) = BF(0)\phi_{0,0}(x) + BF(20)\phi_{0,20}(x) + \text{SURPLUS}\phi_{1,10}(x) \,, \tag{9.7}$$

where $\phi_{0,0}(x)$ is the slide function, $S(x)$, and $\phi_{0,20}(x)$ is the ramp function, $R(x)$, in Fig. 9.2.

It is clear that the name ‘SURPLUS’ (called *hierarchical surplus* in [43]) denotes the excess in function value that the higher-order levels are supposed to accommodate. It’s evaluation at level l requires only one additional blending function to be computed, while using two previously computed at level $l-1$. This is an advantage of the hierarchical structure of the algorithm. If SURPLUS=0, then the interpolator would treat this function as being linear, instead of piecewise linear, in this dimension. Thus, the expansion, (9.6), is reminiscent of a Taylor series, in which the terms corresponding to higher levels of the hierarchy correspond to higher-order polynomial terms in the Taylor series.

The full four-dimensional expansion of (9.6) is given next. The first term is:

$$BF(0,0,0,0)S(x_1)S(x_2)S(x_3)S(x_4) + BF(20,0,0,0)R(x_1)S(x_2)S(x_3)S(x_4) +$$

$$BF(0,20,0,0)S(x_1)R(x_2)S(x_3)S(x_4) + BF(20,20,0,0)R(x_1)R(x_2)S(x_3)S(x_4) +$$

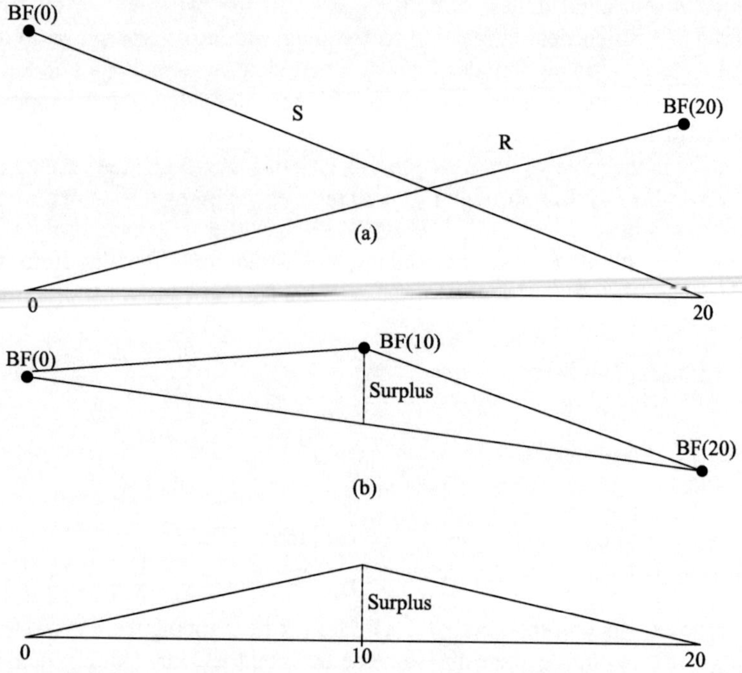

Fig. 9.2 Illustrating a one-dimensional example of the hierarchical expansion at levels 0 (**a**) and 1 (**b**). The blending functions, $BF()$, are labeled at the nodes of the one-dimensional simplex. The 'Surplus' value defines the magnitude of the level-1 spline

$$
\begin{aligned}
&BF(0,0,20,0)S(x_1)S(x_2)R(x_3)S(x_4) + BF(20,0,20,0)R(x_1)S(x_2)R(x_3)S(x_4) +\\
&BF(0,20,20,0)S(x_1)R(x_2)R(x_3)S(x_4) + BF(20,20,20,0)R(x_1)R(x_2)R(x_3)S(x_4) +\\
&BF(0,0,0,20)S(x_1)S(x_2)S(x_3)R(x_4) + BF(20,0,0,20)R(x_1)S(x_2)S(x_3)R(x_4) +\\
&BF(0,20,0,20)S(x_1)R(x_2)S(x_3)R(x_4) + BF(20,20,0,20)R(x_1)R(x_2)S(x_3)R(x_4) +\\
&BF(0,0,20,20)S(x_1)S(x_2)R(x_3)R(x_4) + BF(20,0,20,20)R(x_1)S(x_2)R(x_3)R(x_4) +\\
&BF(0,20,20,20)S(x_1)R(x_2)R(x_3)R(x_4) + BF(20,20,20,20)R(x_1)R(x_2)R(x_3)R(x_4)\,;
\end{aligned}
\tag{9.8}
$$

the second:

$$
\begin{aligned}
&\mathrm{SURPLUS}(10,0,0,0)\phi_{1,10}(x_1)S(x_2)S(x_3)S(x_4)\\
&\quad +\mathrm{SURPLUS}(10,20,0,0)\phi_{1,10}(x_1)R(x_2)S(x_3)S(x_4) +\\
&\mathrm{SURPLUS}(10,0,20,0)\phi_{1,10}(x_1)S(x_2)R(x_3)S(x_4)\\
&\quad +\mathrm{SURPLUS}(10,20,20,0)\phi_{1,10}(x_1)R(x_2)R(x_3)S(x_4) +
\end{aligned}
$$

$$\mathrm{SURPLUS}(10,0,0,20)\phi_{1,10}(x_1)S(x_2)S(x_3)R(x_4)$$
$$+\mathrm{SURPLUS}(10,20,0,20)\phi_{1,10}(x_1)R(x_2)S(x_3)R(x_4)+$$
$$\mathrm{SURPLUS}(10,0,20,20)\phi_{1,10}(x_1)S(x_2)R(x_3)R(x_4)$$
$$+\mathrm{SURPLUS}(10,20,20,20)\phi_{1,10}(x_1)R(x_2)R(x_3)R(x_4)\,; \tag{9.9}$$

the third:

$$\mathrm{SURPLUS}(0,10,0,0)S(x_1)\phi_{1,10}(x_2)S(x_3)S(x_4)$$
$$+\mathrm{SURPLUS}(20,10,0,0)R(x_1)\phi_{1,10}(x_2)S(x_3)S(x_4)+$$
$$\mathrm{SURPLUS}(0,10,20,0)S(x_1)\phi_{1,10}(x_2)R(x_3)S(x_4)$$
$$+\mathrm{SURPLUS}(20,10,20,0)R(x_1)\phi_{1,10}(x_2)R(x_3)S(x_4)+$$
$$\mathrm{SURPLUS}(0,10,0,20)S(x_1)\phi_{1,10}(x_2)S(x_3)R(x_4)$$
$$+\mathrm{SURPLUS}(20,10,0,20)R(x_1)\phi_{1,10}(x_2)S(x_3)R(x_4)+$$
$$\mathrm{SURPLUS}(0,10,20,20)S(x_1)\phi_{1,10}(x_2)R(x_3)R(x_4)$$
$$+\mathrm{SURPLUS}(20,10,20,20)R(x_1)\phi_{1,10}(x_2)R(x_3)R(x_4)\,; \tag{9.10}$$

the fourth:

$$\mathrm{SURPLUS}(0,0,10,0)S(x_1)S(x_2)\phi_{1,10}(x_3)S(x_4)$$
$$+\mathrm{SURPLUS}(20,0,10,0)R(x_1)S(x_2)\phi_{1,10}(x_3)S(x_4)+$$
$$\mathrm{SURPLUS}(0,20,10,0)S(x_1)R(x_2)\phi_{1,10}(x_3)S(x_4)$$
$$+\mathrm{SURPLUS}(20,20,10,0)R(x_1)R(x_2)\phi_{1,10}(x_3)S(x_4)+$$
$$\mathrm{SURPLUS}(0,0,10,20)S(x_1)S(x_2)\phi_{1,10}(x_3)R(x_4)$$
$$+\mathrm{SURPLUS}(20,0,10,20)R(x_1)S(x_2)\phi_{1,10}(x_3)R(x_4)+$$
$$\mathrm{SURPLUS}(0,20,10,20)S(x_1)R(x_2)\phi_{1,10}(x_3)R(x_4)$$
$$+\mathrm{SURPLUS}(20,20,10,20)R(x_1)R(x_2)\phi_{1,10}(x_3)R(x_4)\,; \tag{9.11}$$

and the fifth:

$$\mathrm{SURPLUS}(0,0,0,10)S(x_1)S(x_2)S(x_3)\phi_{1,10}(x_4)$$
$$+\mathrm{SURPLUS}(20,0,0,10)R(x_1)S(x_2)S(x_3)\phi_{1,10}(x_4)+$$

$$
\begin{aligned}
&\mathrm{SURPLUS}(0,20,0,10)S(x_1)R(x_2)S(x_3)\phi_{1,10}(x_4)\\
&\quad+\mathrm{SURPLUS}(20,20,0,10)R(x_1)R(x_2)S(x_3)\phi_{1,10}(x_4)+\\
&\mathrm{SURPLUS}(0,0,20,10)S(x_1)S(x_2)R(x_3)\phi_{1,10}(x_4)\\
&\quad+\mathrm{SURPLUS}(20,0,20,10)R(x_1)S(x_2)R(x_3)\phi_{1,10}(x_4)+\\
&\mathrm{SURPLUS}(0,20,20,10)S(x_1)R(x_2)R(x_3)\phi_{1,10}(x_4)\\
&\quad+\mathrm{SURPLUS}(20,20,20,10)R(x_1)R(x_2)R(x_3)\phi_{1,10}(x_4)\,.
\end{aligned}
\tag{9.12}
$$

The arguments of the BF and SURPLUS functions are the depth parameters of the corresponding slabs in Fig. 6.3. Thus, $BF(0, 20, 20, 20)$ is the blending function computed by **VIC-3D®** when slab 1 has a depth of 0, and the other three slabs are at a full depth of 20. In (9.9)–(9.12), we have

$$
\begin{aligned}
\mathrm{SURPLUS}(10,a,b,c) &= BF(10,a,b,c) - 1/2(BF(20,a,b,c) + BF(0,a,b,c))\\
\mathrm{SURPLUS}(a,10,b,c) &= BF(a,10,b,c) - 1/2(BF(a,20,b,c) + BF(a,0,b,c))\\
\mathrm{SURPLUS}(a,b,10,c) &= BF(a,b,10,c) - 1/2(BF(a,b,20,c) + BF(a,b,0,c))\\
\mathrm{SURPLUS}(a,b,c,10) &= BF(a,b,c,10) - 1/2(BF(a,b,c,20) + BF(a,b,c,0))
\end{aligned}
\tag{9.13}
$$

9.3 Clenshaw-Curtis Grids

The Clenshaw-Curtis family of grids [57, 58] are, for the most part, superior to others of the genre that we have just described, in the sense that the number of points in C-C grids increases more slowly with the level of refinement, while retaining the same asymptotic error decay rate. Our presentation follows [58].

The points, x^i_j, of the C-C grid in one dimension are defined as

$$
m_i = \begin{cases} 1, & \text{if } i = 1,\\ 2^{i-1}+1, & \text{if } i > 1, \end{cases}
$$

$$
x^i_j = \begin{cases} (j-1)/(m_i-1) & \text{for } j = 1,\ldots,m_i \text{ if } m_i > 1,\\ 0.5 & \text{for } j = 1 \text{ if } m_i = 1. \end{cases}
\tag{9.14}
$$

The index, i, is used to indicate a level of refinement of the grid in the appropriate dimension. It is clear from (9.14) that the set of points, X^i, generated at the ith level, is a subset of X^{i+1}: $X^i \subset X^{i+1}$. Furthermore, this implies that the multidimensional sparse grid generated by the tensor product of the one-dimensional grids

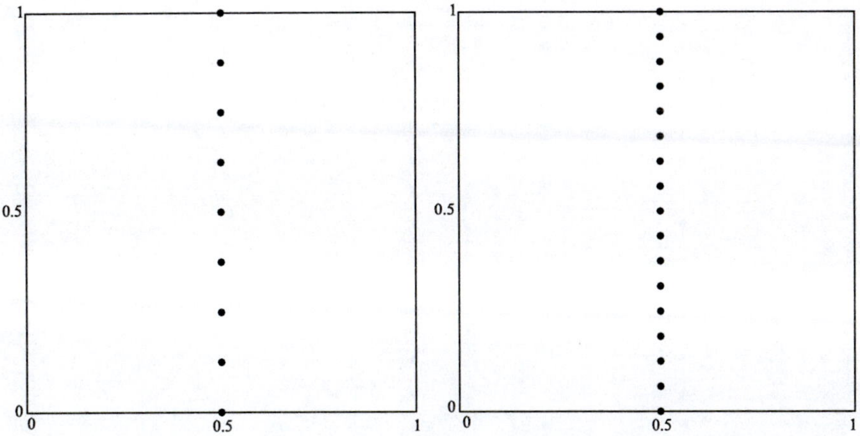

Fig. 9.3 Example of two point sets of the C-C grid. Left: $X^1 \otimes X^4$. Right: $X^1 \otimes X^5$. Note that $X^1 \otimes X^4$ is a subset of $X^1 \otimes X^5$

$$H_{q,d} = \cup_{q-d+1 \le |\mathbf{l}|_1 \le q} (X^{i_1} \otimes \cdots \otimes X^{i_d}) \,, \tag{9.15}$$

also satisfies the inclusion property: $H_{q,d} \subset H_{q+1,d}$. The index, q, is associated with the level of the dth dimension. The importance of this last relationship is that it implies that those blending functions that have already been computed on the set $H_{q,d}$ can be used in $H_{q+1,d}$, with only the surplus blending-functions being needed to be computed in $H_{q+1,d}$. When these surpluses are smaller than a threshold, we can stop refining the grid. Figure 9.3 illustrates two point sets of the C-C grid, $X^1 \otimes X^4$ and $X^1 \otimes X^5$. The former is a subset of the latter, which is at the next level of refinement.

An example of (9.15) for the case $q = 6,\ d = 4$ is given here:

$$\begin{aligned} H_{6,2} &= \cup_{5 \le i_1 + i_2 \le 6} X^{i_1} \otimes X^{i_2} \\ &= X^1 \otimes X^4 \cup X^1 \otimes X^5 \cup X^2 \otimes X^3 \cup X^2 \otimes X^4 \cup X^3 \otimes X^2 \cup X^3 \otimes X^3 \\ &\cup X^4 \otimes X^1 \cup X^4 \otimes X^2 \cup X^5 \otimes X^0 \cup X^5 \otimes X^1 \,, \end{aligned} \tag{9.16}$$

with $X^0 = \emptyset$. The C-C grid corresponding to (9.16) is shown in Fig. 9.4.

With $X^i_\Delta = X^i \backslash X^{i-1}$, namely those points in X^i that are not in X^{i-1} (the 'excess' points), (9.15) can be expanded as

$$H_{q,d} = \cup_{|\mathbf{i}|_1 \le q} (X^{i_1}_\Delta \otimes \cdots \otimes X^{i_d}_\Delta) = H_{q-1,d} \cup \Delta H_{q,d} \,, \tag{9.17}$$

where

$$\Delta H_{q,d} = \cup_{|\mathbf{i}|_1 = q} (X^{i_1}_\Delta \otimes \cdots \otimes X^{i_d}_\Delta) \,, \tag{9.18}$$

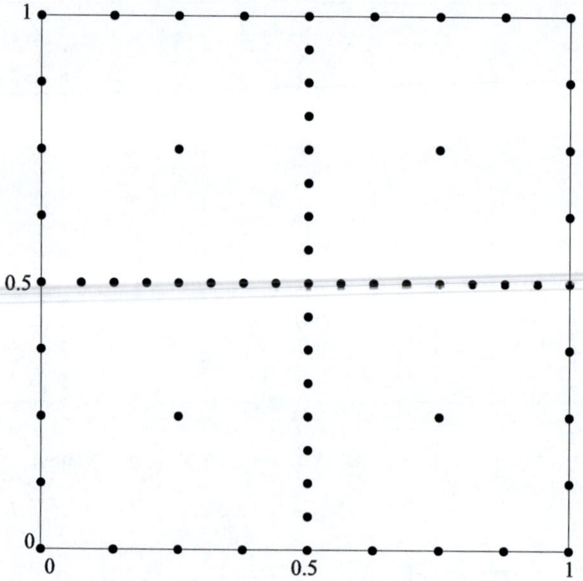

Fig. 9.4 Sparse grid $H_{6,2}^{CC}$ at level 4. There are a total of 65 points

Table 9.3 Number of grid points at level $n = q - d$ for Clenshaw-Curtis grids

n	$d = 2$	$d = 4$	$d = 8$	$d = 16$
0	1	1	1	1
1	5	9	17	33
2	13	41	145	545

and $H_{d-1,d} = \emptyset$. This is more convenient for expansion of the interpolant using successive refinements of the grid with increasing parameter, q.

Table 9.3 shows the number of grid points at level $n = q - d$ for Clenshaw-Curtis grids of up to 16 dimensions. We see that the increase in grid points with dimension is rather slow, which is a significant advantage of C-C grids. If we argue, as we did earlier, that the complex flaw of Fig. 6.3 requires only three grid points per dimension, then it follows from Table 9.3 that we may be able to get by with a total of 33 points for 16 dimensions at level 1! This is due to the fact that the midpoint is used at the 0th level, and the two end points at the first level, and is an incredible savings. If we are conservative, and decide that we need to go to level 2, we can go up to eight dimensions and require only 145 blending functions. If we used a full grid for an eight-dimensional problem with three points per dimension, we would need a total of 6561 points! The curse of dimensionality strikes again.

Klimke, [57, 58], allows one to compute the coordinates of a C-C grid, which then allows the user to determine *a priori* what blending functions to compute for an interpolation table. Table 9.4 lists the coordinates of hierarchical C-C grid points for $d = 4$, $n = 0 : 2$. The table is arranged to show the grid points that are added

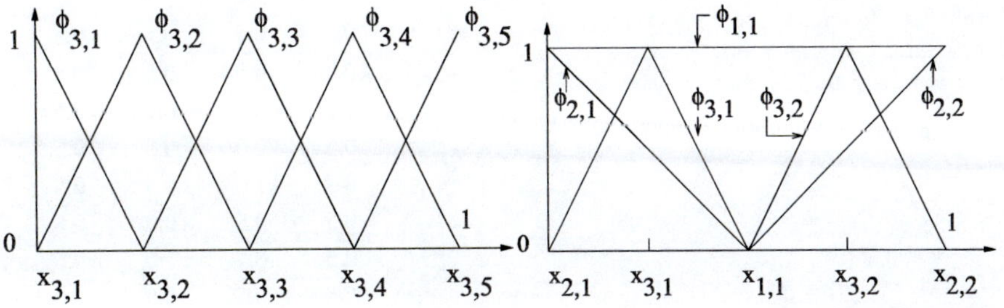

Fig. 9.5 Illustrating nodal basis functions, $\phi_{3,j}$, $x_{3,j} \in X^3$ (left) and hierarchical basis functions, $\phi_{i,j}$, with their support nodes, $x_{i,j} \in X^i_\Delta$, $i = 1, \ldots, 3$ (right) for the Clenshaw-Curtis grid (see [58])

to the interpolant at level n. The total number of points is 41, and agrees with the corresponding entry in Table 9.3.

Now that we have discussed the nature of C-C grids, it is time to address the expansion of the interpolant in function space. Figure 9.5 illustrates the typical piecewise nodal basis functions used in conventional interpolation schemes, as well as the hierarchical basis functions used with Clenshaw-Curtis grids. Note that in the C-C basis, the lowest level function in the hierarchy is simply a constant, whereas the basis system in the next level are a slide (left) and ramp (right) which do not span the entire line, but only one-half. Finally, the basis functions at level 2 are the usual disjoint tent functions that we have seen before in Fig. 9.1.

Figure 9.6 illustrates how the two sets of basis functions are used in interpolations. Our interest is in the hierarchical system shown at the bottom of the figure. The $w_{i,j}$ denote the hierarchical surpluses that are the expansion coefficients for the hierarchical interpolation formula:

$$U(x) = w_{1,1}\phi_{1,1}(x) + w_{2,1}\phi_{2,1}(x) + w_{2,2}\phi_{2,2}(x) + w_{3,1}\phi_{3,1}(x) + w_{3,2}\phi_{3,2}(x)\,. \tag{9.19}$$

They are given by: $w_{1,1} = f(0.5)$, $w_{2,1} = f(0) - f(0.5)$, $w_{2,2} = f(1) - f(0.5)$, $w_{3,1} = f(0.25) - 1/2(f(0) + f(0.5))$, and $w_{3,2} = f(0.75) - 1/2(f(1) + f(0.5))$, where $f(x)$ is shown as the dashed curve in the figure. (The surpluses are the expansion coefficients because the basis functions all have unit amplitude, and do not overlap with each other at a given level.) Note the efficiency in the hierarchical algorithm, in that function values computed at one level are reused in the next higher level.

In reality, the nodal values in Fig. 9.6 are blending functions, comprising an entire 1- or 2-D scan impedance response. Here's how we use the coordinate information in Table 9.4 to determine the computation of the associated blending function in Fig. 6.3. For $n = 0$, we set the boundary of each slab in Fig. 6.3 to be 10 mils (recall that we are scaling the physical dimensions to fit into a unit hypercube in 4-space), and use **VIC-3D®** to compute the response. This, then, is the blending function associated with the midpoint of the hypercube.

Table 9.4 Coordinates of hierarchical C-C grid points for $d = 4,\ n = 0:2$

$n=0$	0.5	0.5	0.5	0.5
$n=1$	0.0	0.5	0.5	0.5
	1.0	0.5	0.5	0.5
	0.5	0.0	0.5	0.5
	0.5	1.0	0.5	0.5
	0.5	0.5	0.0	0.5
	0.5	0.5	1.0	0.5
	0.5	0.5	0.5	0.0
	0.5	0.5	0.5	1.0
$n=2$	0.25	0.5	0.5	0.5
	0.75	0.5	0.5	0.5
	0.0	0.0	0.5	0.5
	1.0	0.0	0.5	0.5
	0.0	1.0	0.5	0.5
	1.0	1.0	0.5	0.5
	0.5	0.25	0.5	0.5
	0.5	0.75	0.5	0.5
	0.0	0.5	0.0	0.5
	1.0	0.5	0.0	0.5
	0.0	0.5	1.0	0.5
	1.0	0.5	1.0	0.5
	0.5	0.0	0.0	0.5
	0.5	1.0	0.0	0.5
	0.5	0.0	1.0	0.5
	0.5	1.0	1.0	0.5
	0.5	0.5	0.25	0.5
	0.5	0.5	0.75	0.5
	0.0	0.5	0.5	0.0
	1.0	0.5	0.5	0.0
	0.0	0.5	0.5	1.0
	1.0	0.5	0.5	1.0
	0.5	0.0	0.5	0.0
	0.5	1.0	0.5	0.0
	0.5	0.0	0.5	1.0
	0.5	1.0	0.5	1.0
	0.5	0.5	0.0	0.0
	0.5	0.5	1.0	0.0
	0.5	0.5	0.0	1.0
	0.5	0.5	1.0	1.0
	0.5	0.5	0.5	0.25
	0.5	0.5	0.5	0.75

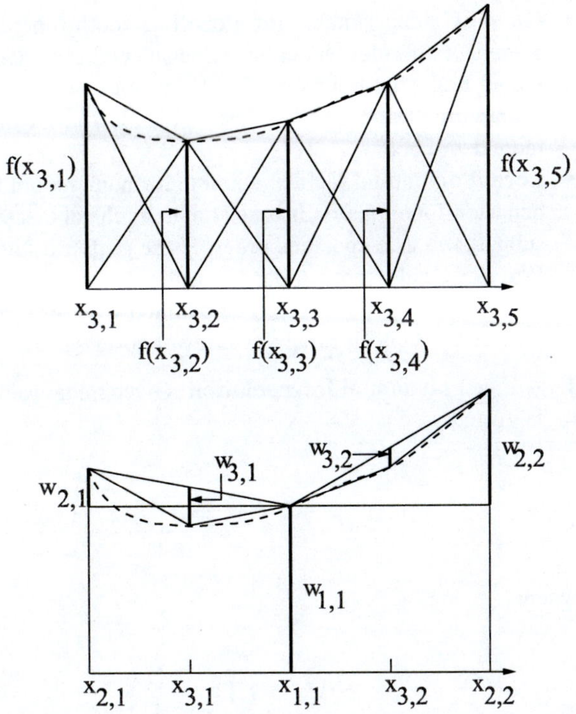

Fig. 9.6 Illustrating nodal (top) versus hierarchical (bottom) interpolation in one dimension (see [58]). The $w_{i,j}$ denote the hierarchical surpluses that are the expansion coefficients for the hierarchical interpolation formula:
$U(x) = w_{1,1}\phi_{1,1}(x) + w_{2,1}\phi_{2,1}(x) + w_{2,2}\phi_{2,2}(x) + w_{3,1}\phi_{3,1}(x) + w_{3,2}\phi_{3,2}(x)$.
They are given by:
$w_{1,1} = f(0.5)$,
$w_{2,1} = f(0) - f(0.5)$,
$w_{2,2} = f(1) - f(0.5)$,
$w_{3,1} = f(0.25){-}1/2(f(0){+}f(0.5))$,
and $w_{3,2} = f(0.75){-}1/2(f(1){+}f(0.5))$,
where $f(x)$ is shown as the dashed curve in the figure

The problem configuration for computing the blending function associated with the first entry at $n = 1$ of Table 9.4 would have the depth of the first slab of Fig. 6.3 equal to zero, and the other three fixed at 10 mils. For the first entry at $n = 2$, the depth of the first slab would be 5 mils, and the other three depths would be 10 mils, and so on.

9.4 The TASMANIAN Sparse Grids Module

The Toolkit for Adaptive Stochastic Modeling and Non-Intrusive ApproximatioN is a robust library for high-dimensional integration and interpolation, as well as parameter calibration. The code consists of several modules that can be used individually or conjointly. The project is sponsored by Oak Ridge National Laboratory Directed Research and Development as well as the Department of Energy Office for Advanced Scientific Computing Research (see tasmanian.ornl.org/about.html).

Sparse Grids is a family of algorithms for constructing multidimensional quadrature and interpolation rules from tensor products of one-dimensional rules. The TASMANIAN Sparse Grid code implements a number of different quadrature rules and basis functions (see [123] for details). The rules are grouped into three categories:

- **Global Grids:** suitable for globally smooth functions. Quadrature is based on a number of rules, including Clenshaw-Curtis, and interpolation is based on global Lagrange polynomials. Nodal point selection follows the same rules as quadrature. These grids are most suitable for our use, and will be discussed in more detail below.
- **Local Polynomial Grids:** suitable for non-smooth functions with locally sharp behavior. Interpolation is based on hierarchical piecewise polynomials with local support and user-specified order. These grids are suitable for local refinement.
- **Wavelet Grids:** are similar to local polynomials, except that it is assumed that the order is either 1 or 3. When coupled with local refinement, these grids often provide the same accuracy with fewer abscissas.

Lagrange Polynomial Interpolation Given nodal values, $f(x_i)$, the LP interpolator is given by

$$f(x) = \sum_{i=0}^{N} l_i(x) f(x_i) , \tag{9.20}$$

where

$$l_i(x) = \prod_{\substack{j=0 \\ j\neq i}}^{N} \frac{x - x_j}{x_i - x_j}, \quad i = 0, \ldots, N . \tag{9.21}$$

The interpolating polynomials, $\{l_i(x)\}$, satisfy $l_i(x_j) = \delta_{ij}$. An example for $N = 3$ is given here:

$$\begin{aligned} l_0(x) &= \frac{x - x_1}{x_0 - x_1} \frac{x - x_2}{x_0 - x_2} \frac{x - x_3}{x_0 - x_3} \\ l_1(x) &= \frac{x - x_0}{x_1 - x_0} \frac{x - x_2}{x_1 - x_2} \frac{x - x_3}{x_1 - x_3} \\ l_2(x) &= \frac{x - x_0}{x_2 - x_0} \frac{x - x_1}{x_2 - x_1} \frac{x - x_3}{x_2 - x_3} \\ l_3(x) &= \frac{x - x_0}{x_3 - x_0} \frac{x - x_1}{x_3 - x_1} \frac{x - x_2}{x_3 - x_2} . \end{aligned} \tag{9.22}$$

The nodal points, or knots, are located at the extrema (maxima or minima) of Chebyshev polynomials. If $m_i > 1$ is the number of knots at the ith level of approximation in a given dimension, then the knots, over the interval $[-1, +1]$, are given by

$$x_j^i = -\cos \frac{\pi(j-1)}{m_i - 1}, \quad j = 1, \ldots, m_i , \tag{9.23}$$

with $x_1^i = 0$ if $m_i = 1$. In order for the knots to be nested at the next level of approximation, we choose $m_1 = 1$ and $m_i = 2^{i-1} + 1$ for $i > 1$ [15]. The Lagrange interpolator is an example of Eq. (1.3) in the TASMANIAN User Manual.

9.5 First TASMANIAN Results

In order to test the accuracy of interpolating high-dimensional models at various levels,[1] we start with a simple example in which we compare the interpolated result at various levels with original data that are computed directly with **VIC-3D®** for the complex flaw with 'coordinates' (20,0,0,0) shown in Fig. 9.7. These results are shown in Figs. 9.8 and 9.9. Additional test results are shown in Figs. 9.10, 9.11, and 9.12. It is clear from these tests that TASMANIAN works well as long as the level is chosen correctly. It seems likely that one could determine a suitable level for problems of a given dimension by using theoretical rates of convergence, but in these tests we used an empirical approach to determine such levels.

9.6 Results for 4D-Level 8

We have done a number of numerical experiments to test TASMANIAN, and learn more about the relationship between the number of dimensions in a grid and the level of the grid. For example, we found that there was reasonable convergence from 8D-Level3(593 points) to 8D-Level4(1953 points), but to use an eight-dimensional grid

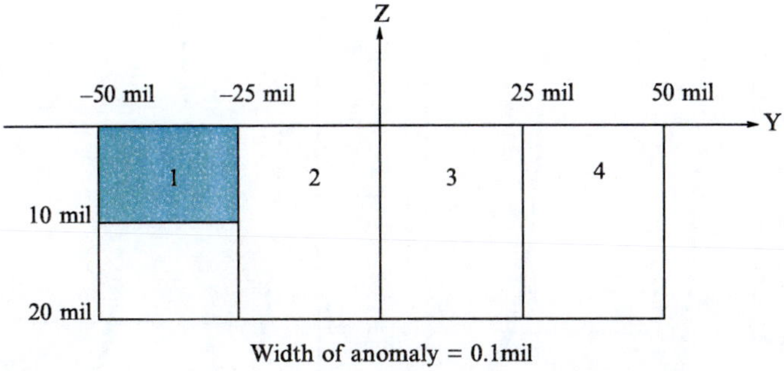

Fig. 9.7 Showing a complex flaw extending over one-half of the first block of Fig. 6.3 and vanishing elsewhere

[1] 'Level', in the context of Lagrange interpolation with the Chebyshev rule, implies the highest order polynomial that can be interpolated exactly.

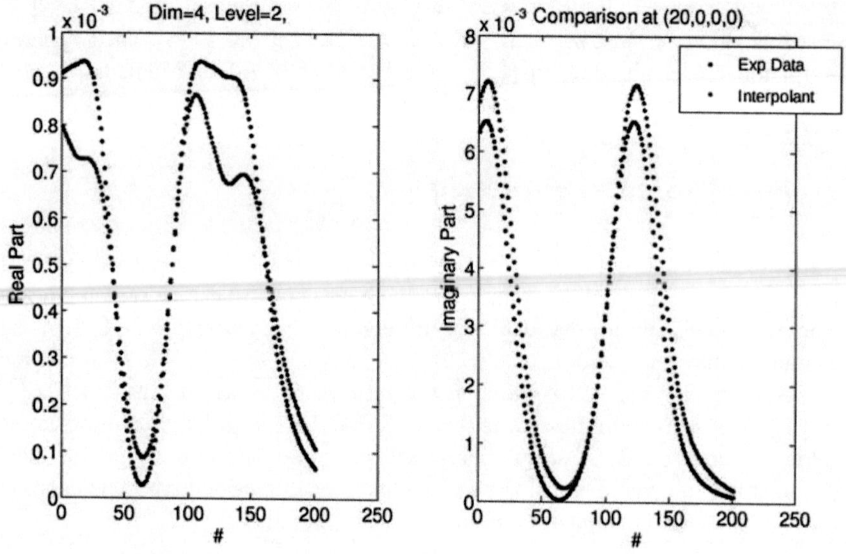

Fig. 9.8 Comparison of interpolated result with original data for the complex flaw with 'coordinates' (20,0,0,0). (See Fig. 9.7.) The four-dimensional TASMANIAN sparse grid was generated at level 2, and required 33 Chebyshev points

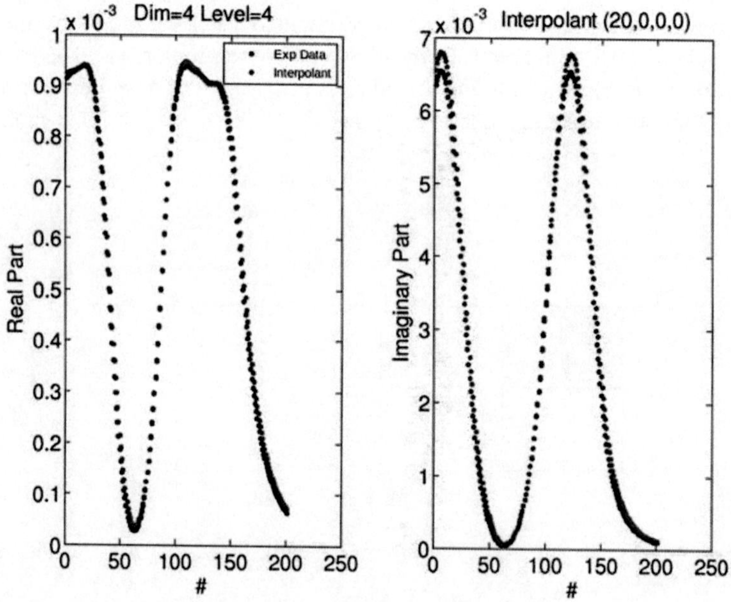

Fig. 9.9 Comparison of interpolated result with original data for the complex flaw with 'coordinates' (20,0,0,0). (See Fig. 9.7.) The four-dimensional TASMANIAN sparse grid was generated at level 4, with 145 points. Note the significant improvement over Fig. 9.8

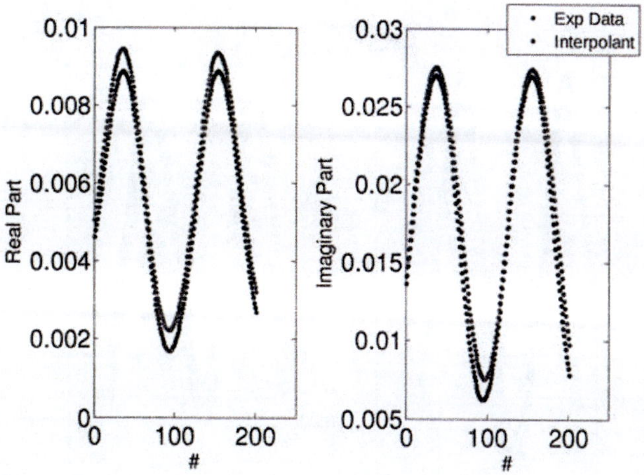

Fig. 9.10 Comparison of interpolated result with original data for the complex flaw with 'coordinates' (20,40,30,10). (See Fig. 6.4.) The four-dimensional TASMANIAN sparse grid was generated at level 2 (33 points)

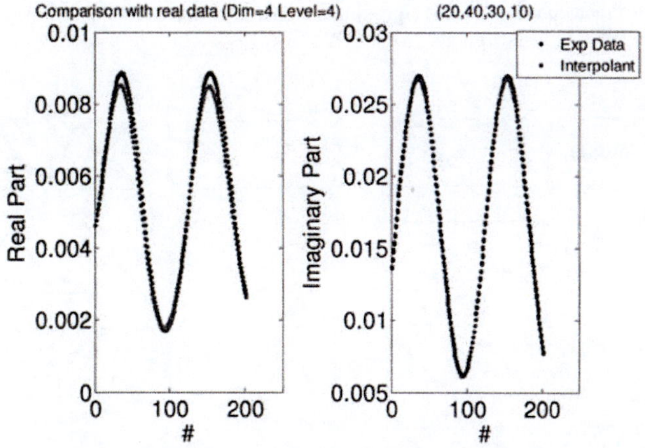

Fig. 9.11 Comparison of interpolated result with original data for the complex flaw with 'coordinates' (20,40,30,10). (See Fig. 6.4.) The four-dimensional TASMANIAN sparse grid was generated at level 4 (145 points). Note the improvement over Fig. 9.10. The relative errors are less than 7% in the real part (resistance), less than 3% in the imaginary part (reactance). The major part of the relative error occurs when the real and imaginary parts, especially the imaginary part, are both small. The relative error is calculated as the ratio of the difference of the 'experimental data' and the interpolated data to the experimental data. The absolute error is the absolute value of the difference between the experimental and interpolated data

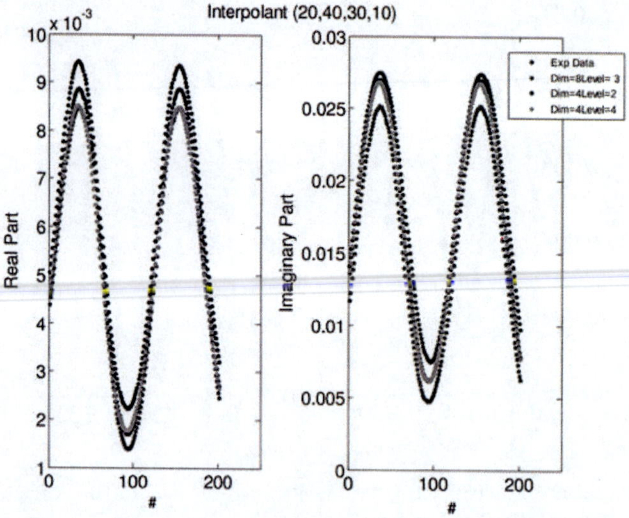

Fig. 9.12 Comparison of interpolated results with original data for the complex flaw with 'coordinates' (20,40,30,10). (See Fig. 6.4.) The coordinates of the flaw in eight dimensions are (20,20,40,40,30,30,10,10), because the length of each of the eight-dimensional slabs is one-half that of the four-dimensional slabs. The eight-dimensional TASMANIAN sparse grid was generated at level 3, with 593 Chebyshev points

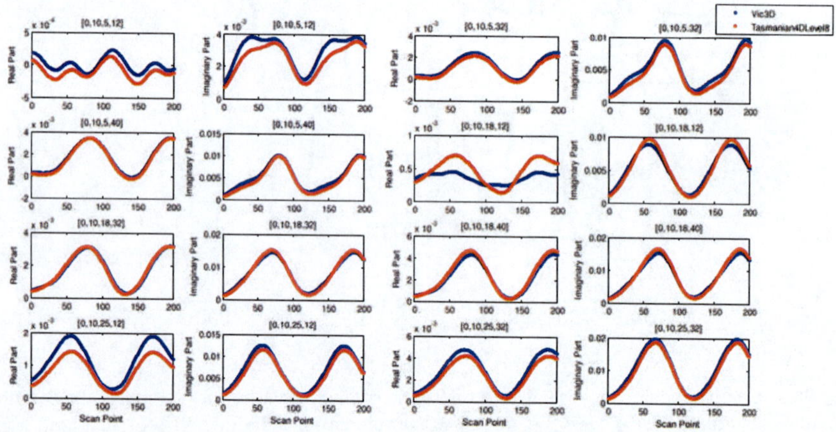

Fig. 9.13 Comparison of interpolated results at 81 arbitrarily selected points with the four-dimensional TASMANIAN grid at level 8, with 1857 Chebyshev points

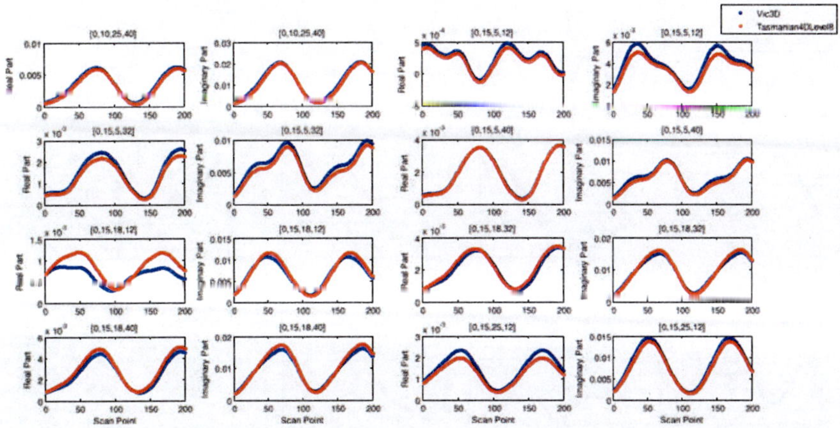

Fig. 9.14 Comparison of interpolated results at 81 arbitrarily selected points with the four-dimensional TASMANIAN grid at level 8 (continued)

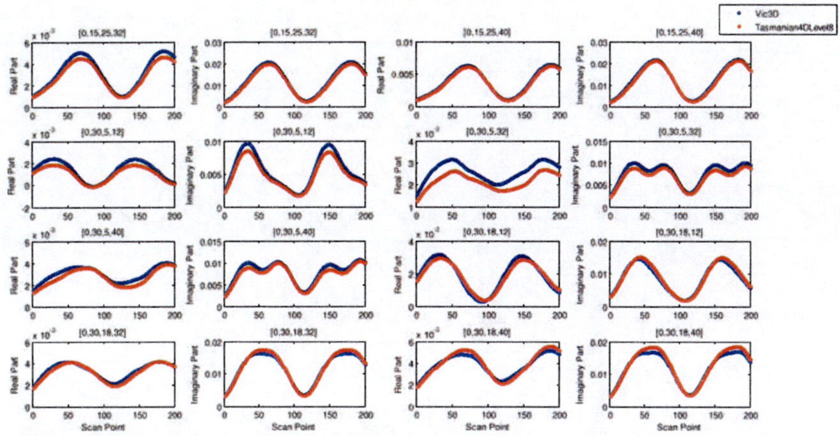

Fig. 9.15 Comparison of interpolated results at 81 arbitrarily selected points with the four-dimensional TASMANIAN grid at level 8 (continued)

accurately would require even higher levels. Since the 8D-Level4 run took 12.5 h, we decided not to go to higher dimensions, until we were absolutely forced to. In fact, it is probably wiser to use voxel-based inverse methods when the number of parameters needed to accurately model the geometry of the problem exceeds, say, 5 or 6.

We also studied the convergence of 4D-Level 6 to 4D-Level 8, and find excellent results. Hence, the remainder of our study at this point will concentrate on 4D-Level 8 situation. As a starter, we show in Figs. 9.13, 9.14, 9.15, 9.16, 9.17, 9.18, 9.19, 9.20, 9.21, 9.22, and 9.23 a comparison of interpolated results at 81 arbitrarily selected points with the TASMANIAN grid at level 8 with 1857 Chebyshev points.

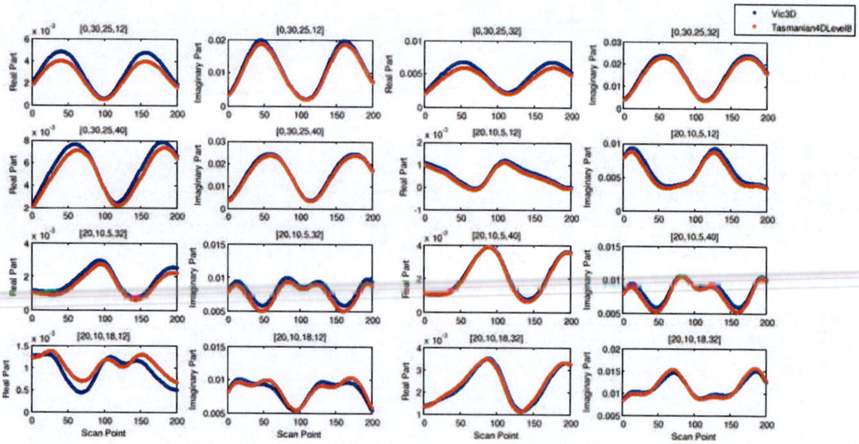

Fig. 9.16 Comparison of interpolated results at 81 arbitrarily selected points with the four-dimensional TASMANIAN grid at level 8 (continued)

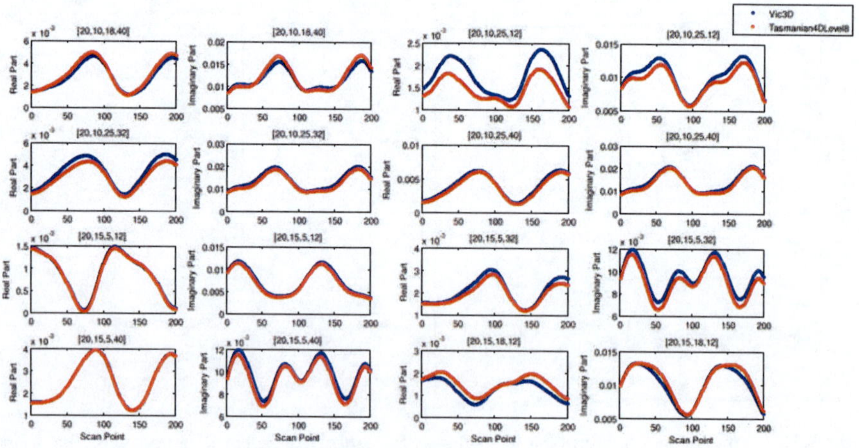

Fig. 9.17 Comparison of interpolated results at 81 arbitrarily selected points with the four-dimensional TASMANIAN grid at level 8 (continued)

Scaling may be a contributing factor to the occasionally poor fit of the real part, since the real part is much smaller than the imaginary, but it is also likely that the real and imaginary parts vary differently over this rather large range of the variables in physical space, which may contribute to the challenge of accurately interpolating each component.

Figure 9.24 shows the **VIC-3D®** model response of Fig. 6.4 when block 2 is varied in depth from 0 to 20 mils in four equal intervals, and the other blocks remain fixed at the values shown. Figure 9.25 shows a fourth-order polynomial fit to the peak values of the impedance data of Fig. 9.24, and Table 9.5 lists the coefficients

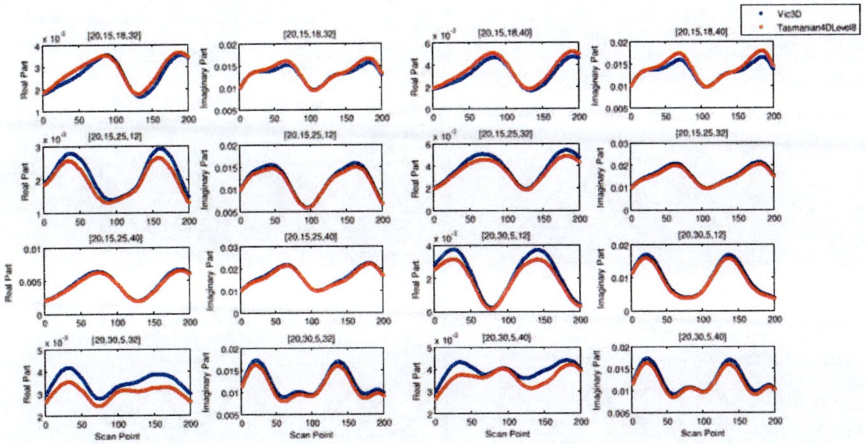

Fig. 9.18 Comparison of interpolated results at 81 arbitrarily selected points with the four-dimensional TASMANIAN grid at level 8 (continued)

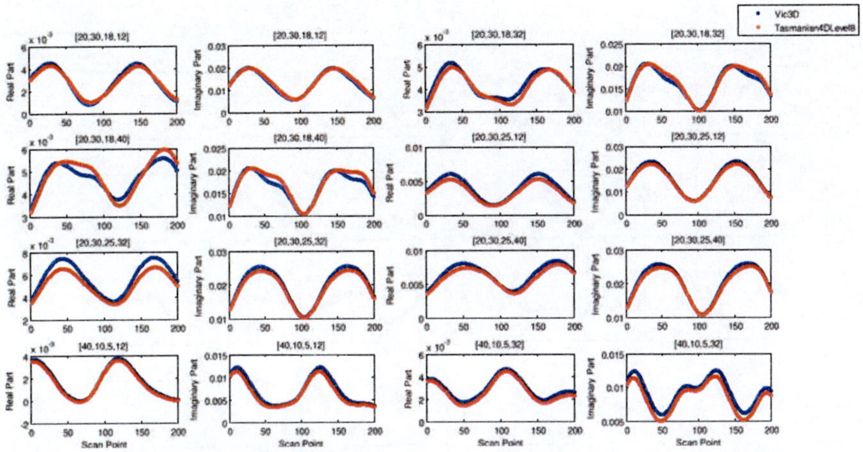

Fig. 9.19 Comparison of interpolated results at 81 arbitrarily selected points with the four-dimensional TASMANIAN grid at level 8 (continued)

of these fits. Clearly, there is a difference in the way that the real and imaginary parts vary with this particular geometric feature of the model. In both Figs. 9.24 and 9.25, the legend indicates depths of 0–40, which is due to the fact that the slabs are represented in **VIC-3D®** as canonical 'blocks,' whose reference coordinate system is at the center of the block. The blocks are defined in **VIC-3D®** by the total length of the unclipped block, so that when the clip plane passes through the origin of the block, as in this case, the 'length' parameter that **VIC-3D®** uses, and which is shown in the figures, is always twice the depth parameter shown in Fig. 6.4.

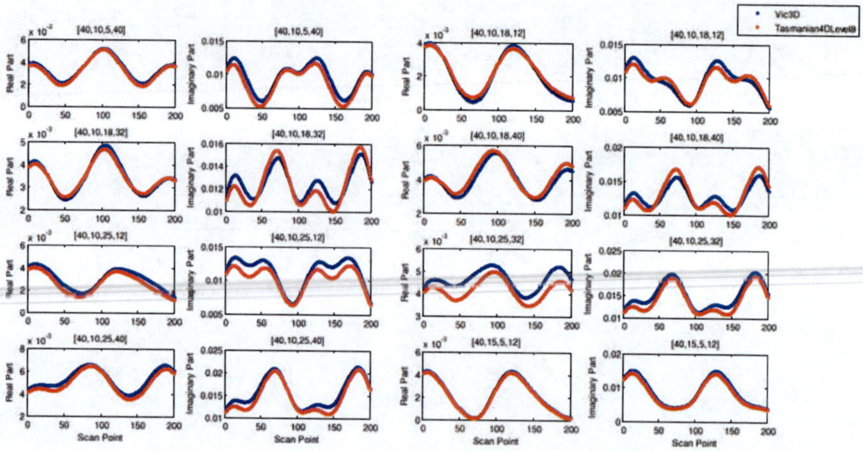

Fig. 9.20 Comparison of interpolated results at 81 arbitrarily selected points with the four-dimensional TASMANIAN grid at level 8 (continued)

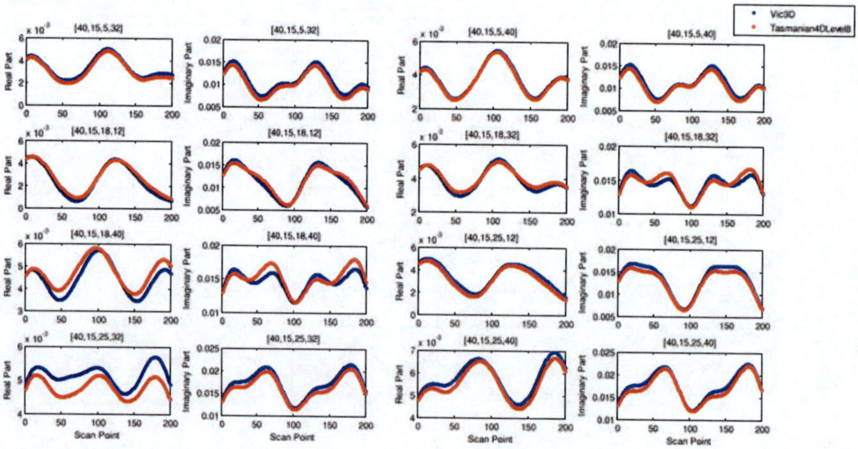

Fig. 9.21 Comparison of interpolated results at 81 arbitrarily selected points with the four-dimensional TASMANIAN grid at level 8 (continued)

9.7 The Geometry of the 4D-Level 8 Chebyshev Sparse Grid

We show in Figs. 9.26, 9.27, 9.28, 9.29, and 9.30 the distribution of points in a number of two-dimensional subspaces of the original four-dimensional grid.

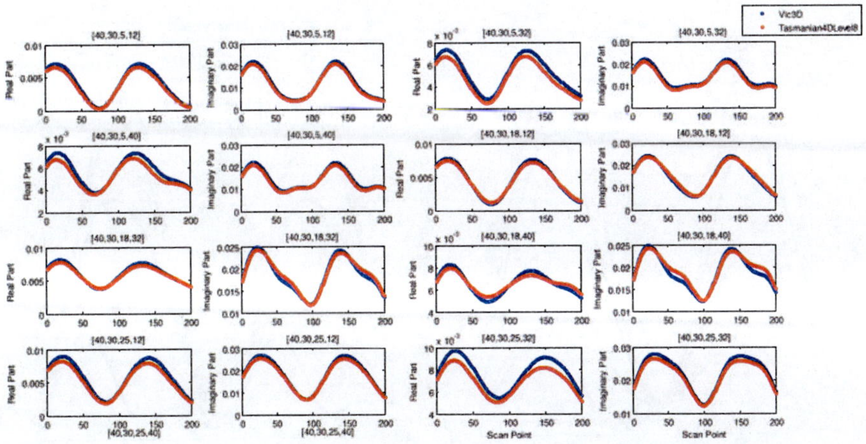

Fig. 9.22 Comparison of interpolated results at 81 arbitrarily selected points with the four-dimensional TASMANIAN grid at level 8 (continued)

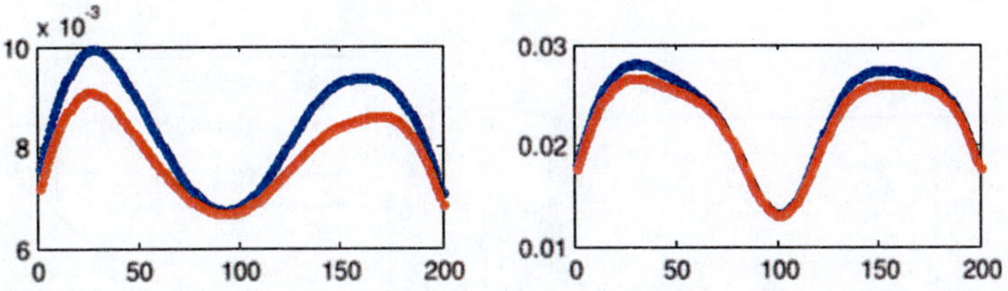

Fig. 9.23 Comparison of interpolated results at 81 arbitrarily selected points with the four-dimensional TASMANIAN grid at level 8 (continued)

Table 9.5 Coefficients of the fourth-order polynomial fits of Fig. 9.25

Order	R	X
4	−2.89708333333331e − 09	9.79583333333349e − 09
3	6.61083333333314e − 08	−1.28541666666668e − 06
2	7.25370833333338e-06	4.77454166666668e − 05
1	−6.38808333333335e − 05	−9.69083333333310e − 05
0	0.00300550000000000	0.0116510000000000

9.8 Searching the Sparse Grid for a Starting Point for Inversion

One of the applications of the sparse grid is as a surrogate for **VIC-3D®** in choosing a starting point for inversion with NLSE. To demonstrate this, we took two 'test data' sets, one with coordinates (0,10,18,32), and the other with coordinates (0,10,5,12). The first corresponds to a 'good' interpolation, as shown in the first column of the third row of Fig. 9.13, and the second to a 'not-so-good' interpolation,

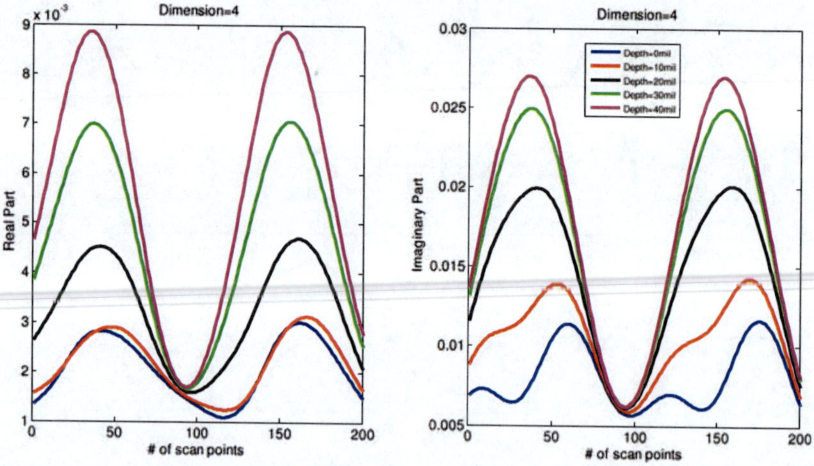

Fig. 9.24 **VIC-3D®** model response of Fig. 6.4 when block 2 is varied in depth from 0 to 20 mils in four equal intervals, and the other blocks remain fixed at the values shown

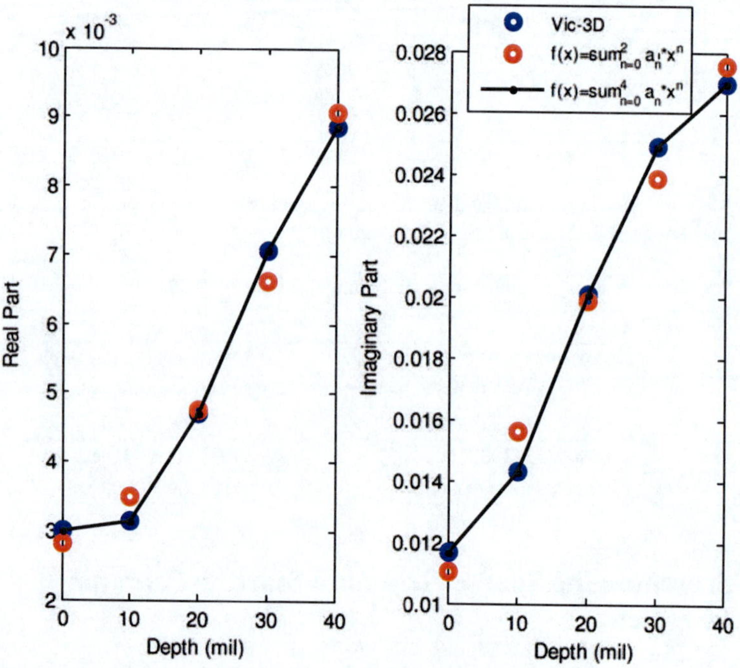

Fig. 9.25 Showing a fourth-order polynomial fit to the peak values of the impedance data of Fig. 9.24

as shown in the first column of the first row of Fig. 9.13. The algorithm for finding the starting point is to determine the 'nearest neighbor' to the test data among the

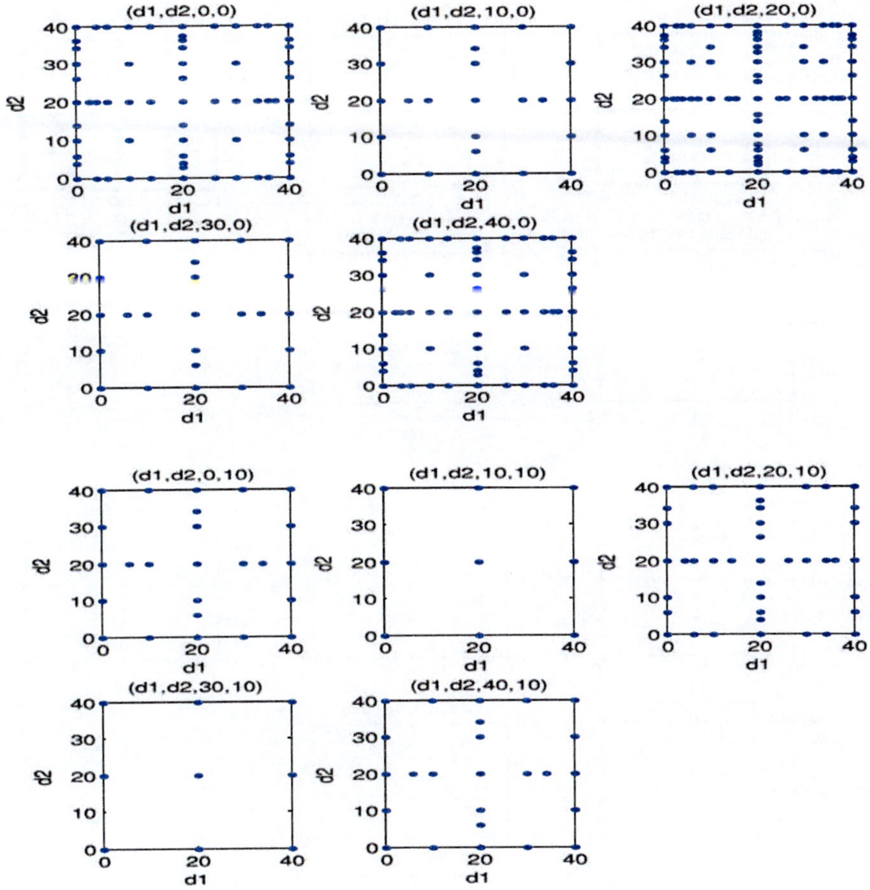

Fig. 9.26 Showing the Chebyshev distribution of points in various two-dimensional subsets of the four-dimensional grid at level 8

Table 9.6 Nearest sparse grid neighbors for two test data sets

Test set	Nearest neighbor/Φ	Second nearest neighbor/Φ
(0, 10, 18, 32)	(0, 5.8579, 20, 30)/9.42(−4)	(5.8579, 5.8579, 20, 20)/1.17(−3)
(0, 10, 5, 12)	(0, 10, 0, 10)/5.43(−4)	(0, 0, 10, 10)/1.08(−3)

1857 sparse grid points by choosing that point with the smallest norm of the residual impedance vector. The result of the experiment is shown in Table 9.6. The second nearest point is also shown for each data vector. The nearest neighbor to (0,10,18,32) lies in the subspace $(d_1, d_2, 20, 30)$ in Fig. 9.27, and the second nearest neighbor lies in the subspace $(d_1, d_2, 20, 20)$ in the same figure. As for (0,10,5,12), its nearest neighbor lies in the subspace $(d_1, d_2, 0, 10)$ shown in Fig. 9.26, and the second nearest neighbor lies in $(d_1, d_2, 10, 10)$ in the same figure.

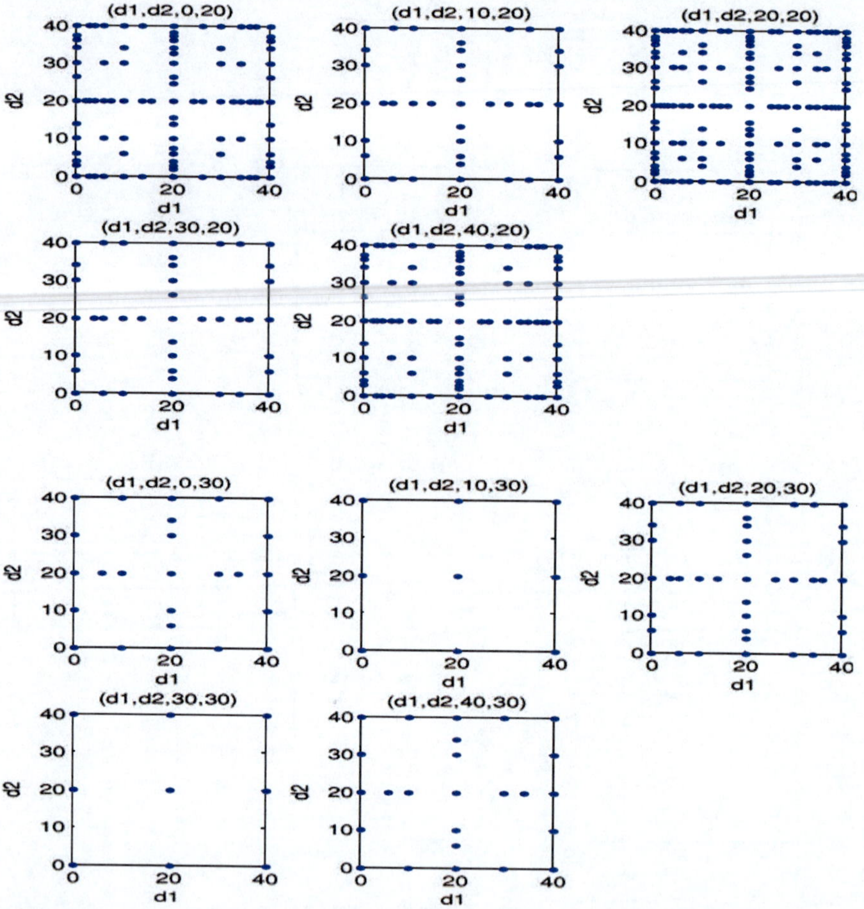

Fig. 9.27 Showing the Chebyshev distribution of points in various two-dimensional subsets of the four-dimensional grid at level 8 (continued)

In this manner we can map the 'most likely' regions of the sparse grid in which the data vector lies, and then compute a compact uniform grid for interpolation in NLSE within these regions. If the uniform grid is much smaller than the sparse grid, we would expect to get tighter estimations of confidence intervals when we perform a stochastic inversion with NLSE.

9.9 A Five-Dimensional Inverse Problem

The proof-of-the-pudding with sparse grids is their ability to simplify the solution of inverse problems that utilize internal interpolation tables as with NLSE. Our hope

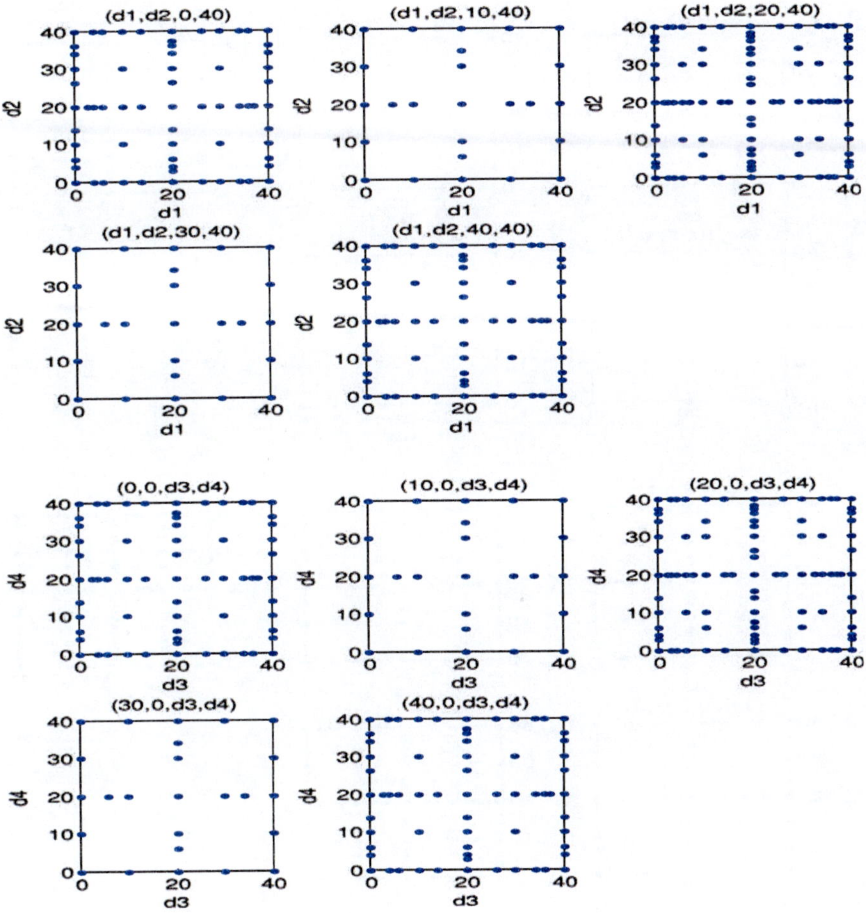

Fig. 9.28 Showing the Chebyshev distribution of points in various two-dimensional subsets of the four-dimensional grid at level 8 (continued)

is that we need to compute far fewer forward solutions with **VIC-3D®**, but then use the resulting sparse-grid solution to compute a much more refined full Cartesian grid which NLSE will then use to complete the inversion problem. This example will demonstrate the validity of this approach.

The problem consists of a split-D probe of the type shown in Fig. 9.31, and which was analyzed in [111, Section 6.6] that is scanned past a rectangular slot whose dimensions are 1 mm × 2 mm × 3 mm. The probe is vertical to the surface of the workpiece, but is rotated about its axis by 22°. The two parameters that define the orientation of the probe are the Euler angles shown in Fig. 9.32. These two angles, together with the liftoff of the probe and the length and depth of the flaw are the five parameters that are to be determined in the inverse problem. The 'true' values of the parameters are: LO = 3.5, $\Theta = 0$, $\Psi = 22$, L = 2, and D = 3.

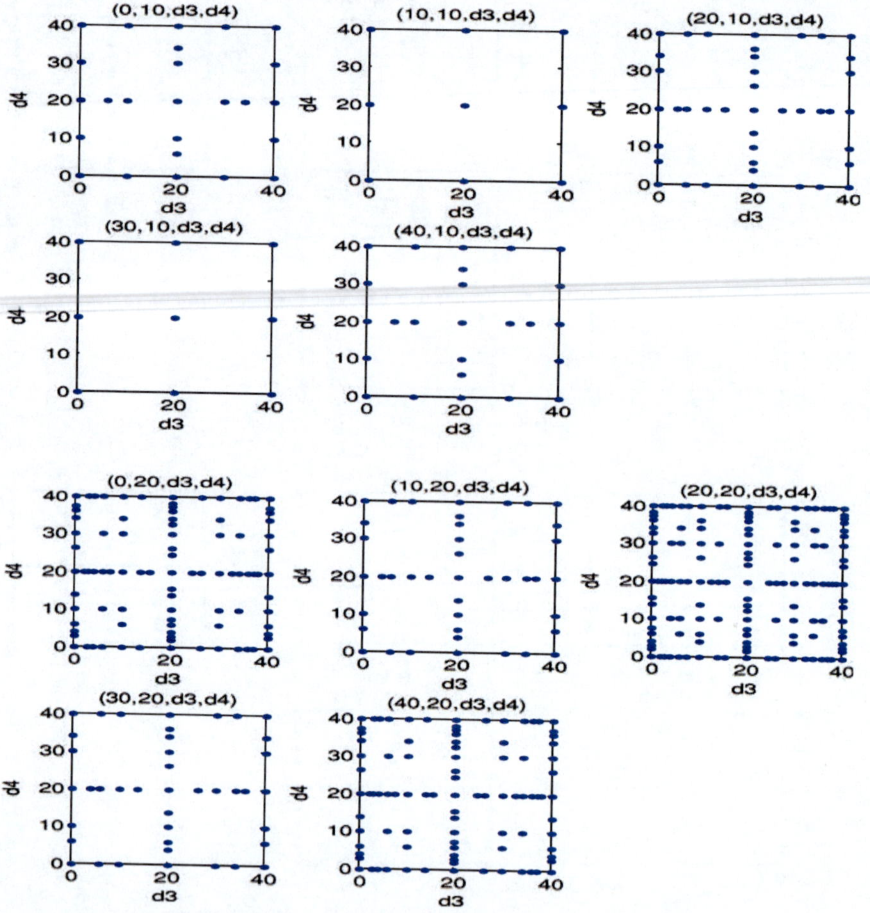

Fig. 9.29 Showing the Chebyshev distribution of points in various two-dimensional subsets of the four-dimensional grid at level 8 (continued)

The five-dimensional parameter space ranges over: LO $=$ [0, 9.99], Θ $=$ [0, 9.99], Ψ $=$ [0, 90], L $=$ [1.7, 2.3], D $=$ [2.7, 3.3], which at level 4 will be covered by 311 Chebyshev points. These are the values that are presented to **VIC-3D®** to generate the sparse interpolation table. The first NLSE full Cartesian table consists of four points in each dimension, uniformly distributed over its range, which yields 1024 nodes for NLSE. The results of the first inversion test are shown in Table 9.7. The final column in the table gives the number of local minima generated by the 500 random starting points that coalesce into the global minimum. In order to improve the accuracy of the inversion of Ψ, we increase the number of NLSE nodes in this parameter to 6, and get the result shown as Test 2 in Table 9.7. Clearly, this parameter benefits from a denser nodal distribution because it covers a large range of [0,90].

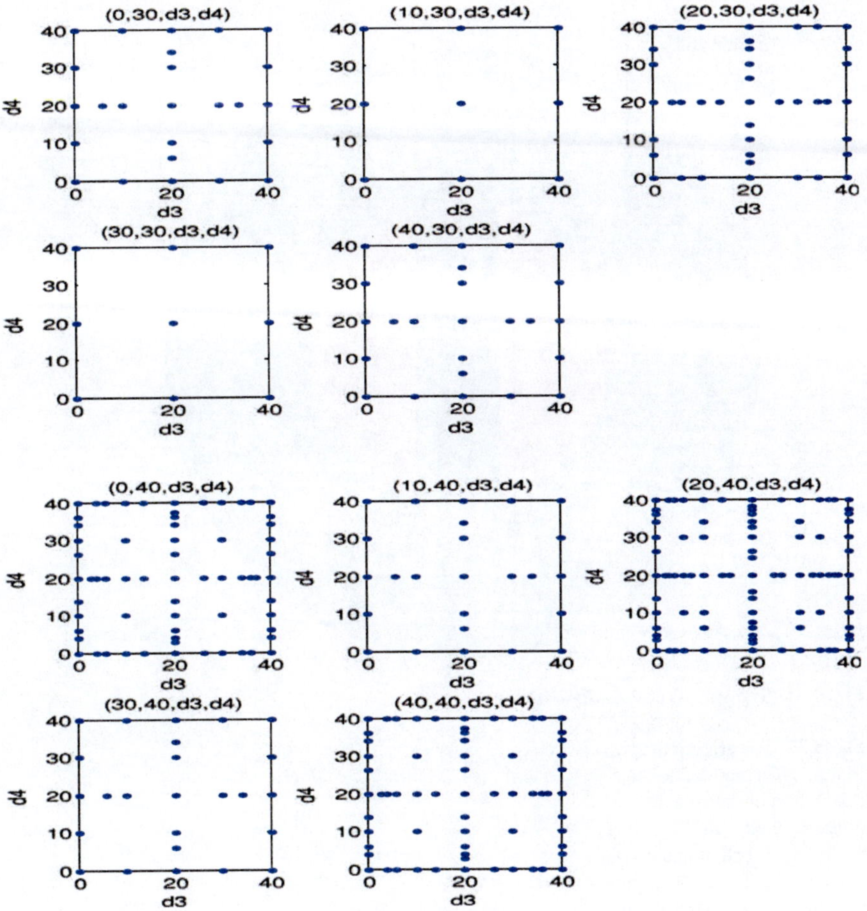

Fig. 9.30 Showing the Chebyshev distribution of points in various two-dimensional subsets of the four-dimensional grid at level 8 (continued)

The five-dimensional, level 4 inversion is quite good, but the important thing to note is the 'leverage-value' of each test in Table 9.7, namely $1024/311 = 3.3$ for the first, and 4.94 for the second. This is significant because the time to compute the 311 points completely dominates the overall inversion process, whereas the computation of function values for NLSE using the full Cartesian grid interpolation table derived from these 311 points is much faster.

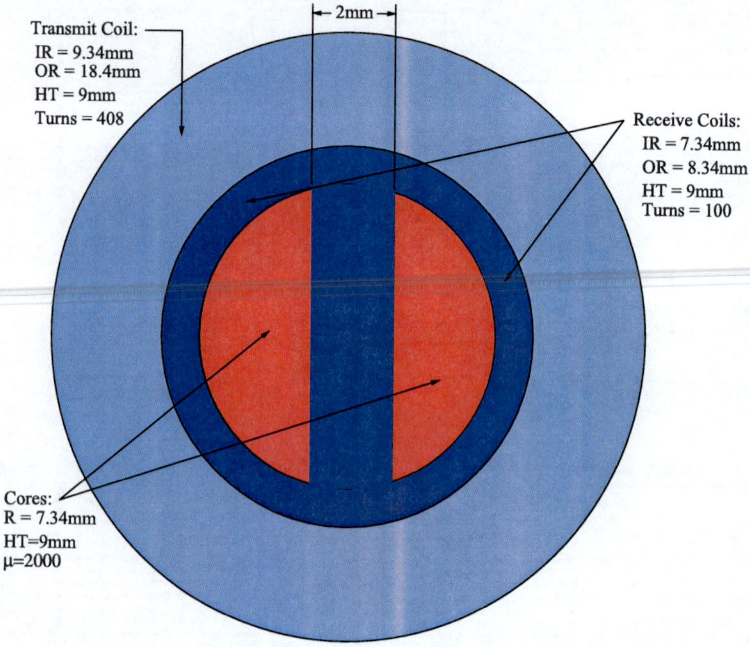

Fig. 9.31 The split-D coil configuration

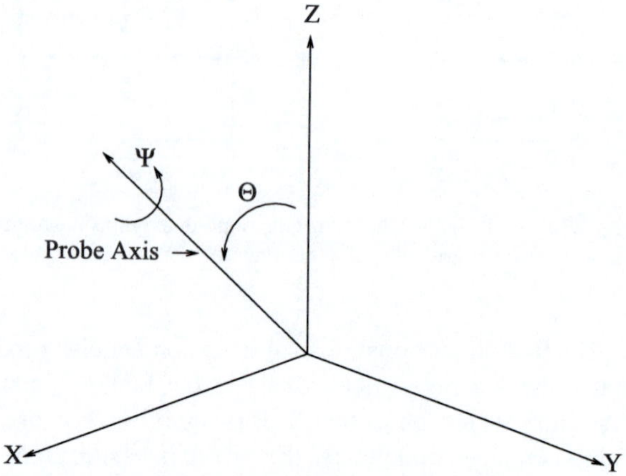

Fig. 9.32 Illustrating the two Euler angles that define the orientation of the split-D probe in the test problem. The probe axis is orthogonal to the plane of Fig. 9.31. The input model data for the inversion are $\Psi = 22°$ and $\Theta = 0°$

9.10 Noisy Data and Uncertainty Propagation

In developing the stochastic inverse model of Chap. 6, we assumed that the input data were given, and that the only stochastic feature of the problem was the random vector of unknown parameters that were to be determined. Each component of the vector was uniformly distributed over a certain range.

Table 9.7 Inversion results for the five-dimensional problem at level 4 with two different interpolation tables for NLSE

Test	Size (NLSE)	Level	No. Cal.	No. Nodes	ϕ	No. Pts.
1	$4 \times 4 \times 4 \times 4 \times 4$	4	311	1024	0.639(−5)	87
2	$4 \times 4 \times 6 \times 4 \times 4$	4	311	1536	0.578(−5)	78

Test	LO/Sensit	Θ/Sensit	Ψ/Sensit	L/Sensit	D/Sensit
1	3.5/2.24(−2)	5.46(−2)/4.17(−2)	20.4/0.47	1.99/4.25(−3)	2.98/0.28
2	3.5/2.02(−2)	0.11/3.71(−2)	21.51/0.36	1.99/3.83(−3)	2.99/0.25

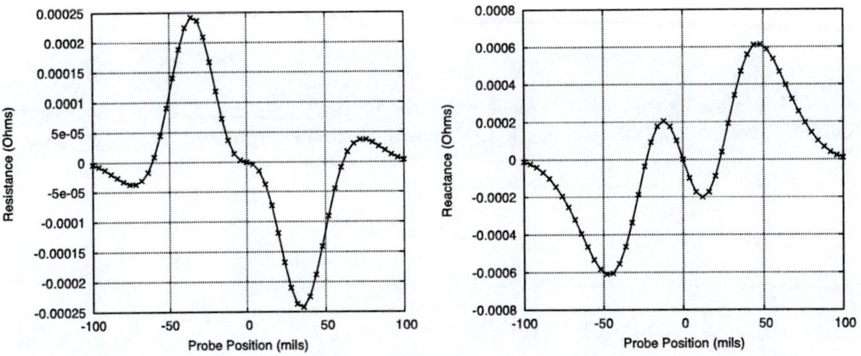

Fig. 9.33 Illustrating the noise-free input to the five-dimensional inverse problem. Left: resistance, Right: reactance

Now we want to extend the model to include the effects of Gaussian random noise that is superimposed on the noise-free input, shown in Fig. 9.33, to the five-dimensional inverse problem discussed in the preceding section. We consider two levels of noise, one with an RMS value of 1×10^{-5} and the other with an RMS value of 3×10^{-5}. Figure 9.34 illustrates a sample function of the former process superimposed on the noiseless data, and Fig. 9.35 illustrates a sample function from the second process superimposed on the noiseless data.

Our interest is in determining how uncertainty in the input data is propagated through the nonlinear least-squares filter into uncertainty in the output parameters. To accomplish this, we do a Monte Carlo analysis, in which the data of Fig. 9.33 are corrupted by ten samples from each noise source, as in Figs. 9.34 and 9.35, and then applied to NLSE using the interpolation table shown as Test 2 in Table 9.7.

The results are shown in Fig. 9.36, which depicts the relative error, defined to be the ratio of the computed value to the 'true' value, except for Θ, which uses an artificial value of 1×10^{-8} for zero. The 'Noise Level' in the figure is the ratio of the RMS value of noise to the peak value of the noiseless resistance, 0.00025, in Fig. 9.33. The ten sample points for each reconstructed parameter are shown as small dots, and the mean of the results is shown as the large red dot. The large black dot is the true value of the parameter. Note that many of the sample points are hidden behind either of the large dots.

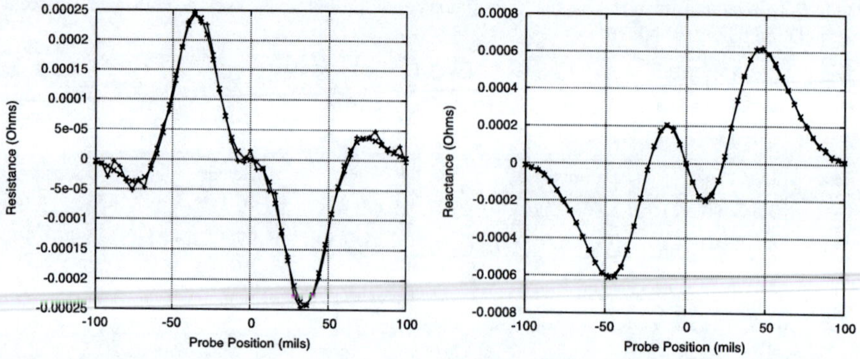

Fig. 9.34 Illustrating the noise-free input to the five-dimensional inverse problem with a sample function of noise at an RMS level of 1×10^{-5} superimposed. Left: resistance, Right: reactance

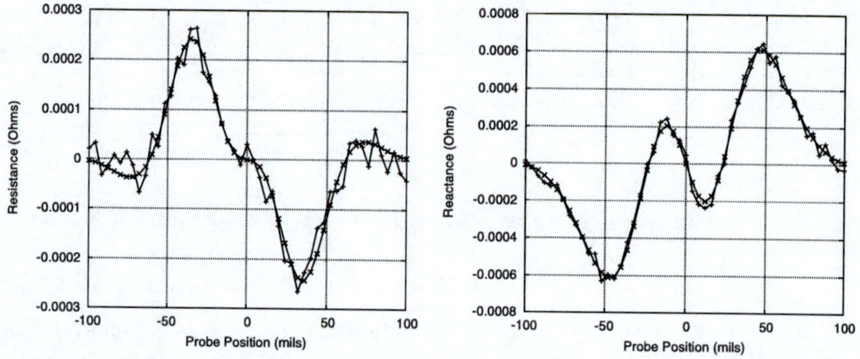

Fig. 9.35 Illustrating the noise-free input to the five-dimensional inverse problem with a sample function of noise at an RMS level of 3×10^{-5} superimposed. Left: resistance, Right: reactance

The mean values are in reasonable agreement with the true values, which suggests that the inversions are reasonable with these levels of input noise. The results for depth, D, may appear strange, in that the error in the mean value for the 1×10^{-5} noise source is greater than that for the 3×10^{-5} source, but keep in mind that the sensitivity coefficient for D in Table 9.7 is large, which, following our discussion in Chap. 6, indicates that the inversion process alone will introduce significant uncertainty in the estimated value of D. Our intuition is restored, however, when we look at the distribution of the errors over the ten samples: it is much larger for all five parameters when the noise level is 0.12. (Note that there is a small dot at ± 0.1 in D for Noise Level$=0.12$.)

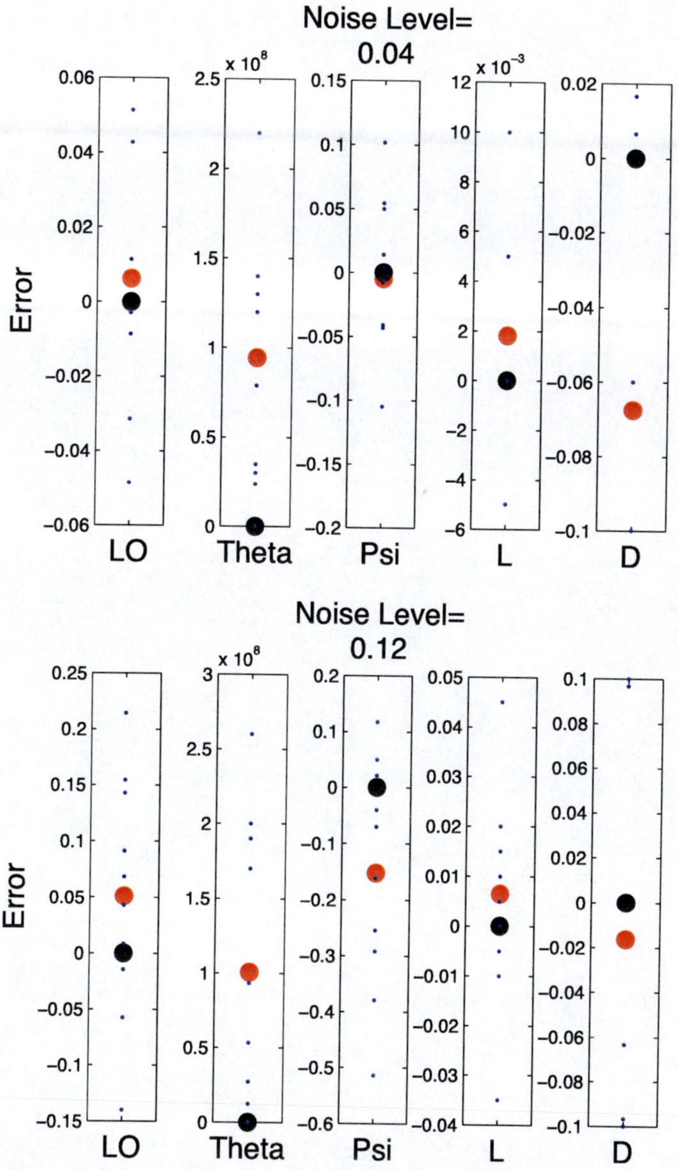

Fig. 9.36 Result of the Monte Carlo test to determine the error in inversion due to random noise at two different RMS values in the input data. Top: RMS $= 1 \times 10^5$; Bottom: RMS $= 3 \times 10^5$. The 'Noise Level' is the ratio of the RMS value of noise to the peak value of the noiseless resistance, 0.00025, in Fig. 9.33

Chapter 10
Characterization of Atherosclerotic Lesions by Inversion of Eddy-Current Impedance Data

10.1 The Model

Figure 10.1 shows a planar model of a type Vb lesion that could be situated in the coronary arteries [122]. This lesion is the largest and most complex stable lesion that can form, and is characterized by the formation of calcium of the outer cap, which gives rise to the expression 'hardening of the arteries.' Because the vulnerability of plaque is not closely correlated to plaque size, but rather to the overall composition, we are interested in determining the amount of fibrous tissue, calcium, lipid core, and smooth muscle that exists in the lesion, and to do this we return to Fig. 10.1.

We excite Eddy-currents in the layered medium by means of the exciting pancake coil, and read the impedance of the coil. We then use NLSE, the nonlinear least-squares parameter estimator in **VIC-3D®** to determine the thickness of each layer. We assume that the conductivity of each layer is known, as shown in the figure. Figure 10.2 illustrates the 'standard conductivity model' for a type Vb lesion. The problem is to determine the widths, $L_1, \cdots, L_5$ of each layer, given the conductivity profile shown. If we assume that the vessel wall's thickness, L_5, is fixed at, say, 0.3 mm, then there are only four unknowns to be determined using NLSE.

The data for the standard model of Fig. 10.2 are shown in Table 10.1, which is taken from [122]. In the model calculations, values for the calcified outer layer were inferred from cancellous bone values (see Table 10.13). In [122] the conductivity values for thrombus (fibrous material) and vessel wall were assumed constant up to 1 MHz. Beyond 1 MHz, however, it was assumed that the conductivity of thrombus and vessel wall increased slightly, and these values were inferred as shown. Additionally, it was assumed that fibrous tissue had relatively low capacitance due to an extracellular matrix of various collagens. The capacitance of the vessel wall was assumed to be mostly due to the smooth muscle, combined with the effects of fat.

H. A. Sabbagh et al., *Advanced Electromagnetic Models for Materials Characterization and Nondestructive Evaluation*, Scientific Computation,
https://doi.org/10.1007/978-3-030-67956-9_10

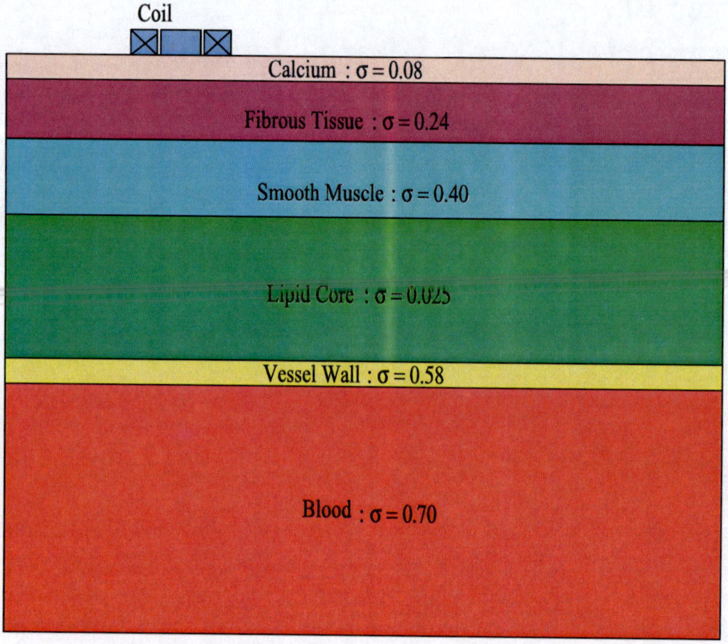

Fig. 10.1 A planar model of a type Vb atherosclerotic lesion, showing the pancake Eddy-current coil for taking impedance data and the conductivities of the various media

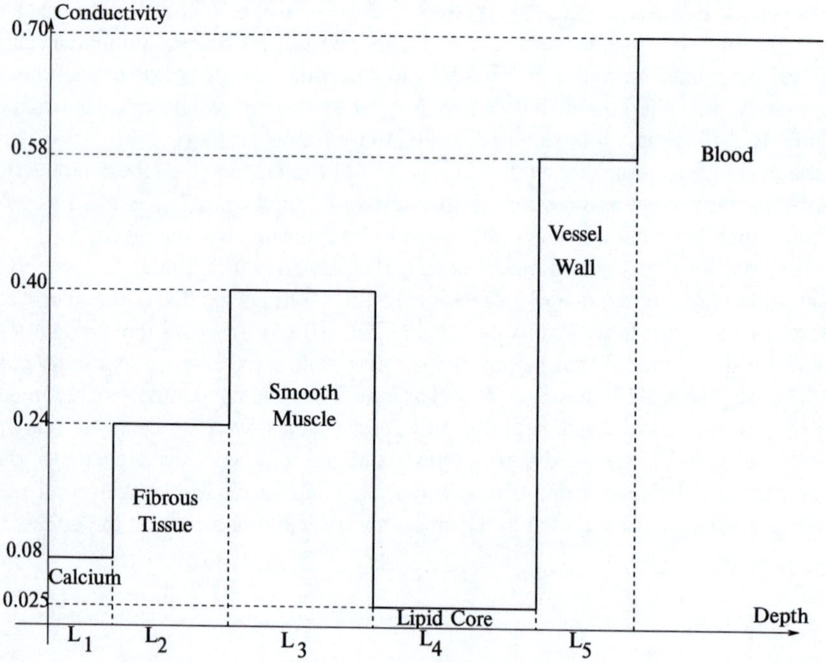

Fig. 10.2 Standard conductivity model of a type Vb lesion

Table 10.1 Values for conductivity and permittivity for type Vb lesions. Values marked with a[a] are inferred

	Conductivity (S/m)				
$\log_{10}$(Freq)	3	5	6	7	8
Blood	0.70	0.70	0.70	1.00	1.49
Fat	0.025	0.025	0.030	0.040	0.060
Muscle	0.40	0.40	0.40	0.40	0.75
Fibrous material	0.24	0.24	0.24	0.29[a]	0.33[a]
Calcium	0.08	0.08	0.10	0.12	0.17
Vessel wall	0.58	0.58	0.58	0.67[a]	0.83[a]
	Relative permittivity				
$\log_{10}$(Freq)	3	5	6	7	8
Blood	4100	4000	2000	300	75
Fat	20,000	100	50	30	12
Muscle	400,000	10,000	8000	200	70
Fibrous material	2000[a]	500[a]	50[a]	5[a]	3[a]
Calcium	10,500	500	250	70	30
Vessel wall	100,000[a]	5000[a]	4000[a]	100[a]	30[a]

Table 10.2 Thickness of each layer in variations of a type Vb lesion. The vessel wall is assumed to be 0.3 mm in all cases

	Thickness of type Vb lesion layers (mm)			
	Calcium	Fibrous	SMC	Lipid
Initial	0.1	0.3	0.3	1.0
Variation 1	0.2	0.2	0.2	1.1
Variation 2	0.2	0.2	0.5	0.8
Variation 3	0.1	0.5	0.3	0.8
Variation 4	0.1	0.1	0.7	0.8

For starters, we will run the arrangements shown in Table 10.2. The coil has 1 turn, with an inner radius of 0.05 mm, outer radius 0.06 mm, a height of 0.01 mm, and is excited over the frequency range of 95–100 GHz, in 21 steps.

Our interest in using Eddy-currents for the intravascular detection and characterization of vulnerable plaque, rather than the more conventional ultrasound, is that the latter can detect calcifications but not the remaining plaque components. We will show that Eddy-currents can do the entire job quite well.

10.2 Sample Impedance Calculations

Impedance calculations for the original lesion and the four variations listed in Table 10.2 are shown in Fig. 10.3. It is clear that each of the configurations is completely resolved over this frequency range. We have found this not to be true at other frequencies, which is the reason that we have chosen this frequency range to

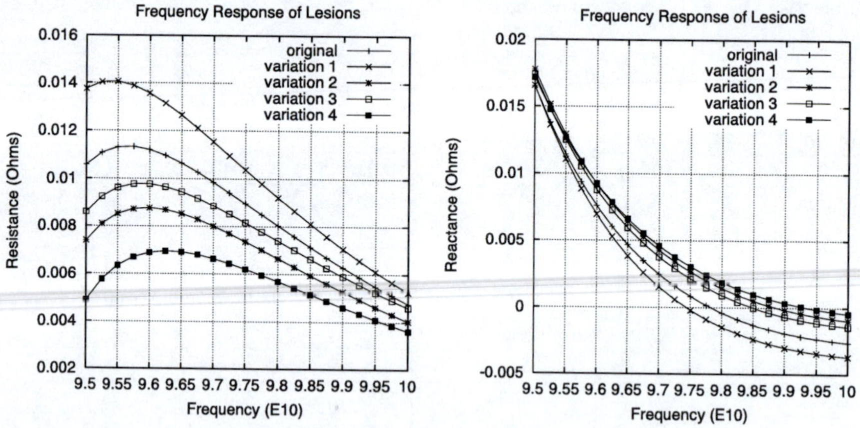

Fig. 10.3 Frequency response of the original lesion and its four variations. Left: resistance; right: reactance

do our inversions. We used the tissue data shown in the standard model of Fig. 10.2, even though they are not correct in the GHz-frequency range. This will be corrected shortly.

In these five models, the total length of the lesion is fixed at 1.7 mm, and we note in Fig. 10.3 that the lesion with the most pronounced response, variation 1, has the longest lipid layer. This means that the higher-conducting layers of variation 1 are more concentrated, having a total length of 0.6 mm, which is smaller than any of the other lesions.

Figure 10.4 shows the freespace frequency response of the same probe that produced the data of Fig. 10.3. This figure, together with Fig. 10.3, indicate the precision with which measurements must be taken. For example, in order to resolve the differences shown in Fig. 10.3, when compared to the freespace response, our instruments must be capable of resolving resistances to two significant digits (40 dB dynamic range), but reactances must be measured to six significant digits or so (120 dB dynamic range). This is typical of Eddy-current measurements on biological tissue.

10.3 The Eight-Layer Inversion Algorithm

The eight-layer inversion algorithm was described in Chapter 20 of [111] in the context of the nondestructive evaluation of coatings. In the present context it starts by assigning a conductivity, σ, to each of the eight layers shown in Fig. 10.5. The height of each layer, which defines the resolution of the algorithm, is $L = 0.2125$ mm, giving a total unknown region of 1.7 mm that contains the calcium,

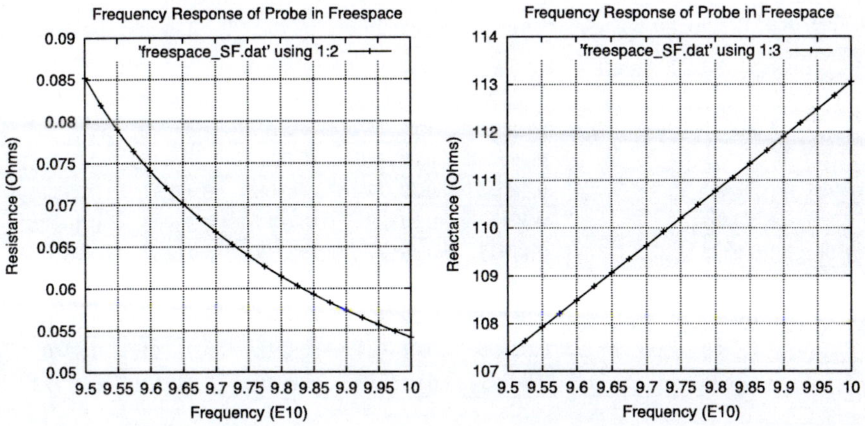

Fig. 10.4 Frequency response of the probe in freespace. Left: resistance; right: reactance

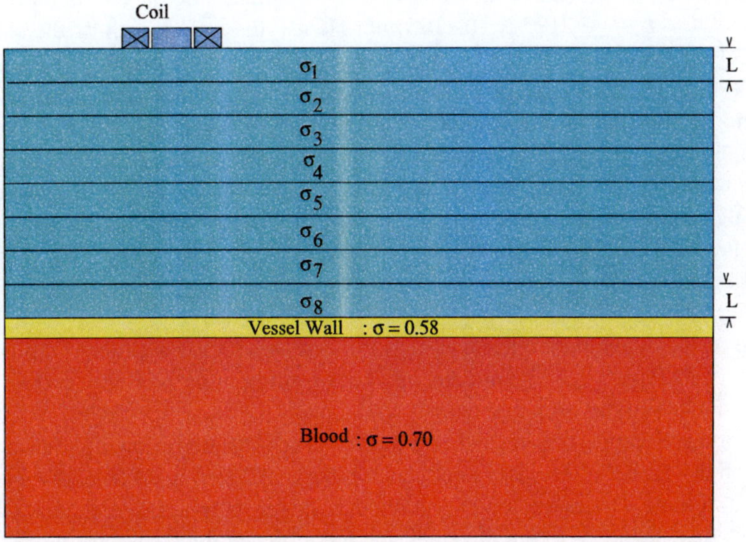

Fig. 10.5 The eight-layer inversion algorithm. The objective is to determine $\sigma_1, \ldots, \sigma_8$, given that the resolution of the algorithm $L = 0.2125$ mm

fibrous tissue, smooth muscle cap, and lipid. Our objective is to determine the size of the layer that contains each of these four materials. NLSE is used for this purpose.

The procedure utilizes simple rules involving average values of conductivities and volume-fractions. Suppose that NLSE produces a value $\sigma_1 = 0.08$ for the first layer. This indicates that the thickness of the calcium layer within the lesion is at least 0.2125 mm. To determine how much more it is, we must go to the second layer and apportion σ_2 between calcium and fibrous tissue. If, for example, $\sigma_2 = 0.15$, then we would calculate L_c, the volume-fraction of calcium in this layer by the

Table 10.3 Results for eight-layer inversion algorithm of Fig. 10.5. The Levenberg–Marquardt parameter is 0.01 for all five cases

		Lesion no.				
Layer	Upper boundary (mm)	Original	1	2	3	4
1	0	0.095828	0.098130	0.075263	0.094353	0.080461
2	−0.2125	0.46058	0.37017	0.48808	0.47237	0.59596
3	−0.425	0.30122	0.27370	0.34817	0.32965	0.40912
4	−0.6375	0.17525	0.13125	0.23358	0.21439	0.26295
5	−0.85	0.089613	0.060255	0.15365	0.13499	0.16217
6	−1.0625	0.041273	0.023305	0.10654	0.089338	0.10104
7	−1.275	0.026568	0.018184	0.089285	0.074301	0.07405
8	−1.4875	0.042713	0.043014	0.099406	0.087321	0.07723

following relationship: $0.15 = 0.08 \times L_c + 0.24 \times (1 - L_c)$. The solution of this equation is $L_c = 0.5625$, which means that the actual amount of calcium in this layer is $l_c = L_c \times 0.2125 = 0.1195$ mm. Hence, the total length of the calcium layer is $0.2125 + 0.1195 = 0.3320$ mm. In carrying out this algorithm, we implicitly assume that no more than two different materials can occupy the same layer of the grid in Fig. 10.5, or we cannot obtain a unique solution. This is a reasonable assumption as long as the layer thickness (or resolution), L, is small.

We can refine the resolution of the calculated results by reapplying the eight-layer algorithm to a modified model, after we have determined the length of some of the layers of material within the lesion. The eight layers would now be applied to the remaining unknown region, and the above volume-fraction algorithm would be repeated.

The results when the eight-layer algorithm of Fig. 10.5 is applied to the lesions are shown in Table 10.3. The numbers in the five columns to the right are the values of the conductivity in S/m.

From these results it is clear that the reconstructions are following the standard conductivity model of Fig. 10.2, in that an initial small step is followed by a longer interval of large values, and ending in a much longer interval of very small values. It is this final sequence of small values that interest us at the outset, for they clearly model the lipid core, whose conductivity is 0.025 S/m.

Layers 5–8 of the Original column and column 1 of Table 10.3 clearly belong to the lipid core, and layers 6–8 of columns 2–4 belong to the lipid core. Hence, we can say that the lipid cores of the reconstructed original lesion and lesion no. 1 are at least $4 \times 0.2125 = 0.85$ mm long.

Assuming that the smooth muscle cap and lipid core share the fifth layer for lesions 2–4, we interpolate within the fifth layer for lesions 2–4 to get a better approximation to the lipid core by using the volume fraction concept. Let L be the volume fraction of the fifth layer that belongs to the lipid core, and $(1 - L)$ be the volume fraction that belongs to the smooth muscle cap. Then, using the data of Table 10.3, we have the following results:

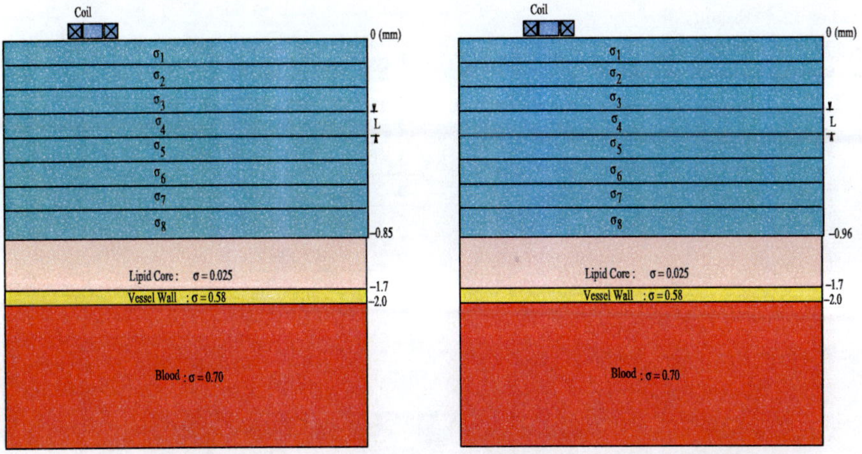

Fig. 10.6 Grid for application of the eight-layer inversion algorithm during the first iteration. Left: $L = 0.10625$ mm for the original lesion and lesion 1. Right: $L = 0.12$ mm for lesions 2–4

$$\begin{aligned} 0.15365 &= 0.0256L + 0.4 \times (1 - L), \text{ or} \\ L &= 0.657 \text{ for lesion 2} \\ 0.13499 &= 0.0256L + 0.4 \times (1 - L), \text{ or} \\ L &= 0.707 \text{ for lesion 3} \\ 0.16217 &= 0.0256L + 0.4 \times (1 - L), \text{ or} \\ L &= 0.634 \text{ for lesion 4}\,. \end{aligned} \tag{10.1}$$

Keeping in mind that the length of each interval is 0.2125 mm, these results yield values of $l = 0.1396,\ 0.1501,\ 0.1348$ mm for the length of the fifth layer occupied by the lipid core in, respectively, the second, third, and fourth lesions. When added to the 'certain' length of 3×0.2125 mm for the lipid core of each of these three lesions, we get estimated values of 0.777, 0.788, and 0.772 mm for the total length of the lipid core for, respectively, lesions 2, 3, and 4. This agrees well with the actual value of 0.8 mm, each, for these three lesions (see Table 10.2). Nevertheless, we are going to be conservative at this stage, and claim that the lipid core of these three lesions is at least 0.74 mm.

Therefore, with the assumptions that the lipid core is at least 0.85 mm long for the original lesion and lesion 1, and is at least 0.74 mm long for the other three lesions, we use the grids shown in Fig. 10.6 for the next (first) refinement of the original calculation.

From this point on, we will reconstruct only the original lesion, starting with the grid on the left of Fig. 10.6. Using a value of the Levenberg–Marquardt parameter, $LM = 0.00025$, we get the results shown in Table 10.4.

Table 10.4 Results of application of eight-layer inversion algorithm using Fig. 10.6. The Levenberg–Marquardt parameter is 0.00025

		Lesion no.
Layer	Upper boundary (mm)	Original
1	0	0.079456
2	−0.10625	0.28272
3	−0.2125	0.34385
4	−0.31875	0.36084
5	−0.425	0.3445
6	−0.53125	0.29836
7	−0.6375	0.22236
8	−0.74375	0.11789

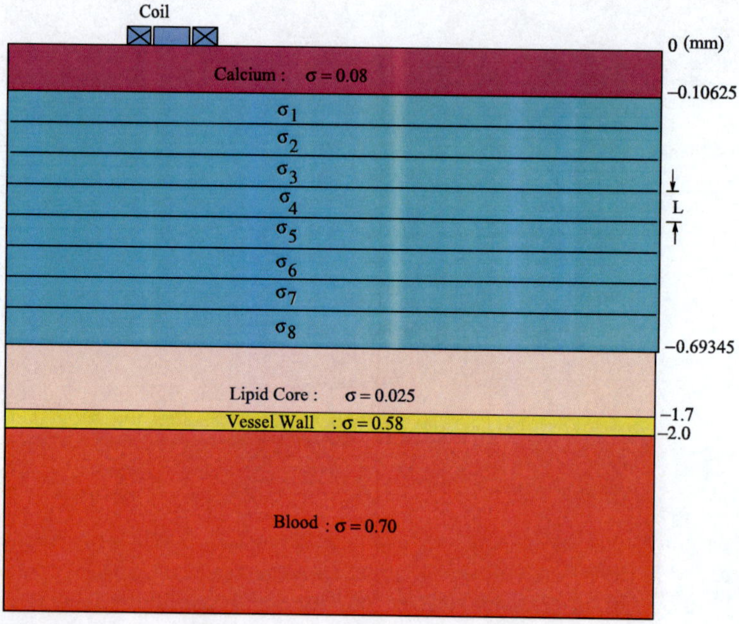

Fig. 10.7 Grid for application of the eight-layer inversion algorithm to the original lesion during the second iteration. $L = 0.0734$ mm

We recognize that the last layer is entirely lipid, and assume that the seventh layer is shared between lipid and smooth muscle. Hence, for this layer we have the volume-fraction relation $0.22236 = 0.025L + 0.40 \times (1 - L)$, or $L = 0.4737$, which means that the length of the seventh layer that is occupied by lipid is $l = 0.4737 \times 0.10625 = 0.0503$ mm. Therefore, the total length of the reconstructed lipid layer is $0.85 + 0.10625 + 0.0503 = 1.00655$ mm, compared with its actual length of 1.0 mm.

Furthermore, we attribute the entire calcium layer to the first layer in Table 10.4, which means that calcium occupies 0.10625 mm in the reconstruction, compared

Table 10.5 Results of application of eight-layer inversion algorithm using Fig. 10.7. The Levenberg–Marquardt parameter is 0.0001

		Lesion No.
Layer	Upper boundary (mm)	Original
1	−0.10625	0.31122
2	−0.17965	0.21456
3	−0.25305	0.26258
4	−0.32645	0.32472
5	−0.39985	0.37471
6	−0.47325	0.40568
7	−0.54665	0.41624
8	−0.62005	0.40678

Table 10.6 Results of application of eight-layer inversion algorithm using a Levenberg–Marquardt parameter 5×10^{-5}. The unknown parameters occupy the region between 0 mm and −0.39985 mm

		Lesion no.
Layer	Upper boundary (mm)	Original
1	0	0.084
2	−0.05	0.083
3	−0.10	0.184
4	−0.15	0.246
5	−0.20	0.279
6	−0.25	0.290
7	−0.30	0.282
8	−0.35	0.258

with 0.10 mm in the original. Using these two results, therefore, we define the new grid, shown in Fig. 10.7, which will be used to generate the second-iteration results.

The results for the second iteration are obtained using a value of the Levenberg–Marquardt parameter, $LM = 0.0001$, and are shown in Table 10.5.

There is a slight oscillation in the results over the last four layers, but the average value of the conductivities is 0.4009, which clearly identifies these layers as belonging to the smooth muscle cap. The average of the first four layers is 0.278, which means that we associate these layers with the fibrous tissue. Hence, we estimate that the smooth muscle cap extends from −0.39985 mm to −0.69345 mm, making it 0.2936 mm long, and that the fibrous tissue extends from −0.10625 mm to −0.39985 mm, making it also 0.2936 mm. The true length of each of these materials is 0.3 mm, so we are in the ballpark.

As one final iteration, however, we put the eight layers extending from 0 mm to −0.39985, and then fixed the smooth muscle cap from −0.39985 to −0.69345 mm, leaving the remainder of the trial lesion as in Fig. 10.7. Now the results, as shown in Table 10.6, are strikingly improved. We used a Levenberg–Marquardt parameter of 5×10^{-5}.

The calcium layer now clearly occupies 0.1 mm, and the average of the remaining six conductivities is 0.2565, which is a better fit to 0.24 than before. Hence, we conclude that the fibrous tissue occupies the next 0.29985 mm, after the calcium. The final results for the original lesion are summarized in Table 10.7.

Table 10.7 Final results for the reconstructed original lesion

	Thickness (mm)	
Material	Computed	Original
Calcium	0.1	0.1
Fibrous Tissue	0.29985	0.3
Smooth Muscle	0.2936	0.3
Lipid	1.00655	1.0

Table 10.8 Results of application of eight-layer inversion algorithm using the right-hand part of Fig. 10.6. The Levenberg–Marquardt Parameter is 0.00025

		Lesion no.
Layer	Upper boundary (mm)	Variation 2
1	0	0.0791
2	−0.12	0.1758
3	−0.24	0.3072
4	−0.36	0.3822
5	−0.48	0.4092
6	−0.60	0.3951
7	−0.72	0.3463
8	−0.84	0.2687

10.4 Lesion 2

Now, we'll go on to reconstruct Lesion 2, starting with the grid on the right-hand side of Fig. 10.6. Using a Levenberg–Marquardt parameter value of 0.00025, the results of the first iteration are shown in Table 10.8.

The first layer is clearly calcium, and the average of the fifth and sixth layer conductivities is 0.402, which suggests that they are muscle. The average of the fourth, fifth and sixth layers also suggest that they are muscle, but we prefer to be conservative at this point, and assign muscle to only the fifth and sixth layers. Hence, our second iteration will use a grid in which calcium is assigned to the region between 0 and −0.12 mm, and muscle to the region between −0.48 and −0.72 mm. The region between −0.12 mm and −0.48 mm will be divided into four layers of 0.09 mm, each, and the region between −0.72 mm and −0.96 mm will be divided into four layers of 0.06 mm, each. The region below −0.96 mm will remain as in the right-hand part of Fig. 10.6.

When we ran the second iteration, using this grid and a Levenberg–Marquardt parameter of 0.0001, we got the results shown in Table 10.9. Layer 8 of Table 10.9 is part of the lipid core, because its conductivity is quite low (in fact it is undershooting the value of 0.025 during its descent from 0.393 in the seventh layer). The average value of layers four through 7 is 0.400, so we are willing to state that the layer of muscle extends from (at least) −0.39 mm to −0.90 mm. Thus, we are ready for the third iteration, using a grid in which calcium is prescribed from 0 mm to −0.12 mm, muscle is prescribed from −0.39 mm to −0.90 mm, and lipid from −0.90 mm to −1.7 mm. The unknown region between −0.12 mm and −0.39 mm is divided into eight layers of 0.03375 mm, each.

Table 10.9 Results of application of eight-layer inversion algorithm during the second iteration for lesion 2. The Levenberg–Marquardt parameter is 0.0001

		Lesion no.
Layer	Upper boundary (mm)	Variation 2
1	−0.12	0.128
2	−0.21	0.223
3	−0.30	0.310
4	−0.39	0.369
5	−0.72	0.427
6	−0.78	0.412
7	−0.84	0.393
8	−0.90	0.0

Table 10.10 Results of application of eight-layer inversion algorithm during the third iteration for lesion 2. The Levenberg–Marquardt parameter is 5×10^{-5}

		Lesion no.
Layer	Upper boundary (mm)	Variation 2
1	−0.12	0.037
2	−0.15375	0.1523
3	−0.1875	0.2086
4	−0.22125	0.2345
5	−0.255	0.2414
6	−0.28875	0.2351
7	−0.3225	0.2209
8	−0.35625	0.1993

Table 10.11 Results of application of eight-layer inversion algorithm during the fourth iteration for lesion 2. The Levenberg–Marquardt parameter is 5×10^{-5}

		Lesion no.
Layer	Upper boundary (mm)	Variation 2
1	−0.12	0.047
2	−0.155	0.148
3	−0.190	0.202
4	−0.225	0.231
5	−0.26	0.244
6	−0.295	0.247
7	−0.33	0.242
8	−0.365	0.230

The results of the third iteration are shown in Table 10.10. It seems clear that layers 1 and 2 are calcium, layer 3 is the transition from calcium to fibrous tissue, and layers 4–8 are fibrous tissue. Note that the average value of the conductivities of layers 4–8 is 0.226, which is not as close to 0.24 as we would like, so we will change the separation interface between fibrous tissue and muscle to -0.4 instead of -0.39, thereby hoping to get an average value closer to 0.24. When we do this, we get the fourth-iteration results of Table 10.11, in which $LM = 5 \times 10^{-5}$.

Now we see that the average of layers 4–8 is 0.239, which is much closer to 0.24, and further, that the variance of the data about the mean is much smaller. Hence, we conclude that layers 4–8 are fibrous tissue, and that the transition between

Table 10.12 Final results for the reconstructed lesion 2

	Thickness (mm)	
Material	Computed	Original
Calcium	0.198	0.2
Fibrous tissue	0.202	0.2
Smooth muscle	0.5	0.5
Lipid	0.8	0.8

calcium and fibrous tissue occurs in layer 3, as before. When we calculate volume-fractions of the different tissue types within layer 3, we conclude that calcium occupies 0.008 mm and that fibrous tissue occupies 0.027 mm. Hence, the total thickness of the calcium layer is $0.190 + 0.008 = 0.198$ mm, and of fibrous tissue $0.027 + (0.40 - 0.225) = 0.202$ mm. The total length of muscle is, of course, 0.90–0.40 = 0.50 mm, and of the lipid core 1.70–0.90 = 0.80 mm. The final results for lesion 2 are shown in Table 10.12.

10.5 Noninvasive Detection and Characterization of Atherosclerotic Lesions

The preceding discussion and inversion example assume that one is using a catheter that has been inserted into the vessel. This, of course, is invasive, though 'minimally' so. Furthermore, the model inversions were used to characterize well-formed, or reasonably well-formed lesions for the purpose of distinguishing them from the surrounding healthy tissue so that they can be properly treated.

A noninvasive scheme for detecting and characterizing atherosclerotic lesions would be highly useful, and the results of a very interesting paper [95] suggest that such a scheme may be quite feasible using our Eddy-current technology. First, a bit of physiology: phagocytes are cells that engulf and digest cells, microorganisms, or other foreign bodies in the bloodstream and tissues, and macrophages are large (very large) molecules that devour things, usually bad things. In [95], rabbits were injected with ultrasmall superparamagnetic particles of iron oxides (USPIOs), and it was observed that these USPIOs were phagocytosed by macrophages in atherosclerotic plaques of the aortic wall in a quantity sufficient to cause susceptibility effects detectable by MRI, and all of this, mind you, before luminal narrowing is present, which means that such lesions might not even show up on an angiogram. Our interest is in detecting the presence of the USPIOs noninvasively by simple Eddy-current means.

Figure 10.8 shows a test setup for modeling. In Fig. 10.9 we show responses to the test setup for varying permeabilities of the sperical lesion, which is centered 1 cm beneath the surface of the serum half-space and has a radius of 1 mm. The coil has one turn, an inner radius of 1 mm, an outer radius of 1.1 mm, a height of 0.1 mm, and is excited at 1 GHz. We see a clear distinction between the response of

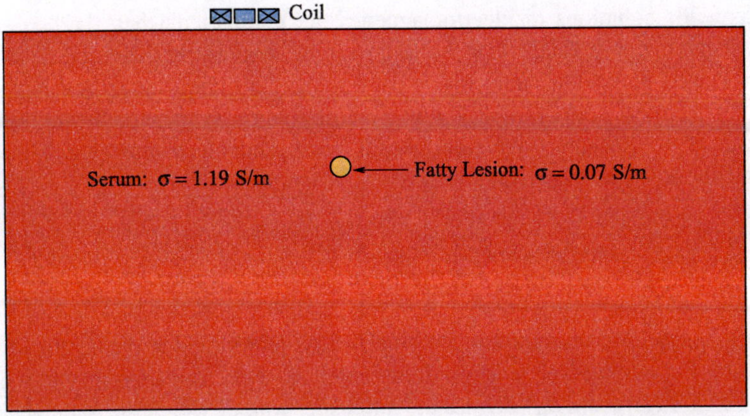

Fig. 10.8 Model of an Eddy-current coil scanned past a fatty lesion embedded in serum

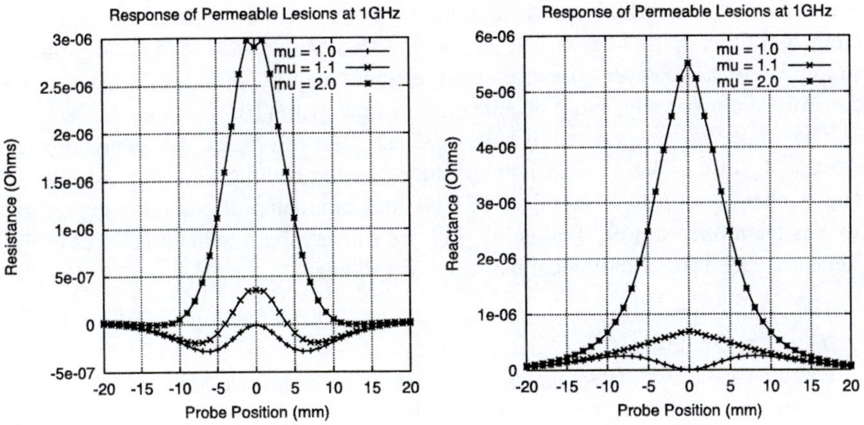

Fig. 10.9 Response of lesions of varying magnetic permeability at 1 GHz. Left: resistance, right: reactance

a nonpermeable lesion, with $\mu = 1.0$, and two permeable lesions with $\mu = 1.1$ and $\mu = 2.0$. The distinction is much clearer in the reactance.

10.6 Electromagnetic Modeling of Biological Tissue

In electromagnetic imaging, our goal is to directly determine the electromagnetic state of the body, from which we infer its physiological state, which could even include determining the change in temperature of tissue noninvasively. This is done by inverting impedance measurements to infer the electromagnetic constitutive properties of the body, from which other state variables can be determined.

Maxwell's equations of electromagnetics are completed when there is a constitutive relationship established between the electric current density, $\mathbf{J}$, and the electric field, $\mathbf{E}$. In the sinusoidal steady-state, the simplest relationship is the linear one, $\mathbf{J}(\mathbf{r}) = j\omega\epsilon_0\hat{\epsilon}(\omega)\mathbf{E}(\mathbf{r})$, which defines the complex generalized (relative) permittivity, $\hat{\epsilon}$. Research over the last 60 years into electromagnetic effects on biological tissue has shown that this permittivity may be written

$$\hat{\epsilon}(\omega) = \epsilon_\infty + \sum_n \frac{\Delta\epsilon_n}{1+(j\omega\tau_n)^{(1-\alpha_n)}} + \frac{\sigma_{dc}}{j\omega\epsilon_0}, \qquad (10.2)$$

where ϵ_∞, σ_{dc}, $\Delta\epsilon_n$, τ_n, and α_n, are empirically determined (Cole–Cole) parameters. ϵ_∞ is the infinite-frequency limit of $\hat{\epsilon}$, and σ_{dc} is the zero-frequecy (dc) limit.

The triplet $(\Delta\epsilon_n, \tau_n, \alpha_n)$ defines the nth dispersion, of which there are generally three observed: α, β and γ, going from lowest to highest frequencies. If $\alpha_n = 0$, then each dispersion could be modeled by a simple series connection of a resistor and capacitor, and (10.2) could be modeled by a parallel combination of linear, lumped, bilateral, passive circuit elements. This would not lead to significant simplifications, however, because these element values would still have to be determined empirically, and it is quite easy to work with (10.2) as it is.

The dielectric properties of biological tissue result from the interaction of electromagnetic radiation with its constituents at the cellular and molecular level. The mechanisms of the interaction are well understood and discussed in the review articles mentioned in [39]. Following [39], we can say that the main features of the dielectric spectrum of biological tissue are as follows:

- The relative permittivity of a tissue may reach values of up to 10^6 or 10^7 at frequencies below 100 Hz
- It decreases at high frequencies in three main steps known as the α, β, and γ dispersions. Other dispersions may also be present.
- The γ dispersion, in the gigahertz region, is due to the polarization of water molecules.
- The β dispersion, in the hundreds of kilohertz region, is due mainly to the polarization of cellular membranes which act as barriers to the flow of ions between the intra and extra cellular media. Other contributions to the β dispersion come from the polarization of protein and other organic macromolecules.
- The low frequency α dispersion is associated with ionic diffusion processes at the site of the cellular membrane.
- Tissues have finite ionic conductivities commensurate with the nature and extent of their ionic content and ionic mobility.

Gabriel et al. [39–41] have tabulated the dielectric properties of tissues over the frequency range 10 Hz to 20 GHz and have determined the Cole–Cole parameters that cover the entire range of dispersions, and they are listed in Table 10.13. Table 10.14 lists values of the conductivity and dielectric permittivity of a number of biological tissues, using the data of Table 10.13.

Table 10.13 Parameters of (10.2) used to predict dielectric properties of tissues [41]

Tissue type	ϵ_∞	$\Delta\epsilon_1$	τ_1 (ps)	α_1	$\Delta\epsilon_2$	τ_2 (ns)	α_2
Blood	4.0	56.0	8.38	0.10	5200	132.63	0.10
Bone (cancellous)	2.5	18.0	13.26	0.22	300	79.58	0.25
Bone (cortical)	2.5	10.0	13.26	0.20	180	79.58	0.20
Brain (grey matter)	4.0	45.0	7.96	0.10	400	15.92	0.15
Brain (white matter)	4.0	32.0	7.96	0.10	100	7.96	0.10
Fat (infiltrated)	2.5	9.0	7.96	0.20	35	15.92	0.10
Fat (not infiltrated)	2.5	3.0	7.96	0.20	15	15.92	0.10
Heart	4.0	50.0	7.96	0.10	1200	159.15	0.05
Kidney	4.0	47.0	7.96	0.10	3500	198.94	0.22
Lens cortex	4.0	42.0	7.96	0.10	1500	79.58	0.10
Liver	4.0	39.0	8.84	0.10	6000	530.52	0.20
Lung (inflated)	2.5	18.0	7.96	0.10	500	63.66	0.10
Muscle	4.0	50.0	7.23	0.10	7000	353.68	0.10
Skin (dry)	4.0	32.0	7.23	0.00	1100	32.48	0.20
Skin (wet)	4.0	39.0	7.96	0.10	280	79.58	0.00
Spleen	4.0	48.0	7.96	0.10	2500	63.66	0.15
Tendon	4.0	42.0	12.24	0.10	60	6.37	0.10

Tissue type	$\Delta\epsilon_3$	τ_3 (μs)	α_3	$\Delta\epsilon_4$	τ_4 (ms)	α_4	σ_{dc}
Blood	0.0			0.0			0.7000
Bone (cancellous)	2.0×10^4	159.15	0.20	2.0×10^7	15.915	0.00	0.0700
Bone (cortical)	5.0×10^3	159.15	0.20	1.0×10^5	15.915	0.00	0.0200
Brain (grey matter)	2.0×10^5	106.10	0.22	4.5×10^7	5.305	0.00	0.0200
Brain (white matter)	4.0×10^4	53.05	0.30	3.5×10^7	7.958	0.02	0.0200
Fat (infiltrated)	3.3×10^4	159.15	0.05	1.0×10^7	15.915	0.01	0.0350
Fat (not infiltrated)	3.3×10^4	159.15	0.05	1.0×10^7	7.958	0.01	0.0100
Heart	4.5×10^5	72.34	0.22	2.5×10^7	4.547	0.00	0.0500
Kidney	2.5×10^5	79.58	0.22	3.0×10^7	4.547	0.00	0.0500
Lens cortex	2.0×10^5	159.15	0.10	4.0×10^7	15.915	0.00	0.3000
Liver	5.0×10^4	22.74	0.20	3.0×10^7	15.915	0.05	0.0200
Lung (inflated)	2.5×10^5	159.15	0.20	4.0×10^7	7.958	0.00	0.0300
Muscle	1.2×10^6	318.31	0.10	2.5×10^7	2.274	0.00	0.200
Skin (dry)	0.0			0.0			0.0002
Skin (wet)	3.0×10^4	1.59	0.16	3.0×10^4	1.592	0.20	0.0004
Spleen	2.0×10^5	265.26	0.25	5.0×10^7	6.366	0.00	0.0300
Tendon	6.0×10^4	318.31	0.22	2.0×10^7	1.326	0.00	0.2500

Table 10.14 Values of conductivity and permittivity for various tissues, using parameters of Table 10.13. Entries marked with[a] are inferred

	Conductivity (S/m)								
$\log_{10}$(Freq)	3	4	5	6	7	8	9	10	11
Blood	0.7	0.7	0.7	0.82	1.1	1.23	1.58	13.13	63.35
Bone (cancellous)	0.08	0.08	0.08	0.09	0.12	0.17	0.36	3.86	16.04
Bone (cortical)	0.02	0.02	0.02	0.02	0.04	0.06	0.16	2.14	8.66
Brain (grey matter)	0.1	0.11	0.13	0.16	0.29	0.56	0.99	10.31	53.24
Brain (white matter)	0.06	0.07	0.08	0.1	0.16	0.32	0.62	7.3	38
Fat (infiltrated)	0.04	0.04	0.04	0.04	0.05	0.07	0.12	1.71	10.63
Fat (not infiltrated)	0.02	0.02	0.02	0.03	0.03	0.04	0.05	0.59	3.56
Heart	0.11	0.15	0.22	0.33	0.5	0.73	1.28	11.84	59.86
Kidney	0.11	0.14	0.17	0.28	0.51	0.81	1.45	11.57	57.08
Lens cortex	0.33	0.34	0.34	0.42	0.75	0.92	1.23	9.78	49.55
Liver	0.04	0.05	0.08	0.19	0.32	0.49	0.9	9.39	42.95
Lung (inflated)	0.08	0.09	0.11	0.14	0.23	0.31	0.47	4.21	21.38
Muscle	0.32	0.34	0.36	0.5	0.62	0.71	0.98	10.62	62.52
Skin (dry)	0	0	0	0.01	0.2	0.49	0.9	8.01	39.45
Skin (wet)	0	0	0.07	0.22	0.37	0.52	0.88	8.95	46.11
Spleen	0.1	0.11	0.12	0.18	0.51	0.8	1.32	11.38	57.34
Tendon	0.38	0.39	0.39	0.39	0.41	0.49	0.76	10.34	34.9
Fibrous material	0.24	0.24	0.24	0.24	0.29[a]	0.33[a]	0.36[a]	0.39[a]	0.45[a]
Vessel wall	0.58	0.58	0.58	0.58	0.67[a]	0.83[a]	0.92[a]	1.1[a]	2.0[a]

	Relative Permittivity								
$\log_{10}$(Freq)	3	4	5	6	7	8	9	10	11
Blood	5258	5248	5120	3026	280	77	61	45	8
Bone (cancellous)	12,320	1658	474	249	71	28	21	13	4
Bone (cortical)	2702	522	228	145	37	15	12	8	3
Brain (grey matter)	164,060	22,241	3222	860	320	80	52	38	8
Brain (white matter)	69,811	12,468	2108	480	176	57	39	28	7
Fat (inflitrated)	19,287	912	101	51	30	13	11	9	4
Fat (not infiltrated)	24,105	1085	93	27	13.8	6.1	5.4	4.6	3
Heart	352,860	70,057	9846	1967	293	91	59	42	8
Kidney	212,900	38,747	7651	2251	371	98	58	40	8
Lens cortex	107,550	10,318	4038	2717	339	66	48	36	7
Liver	85,673	28,927	7499	1536	223	69	46	32	7
Lung (inflated)	141,520	17,174	2581	733	124	32	22	16	4
Muscle	434,930	25,909	8089	1836	171	66	55	43	9
Skin (dry)	1136	1134	1119	991	362	73	41	31	6
Skin (wet)	32,135	29,012	15,369	1835	222	66	46	34	7
Spleen	106,840	13,891	4222	2290	441	91	57	41	8
Tendon	302,710	5336	472	160	103	54	46	29	6
Fibrous material	2000[a]	1000[a]	500[a]	50[a]	5[a]	3[a]	2.0[a]	1.2[a]	1.0[a]
Vessel wall	100,000[a]	22,000[a]	5000[a]	4000[a]	100[a]	30[a]	15[a]	3[a]	1[a]

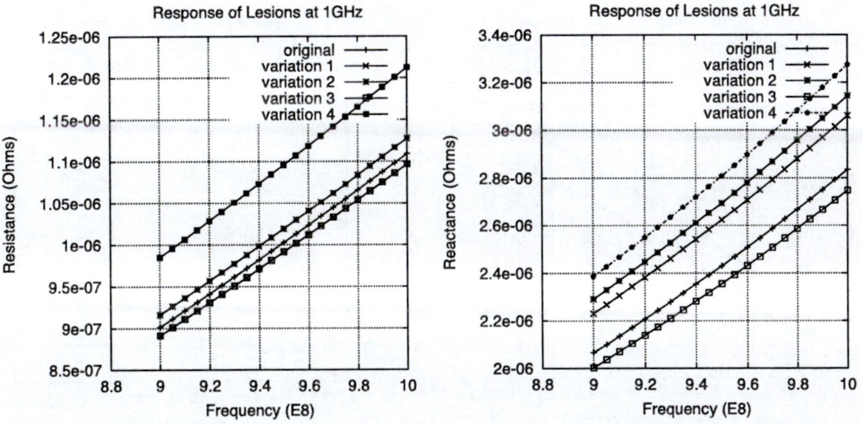

Fig. 10.10 Frequency response of the five lesions in the vicinity of 1 GHz. Both conductivity and permittivity effects are included. Left: resistance; right: reactance

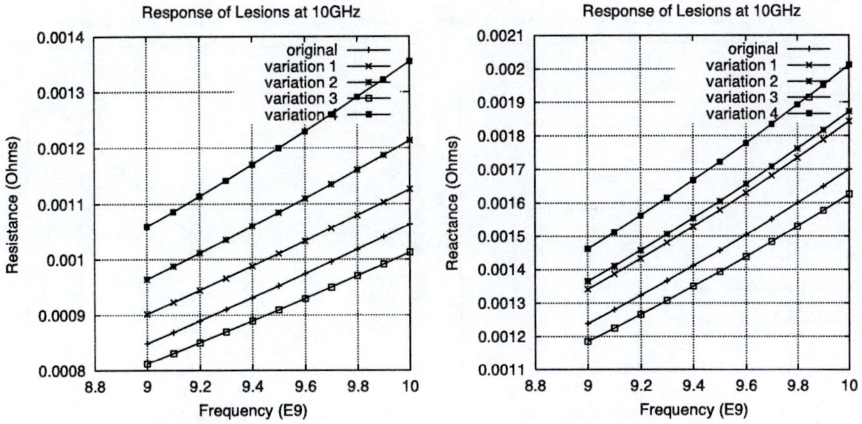

Fig. 10.11 Frequency response of the five lesions in the vicinity of 10 GHz. Both conductivity and permittivity effects are included. Left: resistance; right: reactance

10.6.1 The Lesions Revisited

With the data of Table 10.14 in hand we can redo the calculations of the lesions that were described in Sects. 10.1–10.3, this time correctly accounting for the dispersive properties of the tissue. Earlier, we had used data that were only good in the megahertz range to a model problem set at 100 GHz! Further, we only included conductivity; now we can include dielectric permittivity, though we cannot yet invert both parameters, because **VIC-3D®** is not set up to do so. The correct response for the lesions at 1–100 GHz are shown in Figs. 10.10, 10.11, and 10.12.

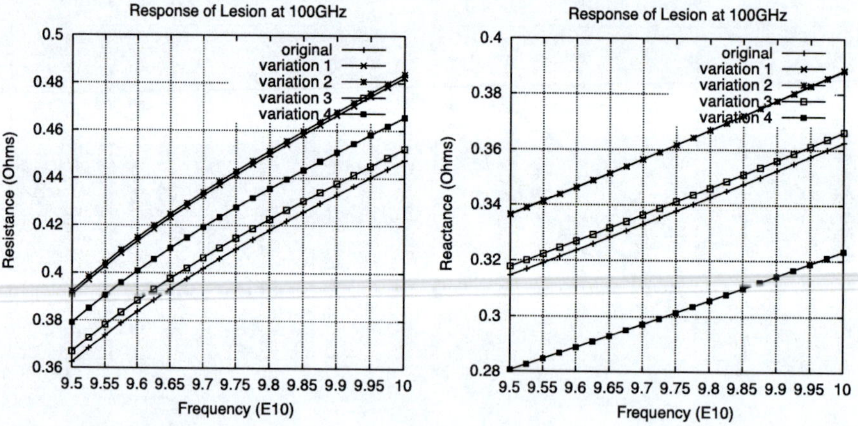

Fig. 10.12 Frequency response of the five lesions in the vicinity of 100 GHz. Both conductivity and permittivity effects are included. Left: resistance; right: reactance

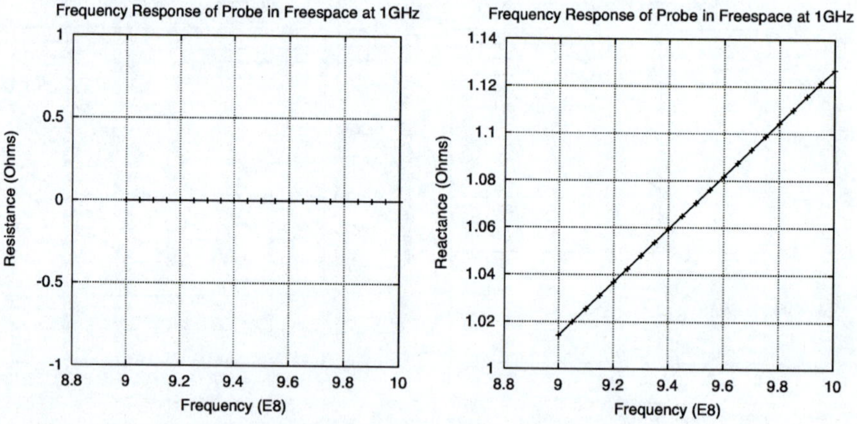

Fig. 10.13 Frequency response of the coil in the vicinity of 1 GHz. Left: resistance; right: reactance

The freespace response of the coil at 100 GHz has already been given in Fig. 10.4. The corresponding results at 1 GHz and 10 GHz are given in Figs. 10.13 and 10.14.

10.7 Determining Coil Parameters

Figure 10.15 is the equivalent circuit of a real coil. **VIC-3D®** can only model the inductor, L_0, and Z_W, the change in the impedance due to the presence of the workpiece. The remaining parameters must be inferred by measurement. Because **VIC-3D®** can model only the right-hand branch, we must subtract the effects of Y_p

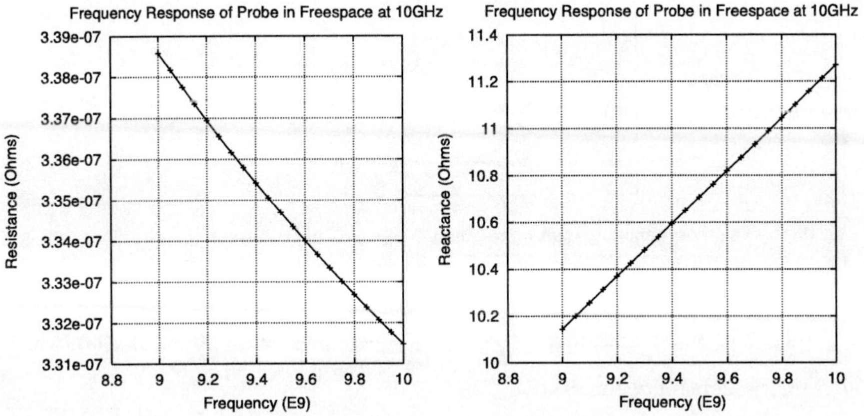

Fig. 10.14 Frequency response of the coil in the vicinity of 10 GHz. Left: resistance; right: reactance

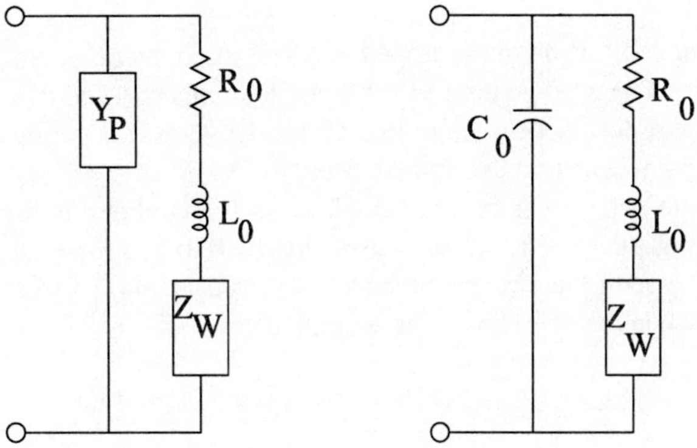

Fig. 10.15 Equivalent circuit of a real coil. It is assumed that $Y_p \to 0$ as $f \to$ on the left. The equivalent capacitance, C_0, in the circuit on the right accounts for the self-resonance of the coil

(or C_0) from the circuit. This is easily done, once we have measured values of the driving-point impedance, Z_{in}, seen at the left-hand terminal-pair.

Consider the situation in which the coil is in air, located well away from the workpiece; then $Z_W = 0$. L_0 is the low-frequency inductance of the coil, and R_0 is the low-frequency resistance of the coil. Each of these parameters is known empirically, with the former perhaps computed by **VIC-3D®** if the coil data are known. Then we have

$$Y_p(\omega) = \frac{1}{Z_{in}^A(\omega)} - \frac{1}{R_0 + j\omega L_0}, \tag{10.3}$$

where Z_{in}^A is the input impedance measured in air.

We assume that Y_p is unchanged in the presence of the workpiece. Therefore, in order to calculate $Z_W(\omega)$, we simply subtract out everything that we know about

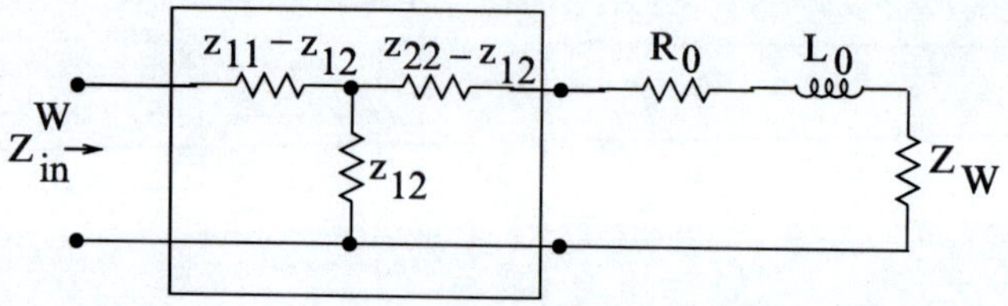

Fig. 10.16 The most general circuit representation of a coil and its connections

the coil and its impedance in air: $\dfrac{1}{R_0 + j\omega L_0 + Z_W(\omega)} = \dfrac{1}{Z_{in}^W(\omega)} - Y_p(\omega)$, which after rearrangement becomes

$$Z_W(\omega) = \frac{Z_{in}^W(\omega)}{1 - Z_{in}^W(\omega)Y_p(\omega)} - R_0 - j\omega L_0\,, \tag{10.4}$$

and Z_{in}^W is the input impedance measured over the workpiece.

The most general equivalent circuit of a coil and its connections includes the two-port network shown in Fig. 10.16. A two-port is defined by its open-circuit driving-point and transfer impedances, $z_{11},\ z_{22},\ z_{21} = z_{12}$, where the latter follows if the two-port is reciprocal. Clearly, the parallel admittance configuration shown in Fig. 10.15 is a special case of Fig. 10.16 when $z_{11} = z_{22} = z_{12} = 1/Y_p$.

With the two-port loaded as shown in Fig. 10.16, the driving-point impedance when the coil is over the workpiece is

$$Z_{in}^W(\omega) = z_{11} - z_{12} + \frac{z_{12}\,(z_{22} - z_{12} + R_0 + j\omega L_0 + Z_W)}{z_{22} + R_0 + j\omega L_0 + Z_W}\,, \tag{10.5}$$

which yields an expression for the change in impedance due to the workpiece:

$$Z_W(\omega) = \frac{z_{12}^2(\omega) - z_{11}(\omega)z_{22}(\omega) + Z_{in}^W(\omega)z_{22}(\omega)}{z_{11}(\omega) - Z_{in}^W(\omega)} - R_0 - j\omega L_0\,. \tag{10.6}$$

If $z_{11} = z_{22} = z_{12} = 1/Y_p$ then (10.6) gives the same result as (10.4).

In order to determine $Y_p(\omega)$ we needed only one measurement in air. Now, we need two additional independent measurements to determine the other two independent impedances of the two-port. These are obtained by measuring $Z_{in}^W(\omega)$ when the coil is over two known workpieces, then using **VIC-3D®** to compute Z_W for these workpieces and finally computing the remaining network parameters from (10.5).

Fig. 10.17 Freespace impedance of a 21.6 mm single-turn loop

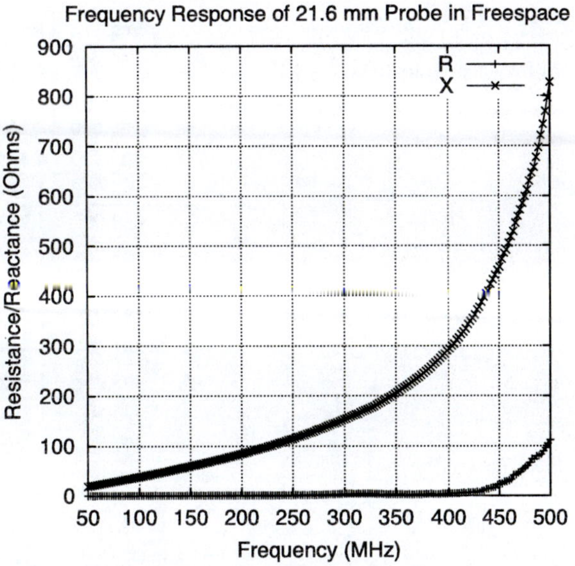

10.7.1 *Application to the 21.6 mm Single-Turn Loop*

Figure 10.17 shows the freespace impedance of a 21.6 mm single-turn loop, measured using the HP8720ET over the frequency range of 50–500 MHz.

The inductance at 50 MHz is 0.0595 μH, and at 68 MHz it is 0.0599 μH, which differ by less than 1%. Hence, we can call 50–68 MHz the 'low-frequency' range for this coil, in that the inductance remains reasonably constant, unaffected by resonance or other effects due to Y_p. Thus, $L_0 = 0.0595 \times 10^{-6}$ in Fig. 10.15. The average of the resistances over the low-frequency range is 0.121 Ω, which we will take to be R_0 in Fig. 10.15. These are the data that we need in order to compute Y_p using (10.3).

Table 10.15 lists the results for a few of the lower frequencies and the highest frequencies. Clearly, Y_p is capacitive because its imaginary part is positive, and it is also lossy, because its real part is positive. It is not important that either the conductance or capacitance remain fixed with frequency, because we do not intend to synthesize Y_p with circuit elements, but it is interesting to note that at the highest frequencies the value of the capacitor remains stable at about 1.32 pF.

10.8 Measuring the Frequency Response of Saline

Using the results of the Y_p calculation in (10.4), we compute $Z_W(\omega)$ when the 21.6 mm single-turn coil is placed above a bag of saline (Fig. 10.18). Impedance data were taken using the same configuration as above with the HP8720ET network

Table 10.15 Y_p for the 21.6 mm single-turn loop. The frequency is in MHz

Freq.	Re	Im
59.000	7.68299869E-05	8.89377407E-05
61.250	6.22287553E-05	1.93887259E-04
63.500	2.16324188E-05	2.83220172E-04
65.750	4.39594660E-05	1.97137473E-04
68.000	1.19457676E-04	2.75384256E-04
482.00	1.71300489E-04	4.02158825E-03
484.25	1.64257421E-04	4.02043480E-03
486.50	1.66601327E-04	4.04098304E-03
488.75	1.59304051E-04	4.05691052E-03
491.00	1.54539463E-04	4.07911092E-03
493.25	1.56246097E-04	4.09630826E-03
495.50	1.56481488E-04	4.11793403E-03
497.75	1.59386982E-04	4.13913047E-03
500.00	1.57347269E-04	4.16233204E-03

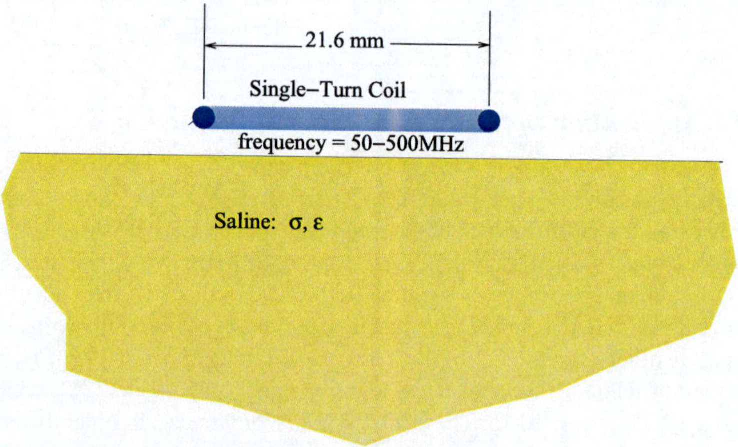

Fig. 10.18 Illustrating a single-turn coil over a saline half-space

analyzer. These data comprise the input impedance, Z_{in}^{W}, of (10.4), but are not listed here. The result of the computation is shown in Fig. 10.19. Note that there are two resonances in the figure, one at about 260 MHz and the other at about 460 MHz. These resonances are not present in the freespace probe response, so they are a property of the saline, but we don't yet know their origin.

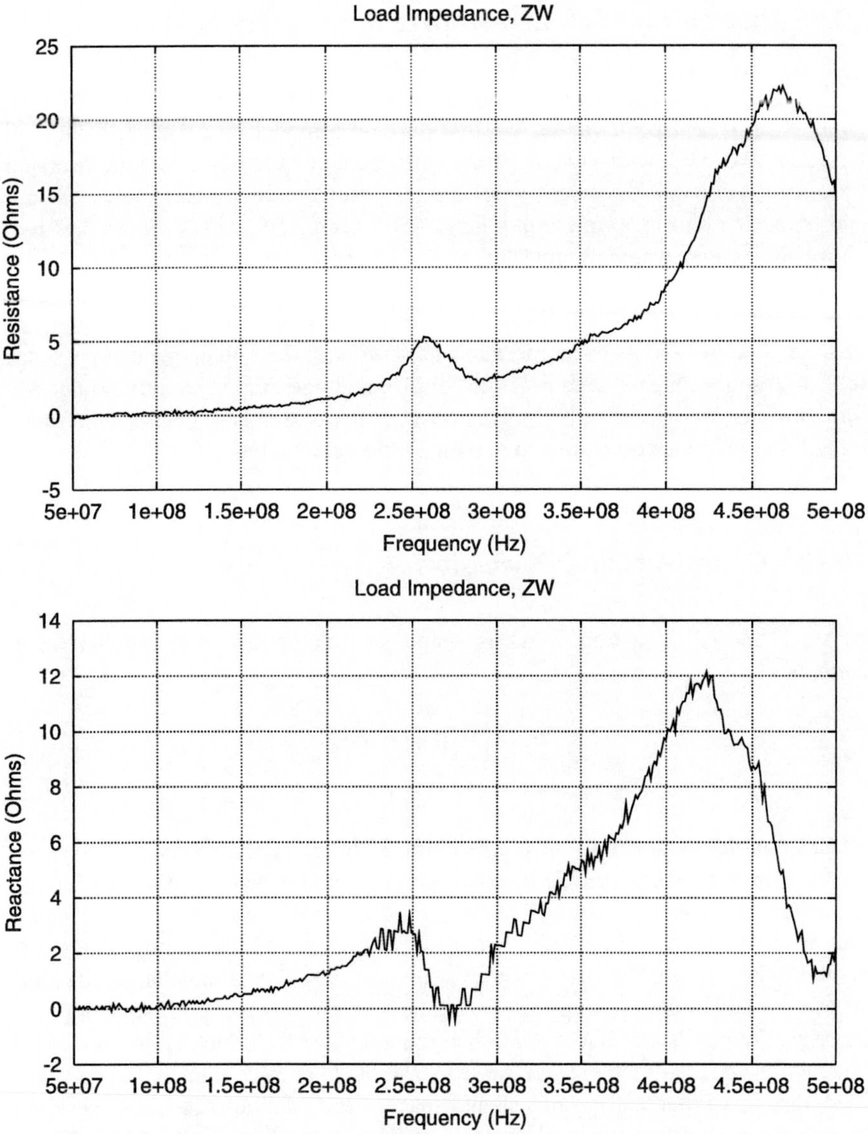

Fig. 10.19 Frequency response of Z_W for the 21.6 mm single-turn loop when placed above a bag of saline. Upper: resistance; lower: reactance

10.9 Determining the Constitutive Parameters of Saline

With the results of Fig. 10.19, we can use **VIC-3D®** to determine the constitutive parameters, σ and ϵ, of saline. We first develop a model of the experiment, and then run **VIC-3D®** with various values of these parameters in a simple inversion process. We model the current loop as a single-turn coil of inner radius 8.8 mm and an outer radius 8.9 mm, with a height of 0.1 mm. The coil is placed 0.55 mm above the saline, which is modeled as a half-space. We then applied NLSE to determine the best values of the electromagnetic parameters to be $\sigma = 1.016$ S/m, $\epsilon = 219.79\epsilon_0$. Figure 10.20 shows the result of this inversion. The result suggests that there are two independent mechanisms at work in the saline, one that produces the two resonances that were described above, and one that apparently produces a third resonance beyond 500 MHz. We now want to discuss these phenomena from a circuit-theoretic viewpoint, mixed in with a little field analysis.

10.10 Comments and Discussion

Maxwell's second equation, which is Ampere's circuital law in differential vector form, is

$$\begin{aligned}\nabla \times \mathbf{H} &= \sigma \mathbf{E} + j\omega\epsilon\mathbf{E} \\ &= \mathbf{J}_c + \mathbf{J}_d\ ,\end{aligned} \tag{10.7}$$

where σ is the conductivity of the medium, ϵ the dielectric constant, **J** the conduction current density, and $\mathbf{J}_d$ the displacement current density. When we induce a current into a workpiece for Eddy-current NDE, we can develop an analogous equivalent circuit, as in Fig. 10.21, to help understand what is happening. This figure shows a simple coupled circuit whose secondary has a load impedance, Z_L, and whose primary is the exciting coil. When the workpiece is weakly conducting, and has a significant dielectric constant, then the total current, which is the sum of the conduction and displacement currents, can be thought of as flowing in the equivalent load impedance that consists of a shunt capacitor and conductance. These elements represent, respectively, the displacement current and conduction current of (10.7).

The equations for the circuit of Fig. 10.21 are:

$$\begin{aligned}E_0 &= I_0(R_0 + j\omega L_0) - j\omega M I_1 \\ 0 &= -j\omega M I_0 + I_1(j\omega L_1 + Z_L)\ ,\end{aligned} \tag{10.8}$$

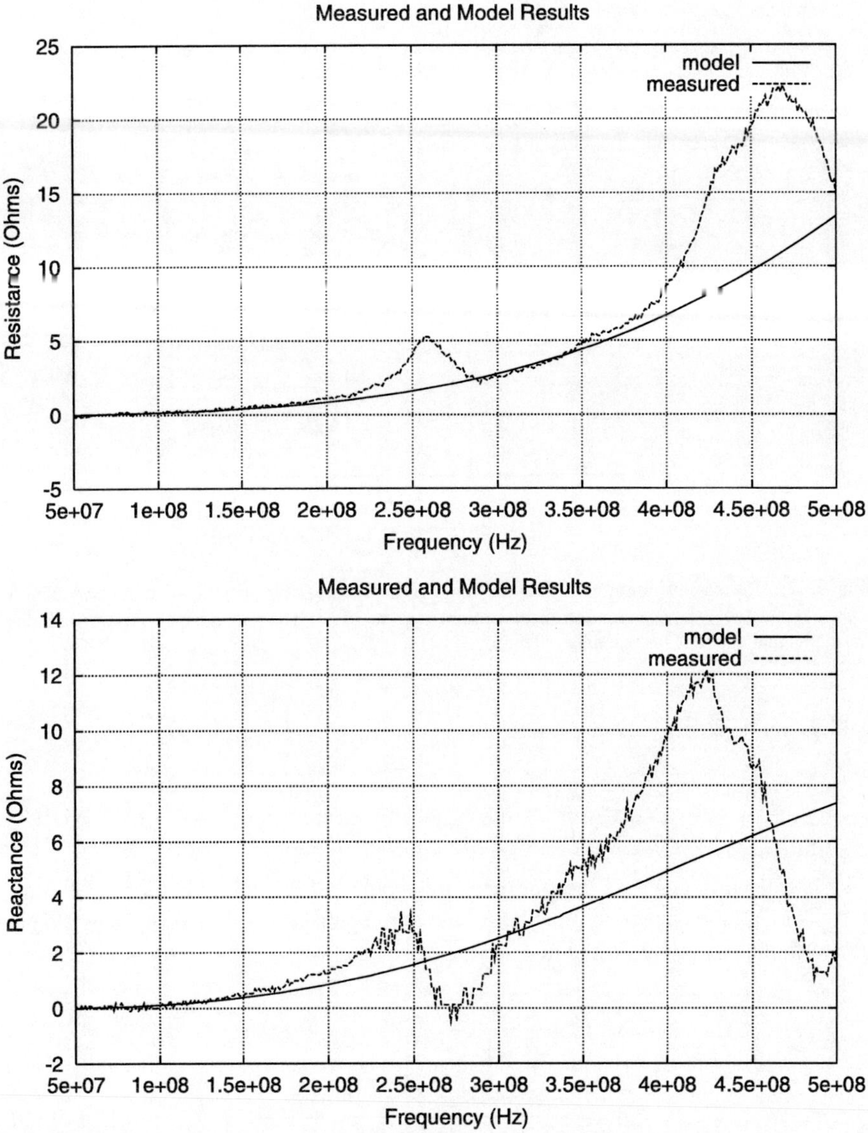

Fig. 10.20 Result of inversion of measured impedance data of saline. The resulting values are $\sigma = 1.016\,\text{S/m}$, $\epsilon = 219.79\epsilon_0$. Upper: resistance; lower: reactance

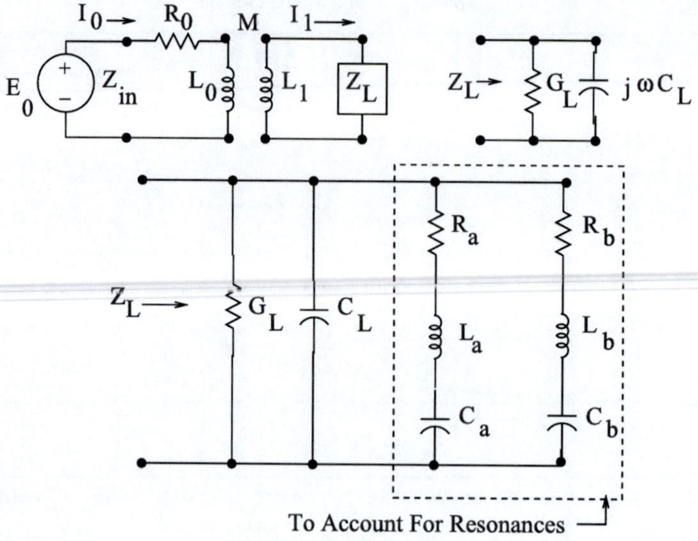

Fig. 10.21 Equivalent circuit for Eddy-current NDE. The resistor accounts for the conduction current, and the capacitor for the displacement current of (). The second load circuit includes elements that account for resonances (if they exist)

which can be easily solved to yield

$$Z_{in} = \frac{E_0}{I_0} = R_0 + j\omega L_0 + \frac{\omega^2 M^2}{j\omega L_1 + Z_L}. \tag{10.9}$$

The first term in (10.9) is the freespace impedance of the coil, whereas the second term is δZ_W, the change in impedance due to the presence of the workpiece. This is the term that we're interested in:

$$\begin{aligned} \delta Z_W &= \frac{\omega^2 M^2}{j\omega L_1 + Z_L} \\ &= \frac{\omega^2 M^2 \left(Z_L^* - j\omega L_1\right)}{|Z_L|^2 + \omega^2 L_1^2 + 2\Re(j\omega L_1 Z_L^*)}, \end{aligned} \tag{10.10}$$

where * denotes the complex conjugate, and $\Re$ denotes the 'real part' of a complex number.

If, as in the upper-right of Fig. 10.21, $Z_L = 1/(G_L + j\omega C_L)$, then

$$\delta Z_W = \frac{\omega^2 M^2 \left[G_L + j\omega C_L \left(1 - Z_0^2 G_L^2 - (\omega/\omega_0)^2\right)\right]}{1 - (\omega/\omega_0)^2 \left(2 - Z_0^2 G_L^2\right) + (\omega^2/\omega_0^2)^2}, \tag{10.11}$$

where $Z_0^2 = L_1/C_L$ is the 'characteristic impedance' and $\omega_0^2 = 1/L_1C_L$ is the resonant frequency of the parallel tank circuit comprising L_1, G_L, and C_L. This result, which is plotted as 'model' in Fig. 10.20, indicates that the displacement current in the host (saline) will eventually resonate with the virtual inductance, L_1, as long as G_L isn't too large. Of course, in a metal G_L is huge, so that the load consists of L_1 and G_L, and there is no resonance. In this case, we can set $\omega_0 = \infty$, and ignore all terms involving ω/ω_0. The result is that

$$\begin{aligned}\delta Z_W &\approx \omega^2 M^2 \left[G_L + j\omega(C_L - L_1 G_L^2)\right] \\ &\approx \omega^2 M^2 G_L(1 - j\omega L_1 G_L) . \end{aligned} \tag{10.12}$$

This shows that in the usual situation when inspecting a metallic workpiece, the change in reactance is always negative, which is an obvious manifestation of Lenz' law, that states that the induced current always produces a flux to oppose the incident flux due to L_0.

Now, let's consider the case in which $\omega L_1 << |Z_L|$, which means that we are well below the resonant frequency, ω_0. Then, from (10.10) we have

$$\begin{aligned}\delta Z_W &\approx \frac{\omega^2 M^2}{Z_L} \\ &= \omega^2 M^2 \left[G_L + j\omega C_L + \frac{\omega^2 R_a C_a^2 + j\omega C_a(1 - \omega^2/\omega_a^2)}{(1-\omega^2/\omega_a^2)^2 + \omega^2 R_a^2 C_a^2} + \frac{\omega^2 R_b C_b^2 + j\omega C_b(1 - \omega^2/\omega_b^2)}{(1-\omega^2/\omega_b^2)^2 + \omega^2 R_b^2 C_b^2}\right] ,\end{aligned} \tag{10.13}$$

where $\omega_a^2 = 1/L_aC_a$, $\omega_b^2 = 1/L_bC_b$ are the resonant frequencies of the two series-resonant branches in Fig. 10.21. If we are well below either of these two resonances, then

$$\delta Z_W \approx \omega^2 M^2 (G_L + j\omega C_L) , \tag{10.14}$$

from which we conclude that at low frequencies, the change in resistance is quadratic in ω, and the change in reactance is cubic, and both changes are positive, exactly as in Fig. 10.20 'model.' Thus, at low frequencies, we see only the effects of the conduction current, as manifested in G_L, and the displacement current, as manifested in C_L. This agrees with the results of the saline experiment in the low-frequency range of 50–150 MHz. The first (small) resonance in the saline experiment occurs around 260 MHz, and the second resonance, which is much larger, occurs around 460 MHz.

When we are at resonance, however, such that $\omega = \omega_a$, or $\omega = \omega_b$, then if these frequencies are reasonably well-separated, (10.13) indicates that the reactance is given simply by $j\omega_a C_L$ or $j\omega_b C_L$, which are the same as if the resonances were not present. Because this appears to closely follow the model and measured results of Fig. 10.20, we conclude that Fig. 10.21 is a reasonably faithful model of the saline experiment. We cannot yet explain the origin of the two resonances, but we are satisfied that they are not due to cavity resonances, because there are no finite boundaries in the model.

Now, when we consider frequencies beyond the second resonance (which we assume to be the last), which is the end of our experimental data, we return to the condition of (10.11). We can only model this condition with **VIC-3D®**, since we have no experimental data for it, and we do that spanning the frequency range 500 MHz to 1 GHZ, with the results shown in Fig. 10.22. We used the same values of $\sigma = 1.016\,\mathrm{S/m}$, $\epsilon = 219.79\epsilon_0$ for the computation. Note that the response is as predicted in (10.11); the resistance continues to increase, but the reactance reaches its peak at about 640 MHz, and then descends through zero at roughly 875 MHz. From (10.11), therefore, we estimate $\omega_0 \approx 875$ MHz.

Eventually, of course, the resistance will go to zero, but the frequency at which this occurs may be so high that the coil may actually begin to radiate, and the equivalent circuit of Fig. 10.21 will no longer be valid. In fact, a careful look at the reactance curve of Fig. 10.22 will show a slight 'glitch' at 620 MHz and a more pronounced one at about 1 GHz. There is a similar glitch in the resistance curve at about 1 GHZ. These indicate that the circuit model may be breaking down in this frequency range.

Finally, we show corresponding results when the saline is assumed to lie within a slab of 38 mm (1.5 in) in Fig. 10.23. The results are virtually indistinguishable from those of a halfspace over the frequency range 50–500 MHz, for which we have experimental data. Beyond 500 MHz, however, there are little wiggles in the computed response, which we believe are actuall cavity resonances that are strongly damped due to the large conductivity of the saline. In any case, the qualitative shape is the same as for the halfspace, so we will continue to rely on the halfspace model, and the equivalent circuit of Fig. 10.21 to anlalyze the results of this experiment.

10.10.1 Summary

The circuit diagram of Fig. 10.21 has four independent parameters: the mutual inductance, M, the virtual secondary inductance, L_1, and the load elements, G_L and C_L. From a field-theoretic viewpoint, which is what **VIC-3D®** looks at, there are also four independent parameters: the lift-off of the coil over the workpiece, the depth of the workpiece, and the electromagnetic parameters of the saline, σ and ϵ. By matching the **VIC-3D®**-model with the measured data, we have found that the lift-off is 0.55 mm, the workpiece extends to infinity in depth, i.e., that it is indeed a half-space, and that $\sigma = 1.016\,\mathrm{S/m}$, $\epsilon = 219.79\epsilon_0$. The reason that we can easily move between circuit models and field models is that the measurable is impedance, which is equally at home in each model.

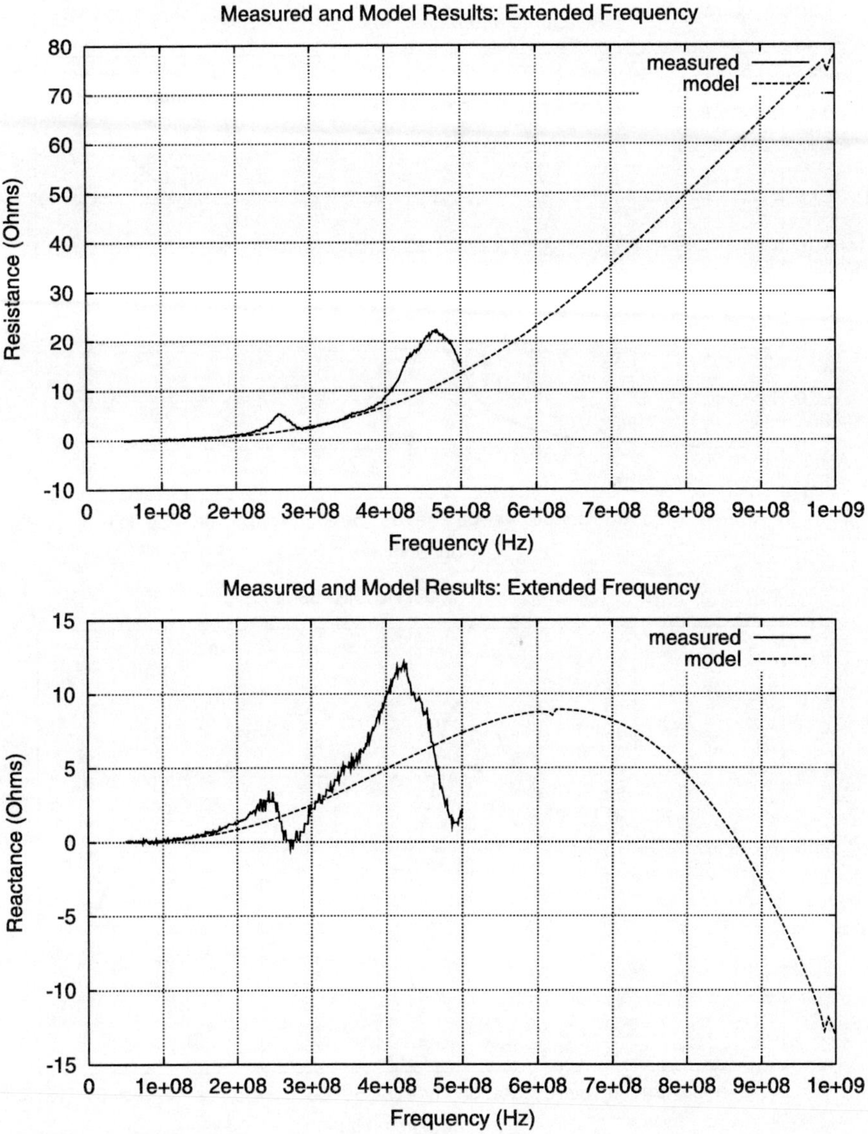

Fig. 10.22 Extending the model results of Fig. 10.20 to 1 GHz. The same values of $\sigma = 1.016\,\mathrm{S/m}$, $\epsilon = 219.79\epsilon_0$ are used in computing the response from 500 MHz to 1 GHz. Upper: resistance; lower: reactance

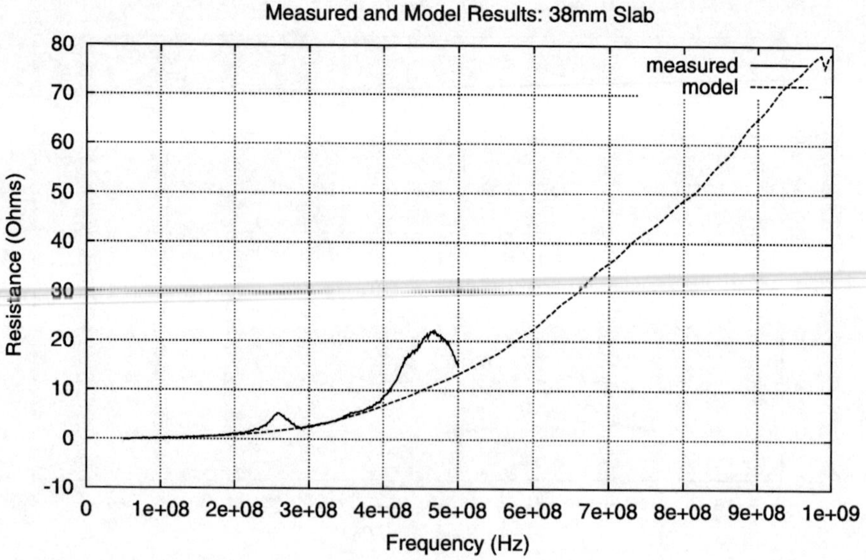

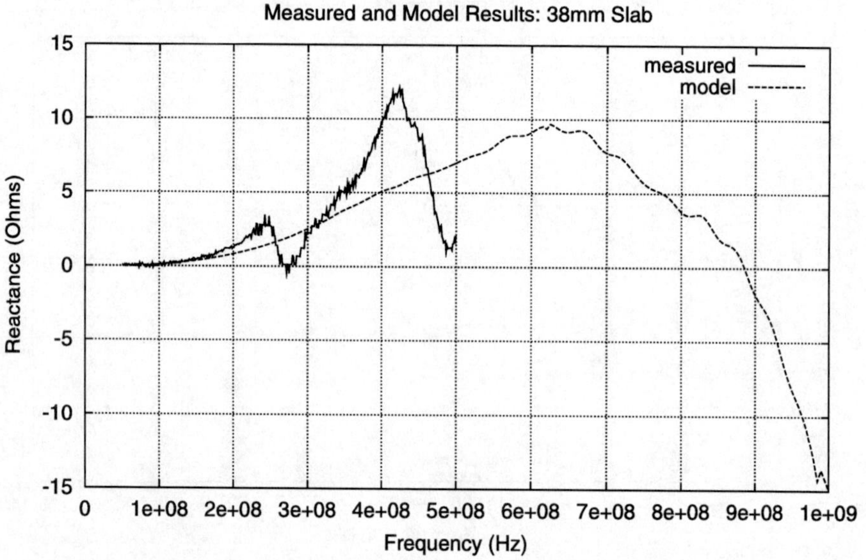

Fig. 10.23 Comparing model and measured results when the half-space is replaced by a slab that is 38 mm high. The same values of $\sigma = 1.016\,\mathrm{S/m}$, $\epsilon = 219.79\epsilon_0$ are used in computing the model response. Upper: resistance; lower: reactance

Appendix: The Levenberg–Marquardt Parameter in Least-Squares Problems

The objective function to be minimized is

$$\Phi(\mathbf{p}|\mathbf{Z}) = |\mathbf{Z} - \mathbf{g}(\mathbf{p})|^2 + \lambda|\mathbf{p} - \overline{\mathbf{p}}|^2 , \tag{10.15}$$

where λ is the Levenberg–Marquardt parameter. It stabilizes the solution by forcing it to remain close to $\overline{\mathbf{p}}$, which is assumed to be known a priori. The degree of 'closeness' is determined by the magnitude of λ. The Levenberg–Marquardt parameter is introduced in [111, p. 263].

Part III
Quantum Effects

Chapter 11
Spintronics

11.1 Introduction

The materials that have been considered thus far in the book are 'classical,' in the sense that their electromagnetic properties are easily stated as parameters that can be easily measured in the laboratory, or can be computed and understood using 'classical' physics. This is true whether the materials are 'structural' or 'biological'. Furthermore, we have shown that materials of these two classes can be characterized using the same classical electromagnetic models.

In order to expand our understanding and application of electromagnetic models to materials, we must consider cases in which the above statements do not hold, and one must resort to more sophisticated physical models that incorporate quantum mechanical principles just to understand the interaction of the electromagnetic field with the material. There are a number of common and novel materials in which this is true. For example, the interaction of an electromagnetic field in a microwave solid-state maser can only be understood through the application of the quantum theory of paramagnetism and electron spin dynamics [117]. Another well-known example is nuclear magnetic resonance (NMR), in which the spin of the proton in the nucleus of atoms provides the interaction that leads to magnetic resonance imaging (MRI).

11.2 Paramagnetic Spin Dynamics and the Spin Hamiltonian

In order to fully understand the possibilities of using paramagnetic phenomena to detect lesions noninvasively, we must review a bit of electron-spin physics. Our interest is in the dynamic response of spins to time-varying fields. These fields are

H. A. Sabbagh et al., *Advanced Electromagnetic Models for Materials Characterization and Nondestructive Evaluation*, Scientific Computation,
https://doi.org/10.1007/978-3-030-67956-9_11

either applied electromagnetic fields or fluctuating fields due to random vibrations of the crystalline surroundings of the spin system.[1]

The system of equations used to describe spin dynamics is derived from Schrödinger's wave equation of quantum mechanics, and is given by

$$\frac{d\rho_{mn}}{dt} = \frac{j}{\hbar}\sum_{k}\left(\rho_{mk}\mathscr{H}_{kn} - \mathscr{H}_{mk}\rho_{kn}\right) + \sum_{pq} R_{mn,pq}\left(\rho_{pq} - \rho_{pq}^{(T)}\right), \quad (11.1)$$

where ρ_{mn} is the density matrix connecting energy states u_m and u_n of the unperturbed system, $R_{mn,pq}$ are real numbers that account for spin-lattice relaxation, and the superscript, T, denotes the thermal equilibrium density matrix. $\mathscr{H}_{jk} = \mathscr{H}_{0jk} + \mathscr{H}_{1jk}(t)$, where $\mathscr{H}_{0jk}$ is the unperturbed, time-independent spin-Hamiltonian associated with the crystalline field, and $\mathscr{H}_{1jk}(t) = gh\beta\,[\mathbf{H}(t)\cdot\mathbf{S}]_{jk}$ is the time-dependent perturbation. Here g is a constant, $h\beta$ the Bohr magneton, h Planck's constant, $\hbar = h/2\pi$, $\mathbf{H}(t)$ the time-dependent (rf) magnetic field, and $\mathbf{S} = S_x\mathbf{a}_x + S_y\mathbf{a}_y + S_z\mathbf{a}_z$ is the vector spin operator.

Because $\{u_m\}$ is an orthonormal system of eigenstates of $\mathscr{H}_0$, it follows immediately that $\mathscr{H}_{0mm} = E_m$, and all off-diagonal elements of $\mathscr{H}_{0mn}$ vanish. Furthermore, in order to get a linear (i.e., first-order in $\mathbf{H}(t)$) response for the overall system, we must set the diagonal terms of (11.1) to their thermal equilibrium values, $\rho_{mn}(t) = \rho_{mn}^{(T)}$, and solve the off-diagonal terms to first-order in $\mathbf{H}(t)$:

$$\frac{d\rho_{mn}}{dt} = \left(j\omega_{0mn} - \frac{1}{\tau_{mn}}\right)\rho_{mn} + \frac{j}{\hbar}\left(\rho_{mm}^{(T)} - \rho_{nn}^{(T)}\right)\mathscr{H}_{1mn}(t), \quad (11.2)$$

where $\omega_{0mn} = \dfrac{E_n - E_m}{\hbar}$, and the relaxation times, τ_{mn}, replace the $R_{mn,pq}$ of (11.1).

For a sinusoidally time-varying field, we have $\mathscr{H}_{1mn}(t) = \dfrac{g\beta h}{2}\left(\mathbf{H}e^{j\omega t} + \mathbf{H}^*e^{-j\omega t}\right)\cdot\mathbf{S}_{mn}$. If we assume solutions of (11.2) of the form $\rho_{mn} = A_{mn}e^{j\omega t} + B_{mn}e^{-j\omega t}$, then the coefficients of the positive-frequency terms, A_{mn}, and negative-frequency terms, B_{mn}, are given by

$$A_{mn} = \frac{(j/\hbar)\left(\rho_{mm}^{(T)} - \rho_{nn}^{(T)}\right)\tau_{mn}g\beta h/2}{1 - j\left(-\omega + \omega_{0mn}\right)\tau_{mn}}\mathbf{S}_{mn}\cdot\mathbf{H}$$

$$B_{mn} = \frac{(j/\hbar)\left(\rho_{mm}^{(T)} - \rho_{nn}^{(T)}\right)\tau_{mn}g\beta h/2}{1 - j\left(\omega + \omega_{0mn}\right)\tau_{mn}}\mathbf{S}_{mn}\cdot\mathbf{H}. \quad (11.3)$$

[1]This discussion follows [96], which deals with spin dynamics in the crystalline field of a solid-state maser. Later we will discuss the changes that occur when the spin system is in a noncrystalline environment, such as biological tissue.

The magnetic dipole-moment operator for each spin is $g\beta h\mathbf{S}$, which means that the average dipole-moment for each spin is $\mathbf{m} = \mathrm{Tr}\,[\rho g\beta h\mathbf{S}]$, where Tr is the trace of an operator (sum of the diagonal elements of its matrix representation). The macroscopic dipole-moment per unit volume, $\mathbf{M}$, is obtained by multiplying $\mathbf{m}$ by the number density, N, of spins. Upon evaluating the trace, we find

$$\mathbf{M} = \gamma^2 \sum_{j<k} \mathbf{S}_{kj}\mathbf{S}_{jk}\left(N_j^{(T)} - N_k^{(T)}\right)$$

$$\tau_{jk}\left[\left(\frac{j/\hbar}{1 - j(\omega_{0jk} - \omega)\tau_{jk}} - \frac{j/\hbar}{1 + j(\omega_{0jk} + \omega)\tau_{jk}}\right)\mathbf{H}e^{j\omega t} + \left(\frac{-j/\hbar}{1 + j(\omega_{0jk} - \omega)\tau_{jk}} - \frac{-j/\hbar}{1 - j(\omega_{0jk} + \omega)\tau_{jk}}\right)\mathbf{H}^* e^{-j\omega t}\right], \quad (11.4)$$

where we have discarded the time-independent static dipole terms, $\mathbf{S}_{mm}\rho_{mm}^{(T)}$, and have set $\gamma^2 = g^2h^2\beta^2$. $N_j^{(T)}$ is the number of spins per-unit-volume occupying the jth energy level when the system is in thermal equilibrium at temperature T. If N is the total number of spins (or systems) in the crystal, then $N_j^{(T)} = \dfrac{N}{Z}\exp(-E_j/kT)$, where $Z = \sum_{j=1}^{J}\exp(-E_j/kT)$ and J is the total number of energy states. Thus, at thermal equilibrium (at positive temperatures), the lower energy states are more densely populated than the higher energy states.

The absorption spectrum, $A(\omega)$, is given by μ_0 times the imaginary part of the generalized magnetic susceptibility, which is the coefficient of $\mathbf{H}e^{j\omega t}$ in (11.4). In the vicinity of the resonant frequency, ω_{0jk}, the absorption spectrum is

$$A(\omega) \approx \mu_0\frac{\gamma^2}{2}\sum_{j<k}|S_{kj}|^2\left(N_j^{(T)} - N_k^{(T)}\right)\frac{\tau_{jk}/\hbar}{1 + (\omega_{0jk} - \omega)^2\tau_{jk}^2}$$

$$= \mu_0\frac{\gamma^2}{2}\frac{N}{Z}\sum_{j<k}|S_{kj}|^2\left(e^{-E_j/kT} - e^{-E_k/kT}\right)\frac{\tau_{jk}/\hbar}{1 + (\omega_{0jk} - \omega)^2\tau_{jk}^2}. \quad (11.5)$$

This spectrum consists of 'lorentzian' curves (resonant curves) centered at the frequencies ω_{0jk}, with line-width $1/\tau_{jk}$. The peak of each resonance is proportional to τ_{jk}, and this gives us the familiar trade-off between bandwidth and magnitude of absorption (or magnitude of gain). The term $N_j^{(T)} - N_k^{(T)}$ yields the population difference per unit volume of the jth and kth energy levels when the system is in thermal equilibrium at temperature T. This population difference will be small if the energy differential, $E_k - E_j$, is small compared to the thermal energy, kT, as is the usual case for paramagnetic spin systems at normal temperatures. In addition to τ_{jk}, an important parameter is the 'line-strength', $|S_{kj}|^2$, or the transition matrix element connecting the jth and kth states. It determines the ease with which pump power is absorbed by the spins, or it determines the gain at signal frequencies.

11.2.1 Application to Fe^{3+} : TiO_2

The five unpaired electrons in Fe^{3+} are each in the 3d state, meaning that the ion is in an S-state, with a spin, $S = 5/2$. The total number of spin-states, therefore, is $N_s = 2S + 1 = 6$. The spin-Hamiltonian, $\mathscr{H}_0$, for the Fe^{3+} : TiO_2 complex is [96]

$$\mathscr{H}_0 = g\beta \mathbf{H}_0 \cdot \mathbf{S} + D\left(S_z^2 - 35/12\right) + E\left(S_x^2 - S_y^2\right) + (a/6)\left(S_x^4 + S_y^4 + S_z^4 - 707/16\right) \\ + (7/36)F\left(S_z^4 - (95/14)S_z^2 + 81/16\right) , \tag{11.6}$$

where the nominal values of the derived constants are $g = 2.0$, $D = 20.35\,\text{GHz}$, $E = 2.21\,\text{GHz}$, $a = 1.1\,\text{GHz}$, $F = -0.5\,\text{GHz}$, and H_0 is the dc magnetic field. S_x, S_y, and S_z are 6×6 Pauli spin-matrices. This Hamiltonian gives us frequency directly, rather than energy. The D term has axial symmetry (about the z-axis), and corresponds to the ion having an electric quadrupole moment, that is acted upon by the crystalline electric fields. The E term represents and additional nonaxially symmetric anisotropy in the xy plane, and corresponds to the ion's possessing an electric moment of higher order than quadrupolar. These are the main terms, as the size of D and E would suggest; the remaining terms are due to the fact that $S > 2$ and that the crystal symmetry is complicated. Clearly, these latter terms are less important, but must be included for completeness.

The eigenvalue equation that determines the unperturbed energy levels (or frequencies in this case) is

$$\mathscr{H}_0 u = E u , \tag{11.7}$$

and when this equation is solved as a function of $\mathbf{H}_0 = H\mathbf{a}_z$, we get the six curves shown in Fig. 11.1. The zero-field energies occur in pairs (Kramers' doublets), as is typical of a system with an odd number of electrons in an electric field (the crystalline field).

Consider the system at $H = 1.78$ kilogauss; the eigenvalues of $\mathscr{H}_0$ are

$$\begin{matrix} E_1 = -58.20 \times 10^9 h & E_2 = -54.15 \times 10^9 h & E_3 = -19.60 \times 10^9 h \\ E_4 = -5.64 \times 10^9 h & E_5 = 56.14 \times 10^9 h & E_6 = 81.05 \times 10^9 h \end{matrix} , \tag{11.8}$$

from which we derive the resonant frequencies (in GHz)

$$\begin{matrix} \omega_{012} = 4.05 & \omega_{023} = 34.55 & \omega_{034} = 13.96 & \omega_{045} = 61.78 & \omega_{056} = 24.91 \\ \omega_{013} = 38.60 & \omega_{024} = 48.51 & \omega_{035} = 75.74 & \omega_{046} = 86.69 & \\ \omega_{014} = 52.56 & \omega_{025} = 110.29 & \omega_{036} = 100.65 & & \\ \omega_{015} = 114.34 & \omega_{026} = 135.20 & & & \\ \omega_{016} = 139.25 & & & & \end{matrix} , \tag{11.9}$$

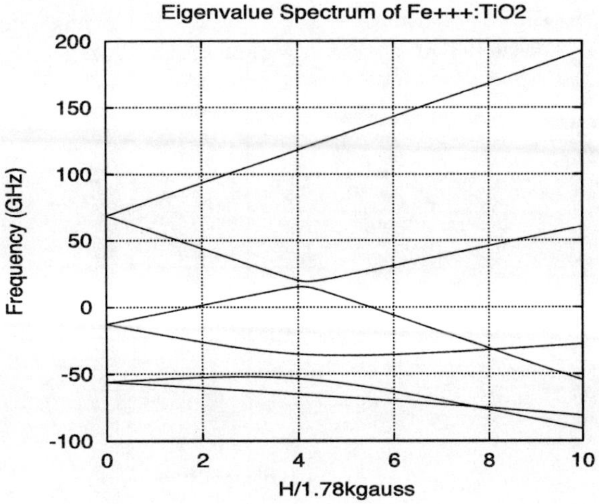

Fig. 11.1 Six-fold energy levels (in frequency units) for $Fe^{3+} : TiO_2$, as a function of the z-directed magnetic field, H

and the transition-matrix elements

$$
\begin{array}{lllll}
S_{x12} = 0.8 & S_{x23} = 0 & S_{x34} = -0.69 & S_{x45} = 0 & S_{x56} = 0 \\
S_{x13} = 1.66 & S_{x24} = -1.50 & S_{x35} = 1.08 & S_{x46} = 1.10 & \\
S_{x14} = 0 & S_{x25} = -0.31 & S_{x36} = 0 & & \\
S_{x15} = 0 & S_{x26} = 0 & & & \\
S_{x16} = 0 & & & & \\
 & & & & \\
S_{y12} = -j0.8 & S_{y23} = 0 & S_{y34} = -j0.53 & S_{y45} = 0 & S_{y56} = 0 \\
S_{y13} = -j1.66 & S_{y24} = -j1.02 & S_{y35} = -j1.08 & S_{y46} = j1.10 & \\
S_{y14} = 0 & S_{y25} = j0.31 & S_{y36} = 0 & & \\
S_{y15} = 0 & S_{y26} = 0 & & & \\
S_{y16} = j0.19 & & & & \\
 & & & & \\
S_{z12} = 0 & S_{z23} = 0.54 & S_{z34} = 0 & S_{z45} = 0 & S_{z56} = 0 \\
S_{z13} = 0 & S_{z24} = 0 & S_{z35} = 0 & S_{z46} = 0 & \\
S_{z14} = 0.33 & S_{z25} = 0 & S_{z36} = 0 & & \\
S_{z15} = 0 & S_{z26} = 0 & & & \\
S_{z16} = 0 & & & &
\end{array}
\qquad (11.10)
$$

From these results we can conclude, for example, that a transition between states 1 and 2 (4.05 GHz) has a 'strength' of $(0.8)^2$ for either x- or y-polarized radiation at that frequency, but cannot occur for z-polarized radiation. Similarly, we can answer the question of pump transitions. There are only two possible pump transitions, the 1–4 transition at 52.56 GHz and the 2–3 transition at 34.55 GHz, if one uses z-polarized pump radiation. If, however, we wish to amplify a signal at 4.05 GHz

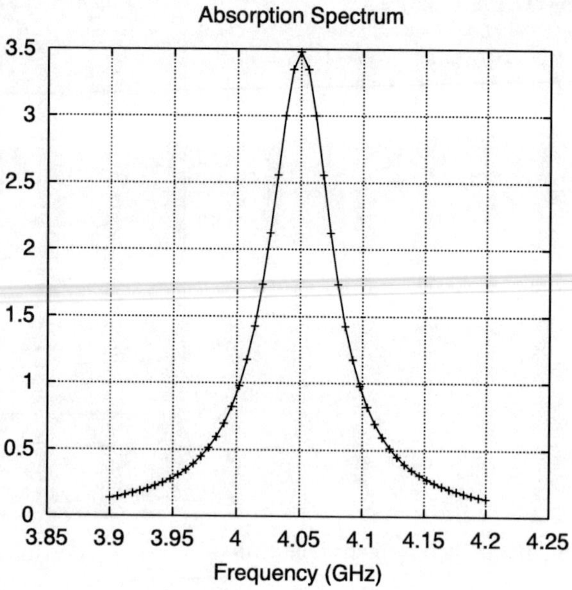

Fig. 11.2 Absorption spectrum in vicinity of 4.05 GHz, with $\tau_{12} = 5.305 \times 10^{-9}$ s. The half-power width is 60 MHz

(1–2 transition) we must pump between the first level and the third or higher level. Hence, we consider pumping only the 1–3 or 1–4 transition if we wish to remain below 100 GHz. The only possible 1–4 transition uses z-polarized radiation and has a strength of $(0.33)^2$. If we pump at 38.60 GHz (the 1–3 transition) we may use x- or y-polarized radiation (or both, as in circular polarization) and improve the absorption strength to $(1.66)^2$.

The width of the absorption curve for the 1–2 (4.05 GHz) transition of $Fe^{3+} : TiO_2$ is 60 MHz. Hence, the spin-lattice relaxation (or simply the *transverse relaxation*) time for the off-diagonal element, ρ_{12}, is $\tau_{12} = 1/2\pi \times 30 \times 10^6 = 5.305 \times 10^{-9}$ s. Figure 11.2 shows the absorption spectrum in the vicinity of 4.05 GHz with this value of τ_{12}.

This example illustrates the utility of the eigenstates in determining the frequency response of a maser. It relies, as we have noted, on knowledge of the crystalline-field environment of the iron ion. It is this information that is lacking when we consider electron-paramagnetic spin systems in biological tissue, and is the basis for one of our research proposals.

11.2.2 $Ho^{++} : CaF_2$

Holmium is a type 4f rare earth, which means that the divalent Holmium ion has its unpaired electrons in the 4f shell where they are effectively screened from their crystalline surroundings by electrons in the outer shells. Therefore, as a reasonable approximation to the effective spin-Hamiltonian we may discard any terms that represent the crystalline field. We must, however, include the spin-spin interaction between the unpaired electrons and the nucleus because these electrons are relatively close to the nucleus. Thus, we use the following spin-Hamiltonian

$$\mathscr{H}_0 = g\beta \mathbf{H}_0 \cdot \mathbf{S} + A\mathbf{I} \cdot \mathbf{S}\,, \tag{11.11}$$

where $g = 5.91$, $\beta = 0.0014$ GHz/gauss, $A = 3.924$ GHz, **S** is the electron spin operator, with effective spin 1/2, and **I** is the nuclear spin operator with spin 7/2.

Because we are dealing with a system of two particles (electron plus nucleus) we cannot simply form matrix products in order to evaluate $\mathscr{H}_0$, but must use the direct product of the appropriate Pauli spin matrices of **I** and **S**. Because there are two possible electron spin-states ("spin up" and "spin down" relative to, say, the axis of $\mathbf{H}_0$) and $2 \times 7/2 + 1 = 8$ possible spin states of the nucleus, we have a composite system of 16 possible states. This means that the combined spin-Hamiltonian, (11.11), will be represented by a 16×16 matrix. When this matrix is written out, and its eigenvalues determined as a function of magnetic field, we get the plot of Fig. 11.3.

A comparison of Figs. 11.1 and 11.3 shows that Ho^{++} has a much more uniform variation of energy (and, hence, resonant frequency) with H than does Fe^{3+}. This follows, as has been mentioned before, because the unpaired electrons in Ho^{++} are screened from the crystalline field, whereas those of Fe^{3+} are not. Hence, Ho^{2+}

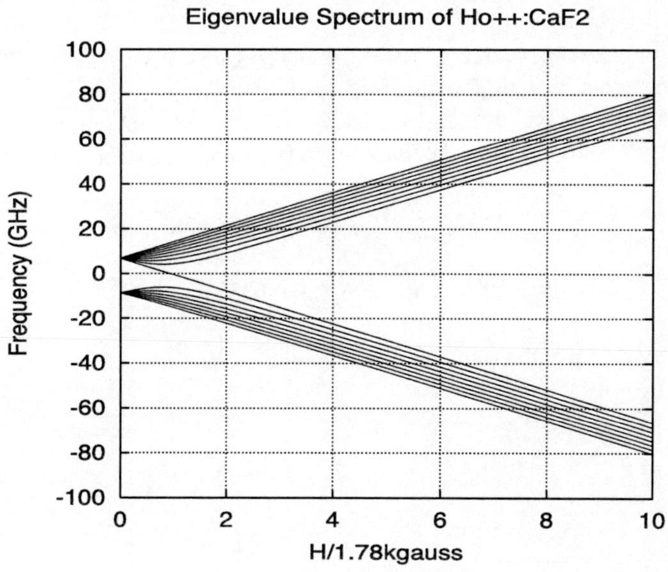

Fig. 11.3 Sixteen-fold energy levels (in frequency units) for Ho^{++} : CaF_2, as a function of the z-directed magnetic field, H

behaves, for H greater than 3 kilogauss, as a free spin in an external magnetic field. We would expect this same qualitative behavior for many of the 4f rare earths, no matter what the crystal lattice is. This suggests that if we use 4f rare-earth spin-systems, it may not be necessary to know anything about the electric-field environment of the biological tissue.

11.3 Superparamagnetic Iron Oxide

This is what started this discussion of paramagnetic spin-systems. Iron oxide, whether it is $\gamma - Fe_2O_3$, called 'maghemite,' or, perhaps magnetite, Fe_3O_4 ([95] is not clear on this), is ferromagnetic. Because of the small size of the particles (~10 nm), their ferromagnetic properties manifest themselves in a single domain, and such single domain particles can behave magnetically in a manner analogous to the paramagnetism of moment-bearing atoms [16]. The main distinction is that the moment of the particle may be 10^5 times the atomic moment, because of the 10^5 atoms ferromagnetically coupled by exchange forces within the single domain.[2] [3]

Two Spins We'll make a simple quantum-mechanical calculation of a system of two electrons coupled through the exchange interaction in a static magnetic field, $\mathbf{H}_0$. The Hamiltonian is

$$\mathscr{H} = -g\beta\mathbf{H}_0 \cdot \left(\mathbf{S}^{(1)} + \mathbf{S}^{(2)}\right) - 2J_{exch}\mathbf{S}^{(1)} \cdot \mathbf{S}^{(2)}\,, \tag{11.12}$$

where $g\beta = g \times 0.0014\,\text{GHz/gauss} = 2.8\,\text{GHz/kgauss}$, if we take $g = 2$. J_{exch} is the exchange energy, with a nominal value of 2.1×10^{-21} J. Dividing by Plancks constant, h, gives us the result in frequency units: $J_{exch}/h = 2.1 \times 10^{-21}/6.626 \times 10^{-34} = 3169.3\,\text{GHz}$. Hence, the normalized Hamiltonian for the system becomes

$$\begin{aligned}\mathscr{H} &= -2.8\mathbf{H}_0 \cdot \left(\mathbf{S}^{(1)} + \mathbf{S}^{(2)}\right) - 6338.7\mathbf{S}^{(1)} \cdot \mathbf{S}^{(2)} \\ &= -2.8H_0\left(S_z^{(1)} + S_z^{(2)}\right) - 6338.7\mathbf{S}^{(1)} \cdot \mathbf{S}^{(2)}\,,\end{aligned} \tag{11.13}$$

where we assume that the static field is along the z-direction.

[2]Additional References on Superparamagnetic and Ferromagnetic Effects: [6, 23, 25, 49, 67, 75, 76, 116, 119, 124, 136].

[3]By a 'single domain particle,' we mean a particle that is in a state of uniform magnetization at any magnetic field[16].

The Pauli spin-matrices for a single electron in a z-directed magnetic field are

$$s_x = \frac{1}{2}\begin{bmatrix} 0 & 1 \\ 1 & 0 \end{bmatrix} s_y = \frac{1}{2}\begin{bmatrix} 0 & -j \\ j & 0 \end{bmatrix} s_z = \frac{1}{2}\begin{bmatrix} 1 & 0 \\ 0 & -1 \end{bmatrix}, \tag{11.14}$$

and the eigenstates of s_z are $\begin{bmatrix} 1 \\ 0 \end{bmatrix}$, $\begin{bmatrix} 0 \\ 1 \end{bmatrix}$, with the first one corresponding to 'spin up' (parallel to the magnetic field), and the second to 'spin down' (antiparallel to the magnetic field).

Since we have two coupled spins, we must work in the four-dimensional direct-product space of the operators of (11.14):

$$S_x^{(1)} S_x^{(2)} = \frac{1}{4}\begin{bmatrix} 0 & 0 & 0 & 1 \\ 0 & 0 & 1 & 0 \\ 0 & 1 & 0 & 0 \\ 1 & 0 & 0 & 0 \end{bmatrix}$$

$$S_y^{(1)} S_y^{(2)} = \frac{1}{4}\begin{bmatrix} 0 & 0 & 0 & -1 \\ 0 & 0 & 1 & 0 \\ 0 & 1 & 0 & 0 \\ -1 & 0 & 0 & 0 \end{bmatrix}$$

$$S_z^{(1)} S_z^{(2)} = \frac{1}{4}\begin{bmatrix} 1 & 0 & 0 & 0 \\ 0 & -1 & 0 & 0 \\ 0 & 0 & -1 & 0 \\ 0 & 0 & 0 & 1 \end{bmatrix}$$

$$\mathbf{S}^{(1)} \cdot \mathbf{S}^{(2)} = S_x^{(1)} S_x^{(2)} + S_y^{(1)} S_y^{(2)} + S_z^{(1)} S_z^{(2)}$$

$$= \frac{1}{4}\begin{bmatrix} 1 & 0 & 0 & 0 \\ 0 & -1 & 2 & 0 \\ 0 & 2 & -1 & 0 \\ 0 & 0 & 0 & 1 \end{bmatrix}. \tag{11.15}$$

The four-dimensional representations of S_x, S_y, S_z are obtained by taking the left- and right-direct products of the single-electron Pauli spin-matrices, (11.14), with the two-dimensional identity matrix. The results are

$$S_x^{(1)} = \frac{1}{2}\begin{bmatrix} 0\,0\,1\,0 \\ 0\,0\,0\,1 \\ 1\,0\,0\,0 \\ 0\,1\,0\,0 \end{bmatrix} \quad S_y^{(1)} = \frac{1}{2}\begin{bmatrix} 0 & 0 & -j & 0 \\ 0 & 0 & 0 & -j \\ j & 0 & 0 & 0 \\ 0 & j & 0 & 0 \end{bmatrix} \quad S_z^{(1)} = \frac{1}{2}\begin{bmatrix} 1 & 0 & 0 & 0 \\ 0 & 1 & 0 & 0 \\ 0 & 0 & -1 & 0 \\ 0 & 0 & 0 & -1 \end{bmatrix}$$

$$S_x^{(2)} = \frac{1}{2}\begin{bmatrix} 0\,1\,0\,0 \\ 1\,0\,0\,0 \\ 0\,0\,0\,1 \\ 0\,0\,1\,0 \end{bmatrix} \quad S_y^{(2)} = \frac{1}{2}\begin{bmatrix} 0 & -j & 0 & 0 \\ j & 0 & 0 & 0 \\ 0 & 0 & 0 & -j \\ 0 & 0 & j & 0 \end{bmatrix} \quad S_z^{(2)} = \frac{1}{2}\begin{bmatrix} 1 & 0 & 0 & 0 \\ 0 & -1 & 0 & 0 \\ 0 & 0 & 1 & 0 \\ 0 & 0 & 0 & -1 \end{bmatrix} \tag{11.16}$$

for the two particles. Note that these two-particle spin matrices satisfy the general commutation relations $\left[S_x^{(p)}, S_y^{(q)}\right] = j\delta_{pq}S_z^{(p)}$. Note further that the product of these matrices gives the same results that we obtained independently in (11.15).

The eigenvectors of the matrix, $S_z^{(1)} + S_z^{(2)}$, in (11.16) are the direct products of the eigenstates of the two-dimensional Pauli spin-matrix, s_z:

$$\begin{bmatrix} 1 \\ 0 \\ 0 \\ 0 \end{bmatrix} \begin{bmatrix} 0 \\ 1 \\ 0 \\ 0 \end{bmatrix} \begin{bmatrix} 0 \\ 0 \\ 1 \\ 0 \end{bmatrix} \begin{bmatrix} 0 \\ 0 \\ 0 \\ 1 \end{bmatrix} \tag{11.17}$$

The first eigenvector in (11.17) corresponds to both spins in the 'up' position (both parallel to the magnetic field), the second to 'spin up; spin down,' the third to 'spin down; spin up,' and the fourth to 'spin down; spin down.'

With this background, we can now write down the matrix representation of the normalized spin-Hamiltonian (11.13):

$$\mathscr{H} = \begin{bmatrix} -2.8H_0 - 1584.7 & 0 & 0 & 0 \\ 0 & 1584.7 & -3169.4 & 0 \\ 0 & -3169.4 & 1584.7 & 0 \\ 0 & 0 & 0 & 2.8H_0 - 1584.7 \end{bmatrix} \tag{11.18}$$

The eigenspectrum of (11.18) is plotted as a function of H_0 in Fig. 11.4. The left-hand figure shows all four solutions, and the right-hand the bottom three eigenvalues. The two parallel branches have a constant separation of 6338.7, which is exactly $2J_{exch}$, where J_{exch} is the exchange energy. It is important to note that the transition (resonant) frequency between states 2 and 3 is the same between as between 3 and 4, for all values of H_0: $\omega_{0_{23}} = \omega_{0_{34}}$.

The eigenstates corresponding to the spectrum of Fig. 11.4 are, from the largest to the smallest eigenvalue (in magnitude):

$$
\begin{matrix} (1) & (2) & (3) & (4) \\ \begin{bmatrix} 0.0 \\ -0.707 \\ +0.707 \\ 0.0 \end{bmatrix} & \begin{bmatrix} 1.0 \\ 0.0 \\ 0.0 \\ 0.0 \end{bmatrix} & \begin{bmatrix} 0.0 \\ 0.707 \\ 0.707 \\ 0 \end{bmatrix} & \begin{bmatrix} 0.0 \\ 0.0 \\ 0.0 \\ 1.0 \end{bmatrix} \end{matrix}, \tag{11.19}
$$

where state 1 corresponds to the eigenvalue, 4754.1, in the left-hand part of Fig. 11.4, and the other eigenvalues are listed in the right-hand part of the figure.

Clearly, state 2 corresponds to both spins being parallel to the static magnetic field, because this gives the lowest energy level, whereas state 4 corresponds to both spins being anti-parallel to the field. States 1 and 3 correspond to linear combinations of spin 1 being parallel and spin 2 anti-parallel, and the converse. The higher-energy state involves a sum and difference of the parallel-anti-parallel combination, whereas the lower-energy state involves only the sum of two such combinations. We'll see this again in the three-electron calculation.

The transition matrix elements of the lowest three energy levels are

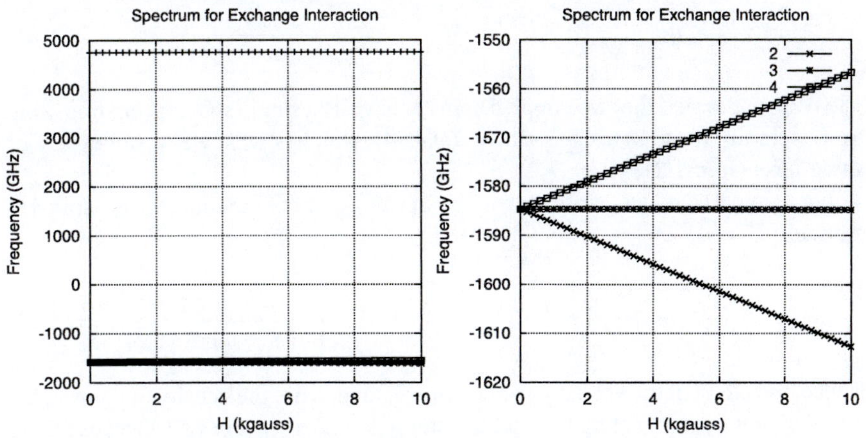

Fig. 11.4 Eigenspectrum of spin-Hamiltonian with exchange interaction. Left: complete spectrum. Right: expanded version of bottom three eigenvalues

$$
\begin{aligned}
S_{x23} &= u_2 \cdot \left(S_x^{(1)} + S_x^{(2)}\right) \cdot u_3 = \quad 0.707 \\
S_{x34} &= u_3 \cdot \left(S_x^{(1)} + S_x^{(2)}\right) \cdot u_4 = \quad 0.707 \\
S_{y23} &= u_2 \cdot \left(S_y^{(1)} + S_y^{(2)}\right) \cdot u_3 = -j0.707 \\
S_{y34} &= u_3 \cdot \left(S_y^{(1)} + S_y^{(2)}\right) \cdot u_4 = -j0.707 \\
S_{z23} &= u_2 \cdot \left(S_z^{(1)} + S_z^{(2)}\right) \cdot u_3 = \quad 0.0 \\
S_{z34} &= u_3 \cdot \left(S_z^{(1)} + S_z^{(2)}\right) \cdot u_4 = \quad 0.0
\end{aligned}, \tag{11.20}
$$

where the orthonormal eigenvectors, $\{u_i\}$, are given in (11.19), and the spin operators are given in (11.16). These results indicate that one cannot induce transitions by using z-directed AC magnetic fields, as we suspected, and that transitions are equally likely with x- or y-directed AC fields (or with circularly polarized AC fields).

When (11.20) is substituted into the expression, (11.5), for the absorption coefficient and use is made of the fact that $\omega_{0_{23}} = \omega_{0_{34}} = \omega_0$ and $\tau_{23} = \tau_{34} = \tau$, we get, after summing over the bottom three energy-states,

$$
A(\omega) = 2 \times \mu_0 \frac{\gamma^2}{4} \frac{1}{Z} \left(e^{-E_2/kT} - e^{-E_4/kT}\right) \frac{\tau/\hbar}{1 + (\omega_0 - \omega)^2 \tau^2} . \tag{11.21}
$$

The response is as if the two coupled spins behave as a single spin-system transiting from 'spin-up' (state 2) to 'spin-down' (state 4), which is what we would expect of a two-level (spin-1/2) system.

For comparison, we write down the result for two non-interacting spin-1/2 particles:

$$
A(\omega) = 2 \times \mu_0 \frac{\gamma^2}{4} \frac{1}{Z} \left(e^{-E_2/kT} - e^{-E_3/kT}\right) \frac{\tau/\hbar}{1 + (\omega_0 - \omega)^2 \tau^2} . \tag{11.22}
$$

Hence, the effect of the exchange interaction is to increase the density of spins in the thermal term by eliminating the middle energy term, $\exp(-E_3/kT)$. Because there is a greater differential in the energies than there was before, we have effectively a greater population difference between the two energy states 2 and 3 that are separated by $\hbar\omega_0$. Clearly, the more interacting spins we have, the greater this population difference becomes, and the greater the absorption spectrum becomes. Because of these two effects, with something of the order of 10^5 spins interacting through the exchange integral, the spectrum becomes significantly larger than in the simple paramagnetic case, giving rise to the name 'superparamagnetism.' We'll give a further example of this next.

Three Spins We'll extend the previous model to include three electrons interacting through the exchange integral. The Hamiltonian now becomes

$$\mathscr{H} = -2.8H_0\left(S_z^{(1)}+S_z^{(2)}+S_z^{(3)}\right)-6338.7\left(\mathbf{S}^{(1)}\cdot\mathbf{S}^{(2)}+\mathbf{S}^{(1)}\cdot\mathbf{S}^{(3)}+\mathbf{S}^{(2)}\cdot\mathbf{S}^{(3)}\right) \tag{11.23}$$

where the various three-particle spin-matrices are obtained by taking three-fold left- and right-direct products of the Pauli spin matrices, s, with the two-dimensional identity matrix, I_2:

$$\mathbf{S}^{(1)} = \mathbf{s}\otimes I_2\otimes I_2 \;\; \mathbf{S}^{(2)} = I_2\otimes\mathbf{s}\otimes I_2 \;\; \mathbf{S}^{(3)} = I_2\otimes I_2\otimes\mathbf{s}\ . \tag{11.24}$$

The results are:

$$\begin{gathered}
S_x^{(1)} = \frac{1}{2}\begin{bmatrix} 0_4 & I_4 \\ I_4 & 0_4 \end{bmatrix} \quad S_x^{(2)} = \begin{bmatrix} S_x^{(1)'} & 0_4 \\ 0_4 & S_x^{(1)'} \end{bmatrix} \quad S_x^{(3)} = \begin{bmatrix} S_x^{(2)'} & 0_4 \\ 0_4 & S_x^{(2)'} \end{bmatrix} \\
S_y^{(1)} = \frac{1}{2}\begin{bmatrix} 0_4 & -jI_4 \\ jI_4 & 0_4 \end{bmatrix} \quad S_y^{(2)} = \begin{bmatrix} S_y^{(1)'} & 0_4 \\ 0_4 & S_y^{(1)'} \end{bmatrix} \quad S_y^{(3)} = \begin{bmatrix} S_y^{(2)'} & 0_4 \\ 0_4 & S_y^{(2)'} \end{bmatrix} \\
S_z^{(1)} = \frac{1}{2}\begin{bmatrix} I_4 & 0_4 \\ 0_4 & -I_4 \end{bmatrix} \quad S_z^{(2)} = \begin{bmatrix} S_z^{(1)'} & 0_4 \\ 0_4 & S_z^{(1)'} \end{bmatrix} \quad S_z^{(3)} = \begin{bmatrix} S_z^{(2)'} & 0_4 \\ 0_4 & S_z^{(2)'} \end{bmatrix}
\end{gathered} \tag{11.25}$$

where the primed,', submatrices refer to the corresponding 4×4 matrices in (11.16), and I_4, 0_4 are the four-dimensional identity and null-matrices, respectively.

Using the results of (11.25), we easily compute the dot-product matrices:

$$\mathbf{S}^{(1)}\cdot\mathbf{S}^{(2)} = \frac{1}{4}\begin{bmatrix}
1 & 0 & 0 & 0 & 0 & 0 & 0 & 0 \\
0 & 1 & 0 & 0 & 0 & 0 & 0 & 0 \\
0 & 0 & -1 & 0 & 2 & 0 & 0 & 0 \\
0 & 0 & 0 & -1 & 0 & 2 & 0 & 0 \\
0 & 0 & 2 & 0 & -1 & 0 & 0 & 0 \\
0 & 0 & 0 & 2 & 0 & -1 & 0 & 0 \\
0 & 0 & 0 & 0 & 0 & 0 & 1 & 0 \\
0 & 0 & 0 & 0 & 0 & 0 & 0 & 1
\end{bmatrix}$$

$$\mathbf{S}^{(1)}\cdot\mathbf{S}^{(3)} = \frac{1}{4}\begin{bmatrix}
1 & 0 & 0 & 0 & 0 & 0 & 0 & 0 \\
0 & -1 & 0 & 0 & 2 & 0 & 0 & 0 \\
0 & 0 & 1 & 0 & 0 & 0 & 0 & 0 \\
0 & 0 & 0 & -1 & 0 & 0 & 2 & 0 \\
0 & 2 & 0 & 0 & -1 & 0 & 0 & 0 \\
0 & 0 & 0 & 0 & 0 & 1 & 0 & 0 \\
0 & 0 & 0 & 2 & 0 & 0 & -1 & 0 \\
0 & 0 & 0 & 0 & 0 & 0 & 0 & 1
\end{bmatrix}$$

$$\mathbf{S}^{(2)} \cdot \mathbf{S}^{(3)} = \frac{1}{4} \begin{bmatrix} 1 & 0 & 0 & 0 & 0 & 0 & 0 & 0 \\ 0 & -1 & 2 & 0 & 0 & 0 & 0 & 0 \\ 0 & 2 & -1 & 0 & 0 & 0 & 0 & 0 \\ 0 & 0 & 0 & 1 & 0 & 0 & 0 & 0 \\ 0 & 0 & 0 & 0 & 1 & 0 & 0 & 0 \\ 0 & 0 & 0 & 0 & 0 & -1 & 2 & 0 \\ 0 & 0 & 0 & 0 & 0 & 2 & -1 & 0 \\ 0 & 0 & 0 & 0 & 0 & 0 & 0 & 1 \end{bmatrix}, \tag{11.26}$$

from which we get the final expression for the Hamiltonian of (11.23):

$$\mathscr{H} = -2.8H_0 \begin{bmatrix} \frac{3}{2} & & & & & & & \\ & \frac{1}{2} & & & & & & \\ & & \frac{1}{2} & & & & & \\ & & & -\frac{1}{2} & & & & \\ & & & & \frac{1}{2} & & & \\ & & & & & -\frac{1}{2} & & \\ & & & & & & -\frac{1}{2} & \\ & & & & & & & -\frac{3}{2} \end{bmatrix}$$

$$-\frac{6338.7}{4} \begin{bmatrix} 3 & 0 & 0 & 0 & 0 & 0 & 0 & 0 \\ 0 & -1 & 2 & 0 & 2 & 0 & 0 & 0 \\ 0 & 2 & -1 & 0 & 2 & 0 & 0 & 0 \\ 0 & 0 & 0 & -1 & 0 & 2 & 2 & 0 \\ 0 & 2 & 2 & 0 & -1 & 0 & 0 & 0 \\ 0 & 0 & 0 & 2 & 0 & -1 & 2 & 0 \\ 0 & 0 & 0 & 2 & 0 & 2 & -1 & 0 \\ 0 & 0 & 0 & 0 & 0 & 0 & 0 & 3 \end{bmatrix}. \tag{11.27}$$

The diagonal matrix in (11.27) is the projection onto the $z-$axis (the magnetic field) of the combined system of particles. It's eigenvectors are

$$\begin{array}{cccccccc}
(3/2) & (1/2) & (1/2) & (1/2) & (-1/2) & (-1/2) & (-1/2) & (-3/2) \\
\begin{bmatrix}1.0\\0.0\\0.0\\0.0\\0.0\\0.0\\0.0\\0.0\end{bmatrix} &
\begin{bmatrix}0.0\\1.0\\0.0\\0.0\\0.0\\0.0\\0.0\\0.0\end{bmatrix} &
\begin{bmatrix}0.0\\0.0\\1.0\\0.0\\0.0\\0.0\\0.0\\0.0\end{bmatrix} &
\begin{bmatrix}0.0\\0.0\\0.0\\0.0\\1.0\\0.0\\0.0\\0.0\end{bmatrix} &
\begin{bmatrix}0.0\\0.0\\0.0\\1.0\\0.0\\0.0\\0.0\\0.0\end{bmatrix} &
\begin{bmatrix}0.0\\0.0\\0.0\\0.0\\0.0\\1.0\\0.0\\0.0\end{bmatrix} &
\begin{bmatrix}0.0\\0.0\\0.0\\0.0\\0.0\\0.0\\1.0\\0.0\end{bmatrix} &
\begin{bmatrix}0.0\\0.0\\0.0\\0.0\\0.0\\0.0\\0.0\\1.0\end{bmatrix} \\
(\bullet\,\bullet\,\bullet) & (\bullet\,\bullet\,\circ) & (\bullet\,\circ\,\bullet) & (\circ\,\bullet\,\bullet) & (\bullet\,\circ\,\circ) & (\circ\,\bullet\,\circ) & (\circ\,\circ\,\bullet) & (\circ\,\circ\,\circ)
\end{array}, \qquad (11.28)$$

where $\bullet$ denotes a spin-up state (parallel to the magnetic field), and $\circ$ denotes a spin-down state.

The eigenspectrum of (11.27), plotted as a function of the static magnetic field, is shown in Fig. 11.5. As is the case with the two-electron problem, the separation between the lowest energy levels is constant and equal to $2.8H_0$ GHz where H_0 is in kGauss. This is identical to the result for a single electron with a spin of 1/2.

The eigenvectors corresponding to the eigenvalues of Fig. 11.5 are:

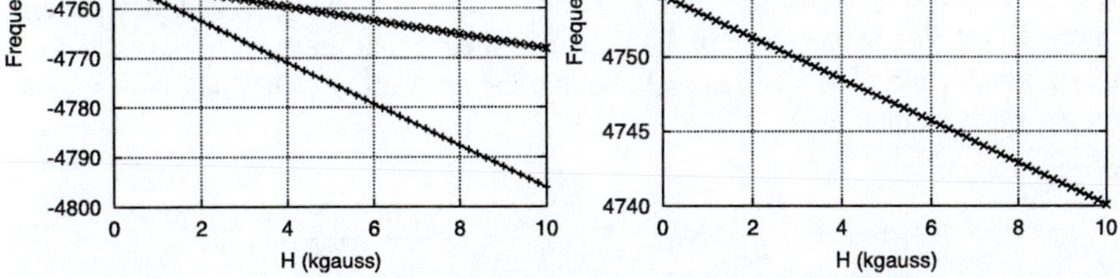

Fig. 11.5 Eigenspectrum of spin-Hamiltonian with exchange interaction for three electrons. Left: Spectrum of bottom four eigenvalues. Right: Spectrum of two largest (degenerate) eigenvalues. The separation of the average value of each spectral cluster is $9508.2 = 3 \times J_{exch}$ for all H_0

$$
\begin{array}{cccccccc}
(1) & (2) & (3) & (4) & (5) & (6) & (7) & (8) \\
\begin{bmatrix} 1.0 \\ 0.0 \\ 0.0 \\ 0.0 \\ 0.0 \\ 0.0 \\ 0.0 \\ 0.0 \end{bmatrix} &
\begin{bmatrix} 0.0 \\ 0.5774 \\ 0.5774 \\ 0.0 \\ 0.5774 \\ 0.0 \\ 0.0 \\ 0.0 \end{bmatrix} &
\begin{bmatrix} 0.0 \\ 0.0 \\ 0.0 \\ 0.5774 \\ 0.0 \\ 0.5774 \\ 0.5774 \\ 0.0 \end{bmatrix} &
\begin{bmatrix} 0.0 \\ 0.0 \\ 0.0 \\ 0.0 \\ 0.0 \\ 0.0 \\ 0.0 \\ 1.0 \end{bmatrix} &
\begin{bmatrix} 0.0 \\ 0.8165 \\ -0.4082 \\ 0.0 \\ -0.4082 \\ 0.0 \\ 0.0 \\ 0.0 \end{bmatrix} &
\begin{bmatrix} 0.0 \\ 0.0 \\ 0.7071 \\ 0.0 \\ -0.7071 \\ 0.0 \\ 0.0 \\ 0.0 \end{bmatrix} &
\begin{bmatrix} 0.0 \\ 0.0 \\ 0.0 \\ 0.8165 \\ 0.0 \\ -0.4082 \\ -0.4082 \\ 0.0 \end{bmatrix} &
\begin{bmatrix} 0.0 \\ 0.0 \\ 0.0 \\ 0.0 \\ 0.0 \\ 0.7071 \\ -0.7071 \\ 0.0 \end{bmatrix}
\end{array}, \tag{11.29}
$$

where we note that vectors 5 and 6 correspond to the same eigenvalue, as do 7 and 8.

When we use (11.25) and (11.29), we compute the following transition matrix elements for the lowest four levels:

$$
\begin{aligned}
S_{x_{12}} &= 3 \times \frac{0.5774}{2} & S_{x_{23}} &= 3 \times \frac{0.6667}{2} & S_{x_{34}} &= 3 \times \frac{0.5774}{2} \\
S_{y_{12}} &= -j3 \times \frac{0.5774}{2} & S_{y_{23}} &= -j3 \times \frac{0.6667}{2} & S_{y_{34}} &= -j3 \times \frac{0.5774}{2} \\
S_{z_{12}} &= 0 & S_{z_{23}} &= 0 & S_{z_{34}} &= 0
\end{aligned} \tag{11.30}
$$

The absorption coefficient for this system is obtained by substituting (11.30) into the general expression, (11.5):

$$
A(\omega) = \frac{\mu_0 \gamma^2}{4Z} \left[3 \left(e^{-E_1/kT} - e^{-E_4/kT} \right) + \left(e^{-E_2/kT} - e^{-E_3/kT} \right) \right] \frac{\tau/\hbar}{1 + (\omega_0 - \omega)^2 \tau^2}. \tag{11.31}
$$

Consider the left-parenthetical term, $3\left(e^{-E_1/kT} - e^{-E_4/kT}\right) = 3e^{-E_1/kT}\left(1 - e^{-(E_4 - E_1)/kT}\right)$, of (11.31), where $E_4 - E_1 = 3\hbar\omega_0$. Under the usual conditions of room (or body) temperature, and a magnetic field of a few kGauss, the exponent, $(E_4 - E_1)/kT$ is of the order of 10^{-3}, which means that the term in parenthesis is approximately equal to $3 \times \hbar\omega_0/kT$, so that the absorption coefficient in (11.31) is approximately equal to

$$
\begin{aligned}
A(\omega) &\approx \frac{9\mu_0\gamma^2}{4Z} e^{-E_1/kT} \frac{\hbar\omega_0}{kT} \frac{\tau/\hbar}{1 + (\omega_0 - \omega)^2 \tau^2} \\
&= \frac{\mu_0 g^2 h^2 (3\beta)^2}{4Z} \frac{e^{-E_1/kT}}{kT} \frac{\omega_0 \tau}{1 + (\omega_0 - \omega)^2 \tau^2},
\end{aligned} \tag{11.32}
$$

where β is the Bohr magneton (the magnetic dipole of a single spin). Therefore, we can conclude from (11.32) that the exchange interaction causes individual spins to align themselves parallel to each other, thereby producing an atomic system of spin-1/2, but with an equivalent dipole three times that of a single spin-1/2 particle. As indicated in (11.31), the effect of the AC field is to cause this single macrospin system to transit from spin-up to spin-down, with all individual spins remaining parallel to each other. In contrast, three noninteracting spins would have an equivalent absorption coefficient of

$$\begin{aligned} A(\omega) &= \frac{3\mu_0\gamma^2}{4Z}\left(e^{-E_1/kT} - e^{-E_2/kT}\right)\frac{\tau/\hbar}{1+(\omega_0-\omega)^2\tau^2} \\ &\approx \frac{3\mu_0\gamma^2}{4Z}\frac{e^{-E_1/kT}}{kT}\frac{\omega_0\tau}{1+(\omega_0-\omega)^2\tau^2}, \end{aligned} \tag{11.33}$$

which is much smaller than (11.32). We can imagine what happens when 10^5 particles interact under exchange effects. This confirms, once again, that the exchange interaction associated with ferromagnetic single-domain particles gives rise to the notion of 'superparamagnetism.'

This model of superparamagnetism results from the large value of J_{exch}, because that isolates the upper energy levels of Figs. 11.4 and 11.5 from the lower levels for all (reasonable) values of H_0. The upper energy levels may be degenerate, as in Fig. 11.5 for the three-spin problem, but the lower levels are always nondegenerate, and have the same number of equal intervals as the number of spins. Furthermore, the large value of the exchange energy ensures that the upper levels will be virtually unpopulated compared to the lower levels. These facts are crucial to the theory.

We can gain further insight into the physics of the problem by considering the combined spin operator $\mathbf{S} = \mathbf{S}^{(1)} + \mathbf{S}^{(2)} + \mathbf{S}^{(3)}$, where the matrices are defined in (11.24) and (11.25). It is straightforward to form $\mathbf{S}^2$, which corresponds to the length-squared of the spin of the composite system of three spin-1/2 particles. The eigenvalues of $\mathbf{S}^2$ give the squares of the lengths when the system is in its allowed states. There are two eigenvalues, 3.750 and 0.750, each four-fold degenerate. Thus, there are two allowed lengths of the composite spin system, $3.750^{1/2}$ and $0.750^{1/2}$. The first corresponds to all three spins being parallel to each other, and the second to two spins being parallel and the third antiparallel. The 'length' of a spin operator is $[S(S+1)]^{1/2}$, so in the first case $S = 3/2$ and in the second $S = 1/2$. Clearly, the first case corresponds to three spin-1/2 particles being aligned in parallel to each other, and the second to two aligned in parallel and the third oppositely aligned, yielding an effective single spin-1/2 particle.

The eigenvectors of $\mathbf{S}^2$ are precisely those shown in (11.29), with the first four corresponding to the eigenvalue, 3.75, and the last four to the eigenvalue 0.75. In the first case, as we stated above, all spins are aligned with each other, yielding a preferred energy state under the effect of the exchange interaction, whereas the second case corresponds to one particle being oppositely aligned to the other two. This results in a significant energy increase due to the large exchange interaction, and this is exactly what we saw in Fig. 11.5. We can further interpret the left-hand spectrum in Fig. 11.5 as being due to the composite system of three parallel spins oriented so that the net component along the magnetic field is maximum (level 1), 1/2 maximum (level 2), 1/2 maximum, but oriented opposite to the field (level 3), and maximum orientation opposite to the field (level 4). This supports our earlier conclusion that under the effects of the exchange interaction the three spins behave as a single large moment, as far as the transitions of the lower energies are concerned.

For the two-spin problem, the eigenvalues of $\mathbf{S}^2$ are 2.000, 2.000, 2.000, and 0.000. The eigenvectors corresponding to the first (degenerate) eigenvalue of 2.000 are the second through fourth eigenvectors listed in (11.19), and the eigenvector corresponding to the eigenvalue 0.000 is the first eigenvector listed in (11.19). The interpretation of the two-spin system follows that of the three-spin system; an eigenvalue of 2.000 means that the two spins are parallel to each other in each of the states shown in the right-hand of Fig. 11.4, whereas the zero eigenvalue means that the two spins are oppositely aligned, thereby cancelling each other, resulting in a zero spin, and no energy variation as H_0 is varied in the top curve of the left-hand of Fig. 11.4. The middle curve on the right-hand part of the figure corresponds to the situation in which the 'macrospin' (both spins aligned with each other) is exactly orthogonal to H_0, meaning that there is no energy variation as H_0 is varied. This explains why states 1 and 3 are constant with respect to H_0. Figure 11.6 summarizes the physics of the problem for the two- and three-spin-1/2 systems.

11.4 Fe^{3+} and Hund's Rules

The models that we have considered so far for superparamagnetism comprised two and three independent electrons, coupled through an exchange interaction only. Real systems contain atoms or ions, which comprise collections of electrons, but whose electrons are not independent. We'll give an example of an important ion, triply-ionized iron [97].

The five unpaired electrons in Fe^{3+} are each in the 3d state. This means that the ion itself is in an orbital S state, i.e., $L = 0$, where L is the orbital angular momentum quantum number. To prove this we use Hund's rules together with the Pauli exclusion principle. Hund's rules are:

1. Assign maximum S (spin) consistent with the Pauli principle.
2. Assign maximum L (orbital angular momentum) consistent with the S. L is defined to be the maximum value of the sum of the z-components of orbital angular momentum for the group of electrons.

Thus, each electron has the same energy quantum number, 3, the same orbital angular momentum quantum number, 2 (corresponding to the d-state), and, if we are to assign maximum spin to the electron group, the same spin quantum number, 1/2. If there is to be no violation of the Pauli principle, therefore, each electron must have a different quantum number, m, corresponding to the z-component of orbital angular momentum. Because $l = 2$ for a d-state, we have $m = 2,\ 1,\ 0,\ -1,\ -2$. Thus, electron number 1 has $m = 2$, number 2 has $m = 1$, etc., to number 5 having $m = -2$. The total $M = m_1 + m_2 + m_3 + m_4 + m_5 = 0$. But since any arrangement

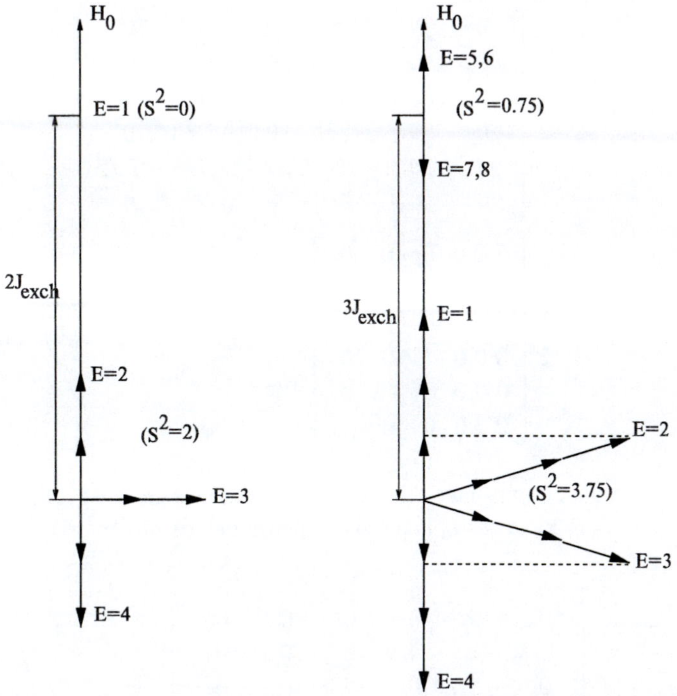

Fig. 11.6 Illustrating the energy levels and states corresponding to two spin-1/2 particles (left) and three spin-1/2 particles (right) in the presence of a magnetic field, H_0. The labeling of the energy eigenstates (E) corresponds to Fig. 11.4 for the two-particle system, and to Fig. 11.5 for the three-particle system. S^2 labels the eigenvalues of the $\mathbf{S}^2$ operator for each of the systems

of the five electrons among the five m-states always yields $M = 0$, we conclude that $L = 0$ (recall that $L = M_{\text{max}}$). Thus, Fe^{3+} is in an S-state ($L = 0$) with a spin equal to 5/2. The fact that $L = 0$ means that the orbital angular momentum is 'quenched,' and cannot contribute to magnetic effects of the ion. Spin is the sole contributor of magnetic effects, and these effects are manifest through the spin-Hamiltonian (Figs. 11.7, 11.8, and 11.9).

The Pauli spin-matrices for a spin-5/2 system are the 6×6 matrices:

$$\sigma_x = \frac{1}{2}\begin{bmatrix} 0 & \sqrt{5} & 0 & 0 & 0 & 0 \\ \sqrt{5} & 0 & \sqrt{8} & 0 & 0 & 0 \\ 0 & \sqrt{8} & 0 & 3 & 0 & 0 \\ 0 & 0 & 3 & 0 & \sqrt{8} & 0 \\ 0 & 0 & 0 & \sqrt{8} & 0 & \sqrt{5} \\ 0 & 0 & 0 & 0 & \sqrt{5} & 0 \end{bmatrix}$$

$$\sigma_y = \frac{1}{2}\begin{bmatrix} 0 & -j\sqrt{5} & 0 & 0 & 0 & 0 \\ j\sqrt{5} & 0 & -j\sqrt{8} & 0 & 0 & 0 \\ 0 & j\sqrt{8} & 0 & -j3 & 0 & 0 \\ 0 & 0 & j3 & 0 & -j\sqrt{8} & 0 \\ 0 & 0 & 0 & j\sqrt{8} & 0 & -j\sqrt{5} \\ 0 & 0 & 0 & 0 & j\sqrt{5} & 0 \end{bmatrix}$$

$$\sigma_z = \frac{1}{2}\begin{bmatrix} 5 & 0 & 0 & 0 & 0 & 0 \\ 0 & 3 & 0 & 0 & 0 & 0 \\ 0 & 0 & 1 & 0 & 0 & 0 \\ 0 & 0 & 0 & -1 & 0 & 0 \\ 0 & 0 & 0 & 0 & -3 & 0 \\ 0 & 0 & 0 & 0 & 0 & -5 \end{bmatrix}. \tag{11.34}$$

The direct-product spin-matrices for the two-ion system are given by $\mathbf{S}^{(1)} = \sigma \otimes I_6$, $\mathbf{S}^{(2)} = I_6 \otimes \sigma$, where I_6 is the six-dimensional identity matrix. Spelled out, these are:

$$S_x^{(1)} = \frac{1}{2}\begin{bmatrix} 0_6 & \sqrt{5}I_6 & 0_6 & 0_6 & 0_6 & 0_6 \\ \sqrt{5}I_6 & 0_6 & \sqrt{8}I_6 & 0_6 & 0_6 & 0_6 \\ 0_6 & \sqrt{8}I_6 & 0_6 & 3 & 0_6 & 0_6 \\ 0_6 & 0_6 & 3 & 0_6 & \sqrt{8}I_6 & 0_6 \\ 0_6 & 0_6 & 0_6 & \sqrt{8}I_6 & 0_6 & \sqrt{5}I_6 \\ 0_6 & 0_6 & 0_6 & 0_6 & \sqrt{5}I_6 & 0_6 \end{bmatrix}$$

$$S_y^{(1)} = \frac{1}{2}\begin{bmatrix} 0_6 & -j\sqrt{5}I_6 & 0_6 & 0_6 & 0_6 & 0_6 \\ j\sqrt{5}I_6 & 0_6 & -j\sqrt{8}I_6 & 0_6 & 0_6 & 0_6 \\ 0_6 & j\sqrt{8}I_6 & 0_6 & -j3I_6 & 0_6 & 0_6 \\ 0_6 & 0_6 & j3I_6 & 0_6 & -j\sqrt{8}I_6 & 0_6 \\ 0_6 & 0_6 & 0_6 & j\sqrt{8}I_6 & 0_6 & -j\sqrt{5}I_6 \\ 0_6 & 0_6 & 0_6 & 0_6 & j\sqrt{5}I_6 & 0_6 \end{bmatrix}$$

$$S_z^{(1)} = \frac{1}{2}\begin{bmatrix} 5I_6 & 0_6 & 0_6 & 0_6 & 0_6 & 0_6 \\ 0_6 & 3I_6 & 0_6 & 0_6 & 0_6 & 0_6 \\ 0_6 & 0_6 & 1I_6 & 0_6 & 0_6 & 0_6 \\ 0_6 & 0_6 & 0_6 & -1I_6 & 0_6 & 0_6 \\ 0_6 & 0_6 & 0_6 & 0_6 & -3I_6 & 0_6 \\ 0_6 & 0_6 & 0_6 & 0_6 & 0_6 & -5I_6 \end{bmatrix}$$

$$S_x^{(2)} = \frac{1}{2}\begin{bmatrix} \sigma_x & 0_6 & 0_6 & 0_6 & 0_6 & 0_6 \\ 0_6 & \sigma_x & 0_6 & 0_6 & 0_6 & 0_6 \\ 0_6 & 0_6 & \sigma_x & 0_6 & 0_6 & 0_6 \\ 0_6 & 0_6 & 0_6 & \sigma_x & 0_6 & 0_6 \\ 0_6 & 0_6 & 0_6 & 0_6 & \sigma_x & 0_6 \\ 0_6 & 0_6 & 0_6 & 0_6 & 0_6 & \sigma_x \end{bmatrix}$$

$$S_y^{(2)} = \frac{1}{2}\begin{bmatrix} \sigma_y & 0_6 & 0_6 & 0_6 & 0_6 & 0_6 \\ 0_6 & \sigma_y & 0_6 & 0_6 & 0_6 & 0_6 \\ 0_6 & 0_6 & \sigma_y & 0_6 & 0_6 & 0_6 \\ 0_6 & 0_6 & 0_6 & \sigma_y & 0_6 & 0_6 \\ 0_6 & 0_6 & 0_6 & 0_6 & \sigma_y & 0_6 \\ 0_6 & 0_6 & 0_6 & 0_6 & 0_6 & \sigma_y \end{bmatrix}$$

$$S_z^{(2)} = \frac{1}{2}\begin{bmatrix} \sigma_z & 0_6 & 0_6 & 0_6 & 0_6 & 0_6 \\ 0_6 & \sigma_z & 0_6 & 0_6 & 0_6 & 0_6 \\ 0_6 & 0_6 & \sigma_z & 0_6 & 0_6 & 0_6 \\ 0_6 & 0_6 & 0_6 & \sigma_z & 0_6 & 0_6 \\ 0_6 & 0_6 & 0_6 & 0_6 & \sigma_z & 0_6 \\ 0_6 & 0_6 & 0_6 & 0_6 & 0_6 & \sigma_z \end{bmatrix}. \quad (11.35)$$

11.5 Crystalline Anisotropy and TiO_2

Now we need to include the effects of crystalline anisotropy. The only spin-Hamiltonian that we have for the interaction of a crystal field with Fe^{3+} is for TiO_2, and is given in (11.6). When we add the D-term of the crystal field to the spin-Hamiltonian for the Zeeman and exchange interactions, and compute the eigenvalues, we get the results shown in Fig. 11.10 for the lowest eleven frequencies, the 'A-section' of Fig. 11.9. The spectrum of the other five sections of Fig. 11.9 are shown in Figs. 11.11, 11.12, 11.13, 11.14, and 11.15. Clearly, the spectrum is altered significantly by the crystal-field interaction, and that is why this interaction is so important to us. In order for us to 'tune' the system to the proper frequency of operation by adjusting the external magnetic field, we must know the nature of the environment of the iron ion, and how this environment reacts with the ion.

In the example just computed, we see that the lowest energy levels, 1 and 2 in the A-spectrum, have a small separation as H_0 is varied. The frequency interval, ω_{012}, varies between 100 MHz and 1.5 GHz, and the transition-matrix element, S_{y12}, is relatively close to $\pm j4.8$ over this range. This is a considerably larger value of the transition-matrix element than those given in (11.10) for a single, noninteracting spin, and, once again, demonstrates the effect of the exchange interaction.

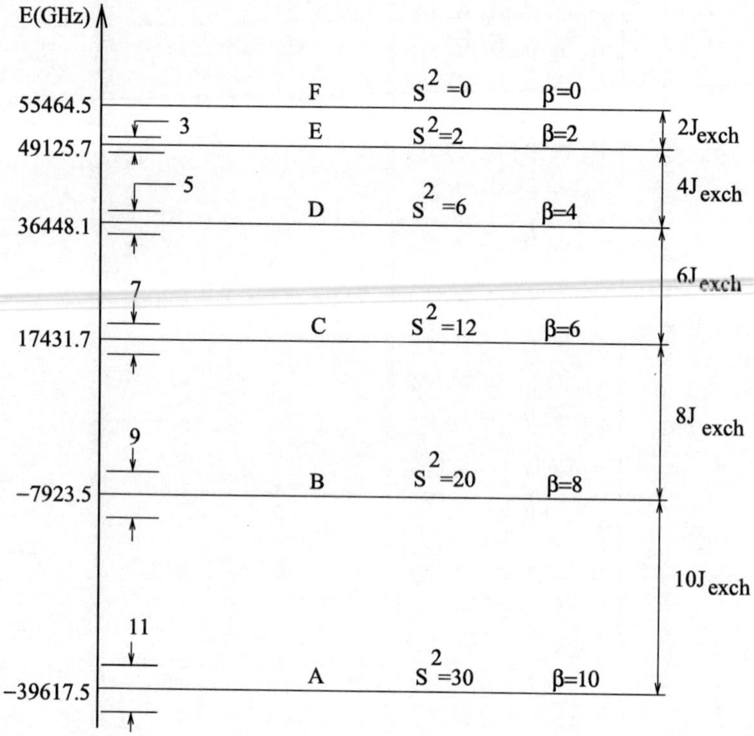

Fig. 11.7 Energy-levels (in GHz) for the system comprising two $Fe^{(3+)}$ ions connected through the exchange interaction $J_{\mathbf{exch}}$. The numbers just inside the vertical line indicate the number of energy levels associated with that branch of the spectrum; the spacing between consecutive pairs of levels is equal to $2.8H_0$, where H_0 is in kGauss. The letters, A-F, correspond to the physical arrangement of the ten electrons in the coupled system, as shown in Table 11.1. S^2 are the eigenvalues of the magnitude-squared spin-operator, $\mathbf{S}^2$, and the β-values indicate the number of Bohr magnetons in the various arrangements of parallel spins

Table 11.1 Spin arrangements for Fig. 11.7. The bullets, •, correspond to spins that are parallel to the z−axis, and the circles, ∘, to spins that are anti-parallel

A	••••••••••	$S = 5 : 2S + 1 = 11$
B	•••••••••∘	$S = 4 : 2S + 1 = 9$
C	••••••••∘∘	$S = 3 : 2S + 1 = 7$
D	•••••••∘∘∘	$S = 2 : 2S + 1 = 5$
E	••••••∘∘∘∘	$S = 1 : 2S + 1 = 3$
F	•••••∘∘∘∘∘	$S = 0 : 2S + 1 = 1$

The eigenvalues (in GHz) corresponding to the lowest eleven energy levels (the 'A-section' spectrum of Fig. 11.10) are listed here in the order, E_1 to E_{11}, left-to-right, for $H_0 = 1$ kGauss:

```
h =  1.000000000000000
A-section spectrum
-39662.7967 -39661.1857 -39580.6189 -39558.0700 -39525.8847 -39453.1469
-39436.4784 -39306.8435 -39284.5805 -39098.4235 -39070.4491
```

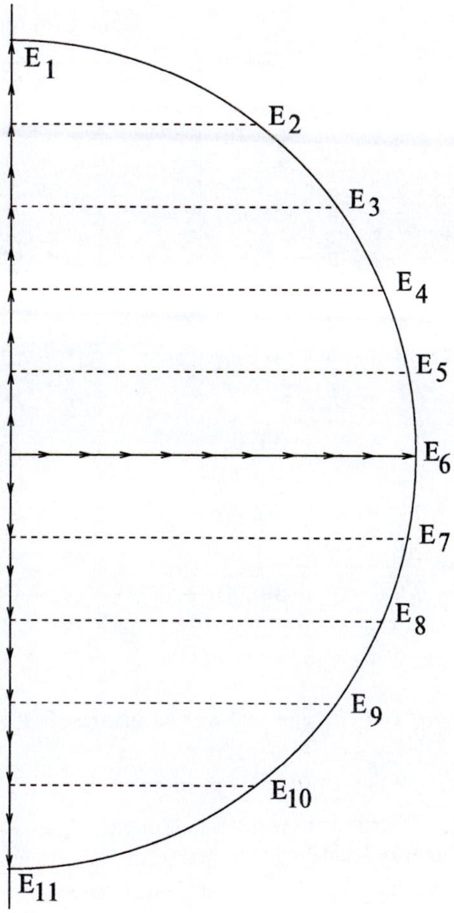

Fig. 11.8 Positions of the 10-spin magnetic dipole that produce the 'A-spectrum' of Fig. 11.7. The numbering of the energy levels, $E_1 \cdots E_{11}$, corresponds to that of Fig. 11.9

The resonant frequencies (in GHz) associated with transitions between these eigenvalues are:

```
h =  1.000000000000000
A-section resonant frequencies
  1.6109   82.1777 104.7267 136.9120 209.6497 226.3183 355.9532
   378.2162 564.3732 592.3476
80.5668 103.1157 135.3010 208.0388 224.7073 354.3423 376.6052
 562.7622 590.7366
22.5489   54.7342 127.4720 144.1405 273.7755 296.0384 482.1954 510.1698
32.1853 104.9231 121.5916 251.2266 273.4895 459.6465 487.6209
72.7377   89.4063 219.0412 241.3042 427.4612 455.4356
16.6685 146.3035 168.5665 354.7234 382.6978
129.6350 151.8979 338.0549 366.0293
 22.2630 208.4200 236.3943
186.1570 214.1314
 27.9744
```

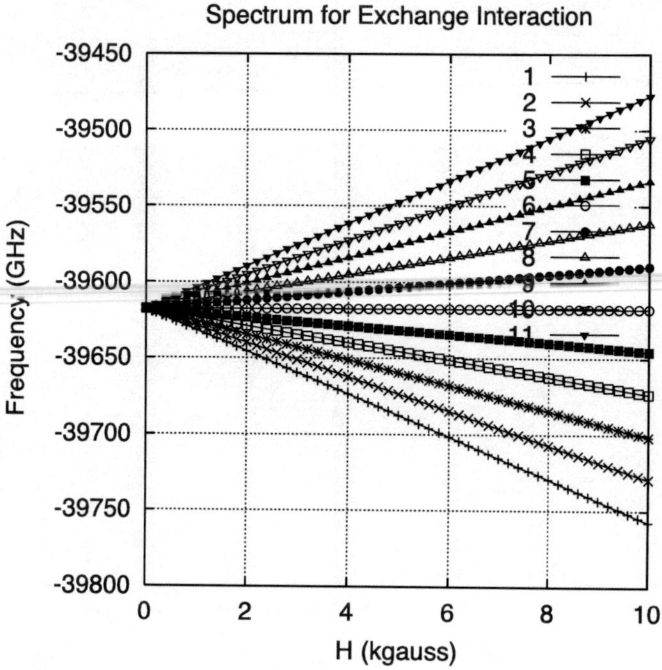

Fig. 11.9 Variation of A-section spectrum of Figs. 11.7 and 11.8 as a function of the magnetic field

The transition-matrix elements, $S_{x_{ij}}$, $S_{y_{ij}}$, $S_{z_{ij}}$, with $i < j$, for the lowest eleven energy levels (the 'A-spectrum') are shown next for $H_0 = 1$ kGauss. The $S_{y_{ij}}$ are complex–in this case pure imaginaries.

```
h =  1.000000000000000
transx
0.0340 2.3641  0.0000 0.0000 -0.3188 -0.3458 0.0000 0.0000 0.0104 0.0083
0.0000 2.2114 -0.2267 0.0000  0.0000 -0.0881 0.0708 0.0000 0.0000
0.4898 3.4479  0.0000 0.0000 -0.2023 -0.1794 0.0000 0.0000
0.0000 2.3182 -1.2585 0.0000  0.0000 -0.0475 0.0315
1.5273 2.5256  0.0000 0.0000 -0.0396 -0.0480
0.0000 2.3001 -0.2578 0.0000  0.0000
0.2658 2.2929  0.0000 0.0000
0.0000 1.6855 -0.0044
0.0044 1.6823
0.0000
transy
0.0000  4.8024 0.0000 -0.0927 0.0000 0.0000 0.0000 0.0000 0.0000 -0.3642
0.0000  0.2450 0.0000  0.0000 0.0000 0.0000 0.0000 0.0096 0.0000 -0.0077
0.0000  0.0000 0.0000  0.1411 0.0000 0.8497 0.0000 0.0000 0.0000  0.0000
0.0000 -0.0777 0.0000 -0.0633 0.0000 0.0000 0.0000 0.0000
0.0000  3.2101 0.0000 -0.5124 0.0000 0.0000 0.0000 0.0000 0.0000 -0.1786
0.0000  0.1571 0.0000  0.0000 0.0000 0.0000
0.0000  0.0000 0.0000  1.3829 0.0000 1.2520 0.0000 0.0000 0.0000  0.0000
0.0000 -0.0435 0.0000 -0.0289
```

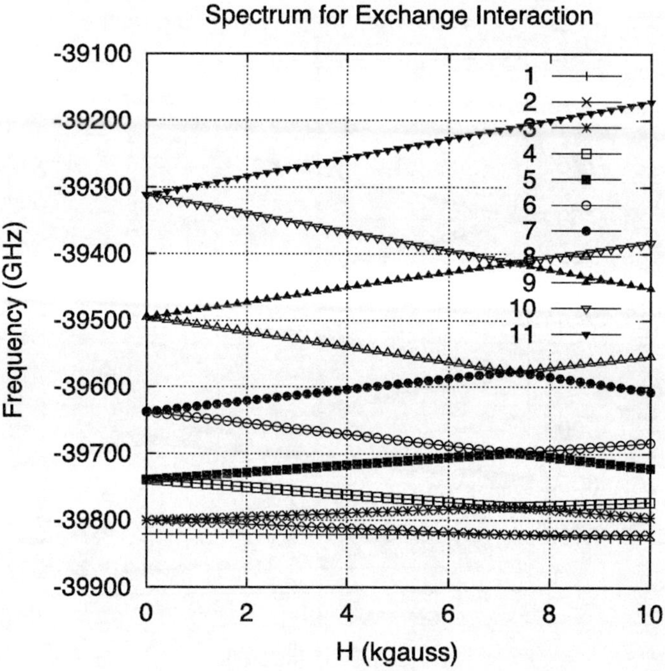

Fig. 11.10 Modification of the A-section spectrum of Fig. 11.9 due to the *D*-term of the crystal field spin-Hamiltonian of TiO_2 shown in (11.6)

```
0.0000  1.5732 0.0000 -1.6175 0.0000 0.0000 0.0000 0.0000 0.0000 -0.0360
0.0000  0.0438
0.0000  0.0000 0.0000  1.9029 0.0000 0.2246 0.0000 0.0000 0.0000  0.0000
0.0000  0.2310 0.0000 -1.9156 0.0000 0.0000 0.0000 0.0000
0.0000  0.0000 0.0000  1.4797 0.0000 0.0039
0.0000  0.0039 0.0000 -1.4829
0.0000  0.0000
transz
0.0000 0.0000 -0.9531  0.1054  0.0000 0.0000 0.0335 -0.0271 0.0000 0.0000
1.0827 0.0000  0.0000 -0.1526 -0.1637 0.0000 0.0000  0.0038 0.0030
0.0000 0.0000 -0.5122  0.3721  0.0000 0.0000 0.0089 -0.0076
1.7013 0.0000  0.0000 -0.1711 -0.1161 0.0000 0.0000
0.0000 0.0000 -0.1487  0.1827  0.0000 0.0000
0.6668 0.0000  0.0000 -0.0964 -0.0108
0.0000 0.0000 -0.0114  0.0935
0.0208 0.0000  0.0000
0.0000 0.0000
0.0002
```

Let's use these results to calculate the peak value of the absorption coefficient for those transitions that occur with a frequency less than 50 GHz. According to the table of 'A-section resonant frequencies,' there are six such transitions: 12, 34, 45, 67, 89, 1011. According to (11.5), the peak value of the absorption coefficient is given by

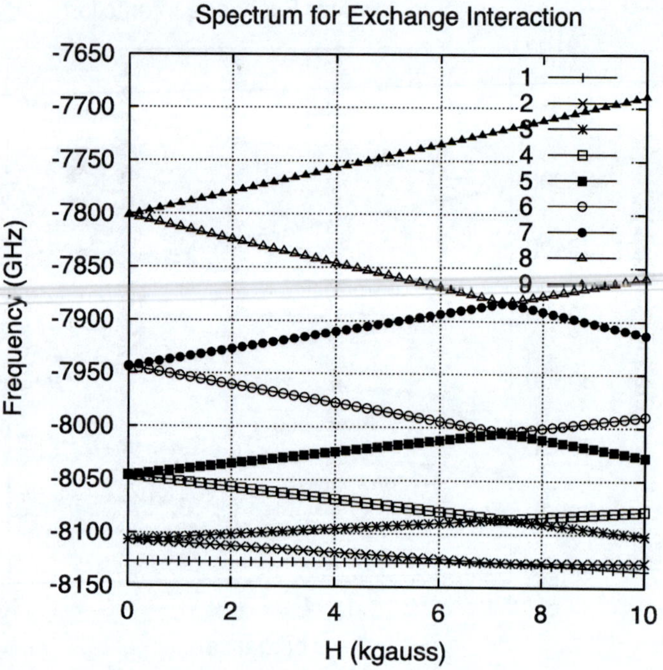

Fig. 11.11 Modification of the B-section spectrum of Fig. 11.9 due to the *D*-term of the crystal field spin-Hamiltonian of TiO_2 shown in (11.6)

$$\begin{aligned} A(\omega_{0_{jk}}) &\propto |S_{kj}|^2 \left(N_j^{(T)} - N_k^{(T)}\right) \tau_{jk}/\hbar \\ &= |S_{kj}|^2 N_j \left(1 - e^{-(E_k - E_j)/kT}\right) \tau_{jk}/\hbar \\ &\approx |S_{kj}|^2 N_j \frac{(E_k - E_j)}{kT} \frac{\tau_{jk}}{\hbar} \\ &= |S_{kj}|^2 N_j \frac{\hbar\omega_{0_{jk}}}{kT} \frac{\tau_{jk}}{\hbar} \\ &= |S_{kj}|^2 N_j \frac{\omega_{0_{jk}} \tau_{jk}}{kT} \,. \end{aligned} \tag{11.36}$$

We will assume that the temperature is 98.6 °F, which is 310 °K, so that $kT = 4.2811 \times 10^{-21}$ J. To express this in terms of frequency, divide by Planck's constant, 6.626×10^{-34} J, and get 6.461×10^{12}Hz $=$ 6461.1 GHz. Since this number is much greater than the resonant frequencies, we are permitted to carry out the approximation in the third line of (11.36). The transition matrix elements in (11.36) include S_x, S_y, and S_z (transx, transy, transz, respectively), and from here on we will assume that the relaxation frequency, τ_{jk}, is fixed for all transitions.

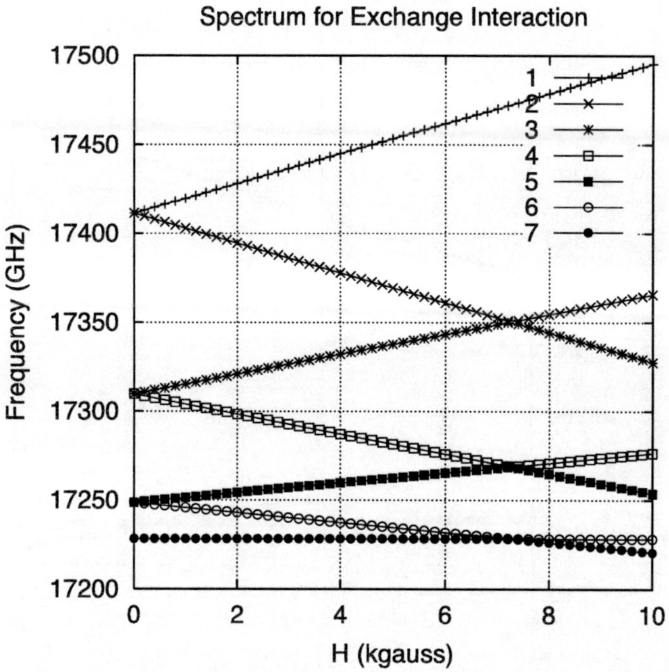

Fig. 11.12 Modification of the C-section spectrum of Fig. 11.9 due to the *D*-term of the crystal field spin-Hamiltonian of TiO_2 shown in (11.6)

The following table lists important data for the calculations:

Transition	ω_{0j} (GHz)	Transx	Transy	Transz	N_j/N_1	ω_{0j}/ω_{01}
12	1.6109	0.0340	j4.8024	0.0	1	1
34	22.5489	0.4898	j3.2101	0.0	0.987	14.0
45	32.1853	0.0	0.0	1.7013	0.984	19.98
67	16.6685	0.0	0.0	0.6668	0.968	10.35
89	22.2630	0.0	0.0	0.0208	0.946	13.82
10,11	27.9744	0.0	0.0	0.0002	0.916	17.37

It turns out that the first three absorption lines are much stronger than the others, and these are plotted in Fig. 11.16. This is an example of how we can ‘tune’ our system to achieve a design feature once we have the physics in the form of a mathematical model. Even though the two higher-frequency lines are stronger, practical considerations would lead us to use the response at 1.6109 GHz for noninvasive probing for lesions.

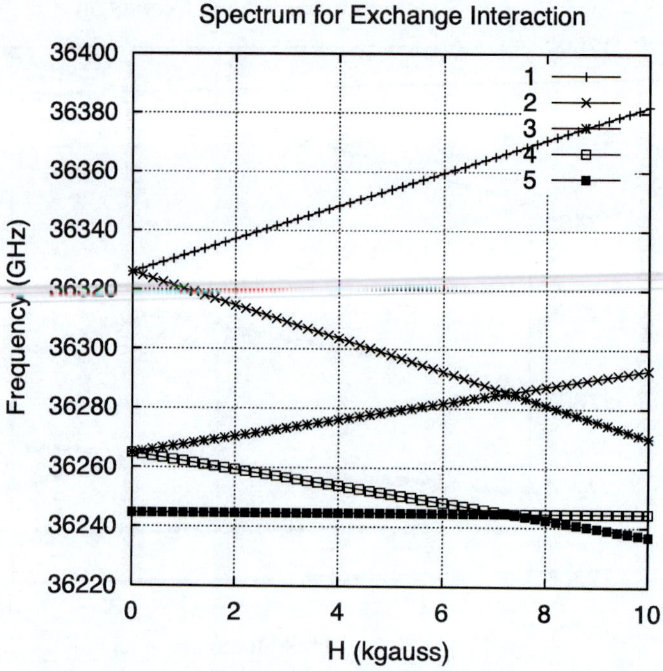

Fig. 11.13 Modification of the D-section spectrum of Fig. 11.9 due to the *D*-term of the crystal field spin-Hamiltonian of TiO_2 shown in (11.6)

11.5.1 Application to a 'Magnetic Lesion'

Consider the model shown in Fig. 11.17, which corresponds to Fig. 10.8, except that the background is tissue with a conductivity that is the average of that found in the body, and the lesion is nonconducting, but is magnetically permeable. If we assume that the same coil is used as in the model of Fig. 10.8, and that the lesion is much deeper, being 5 cm beneath the surface, and that the coil is excited at 1.6 GHz, which corresponds to the lowest transition frequency described above, then the response of the probe to the lesion, whose permeability (at 1.6 GHz) is $\mu = 2,\ 10,$ and 100, is shown in Fig. 11.18.

11.6 Static Interaction Energy of Two Magnetic Moments

The spin-Hamiltonian that we have worked with so far includes only the Zeeman term and the exchange interaction. There are other terms that reflect certain physical processes that need to be included, as well. One such term corresponds to the static interaction energy of two magnetic moments [72, p. 412]: $H' =$

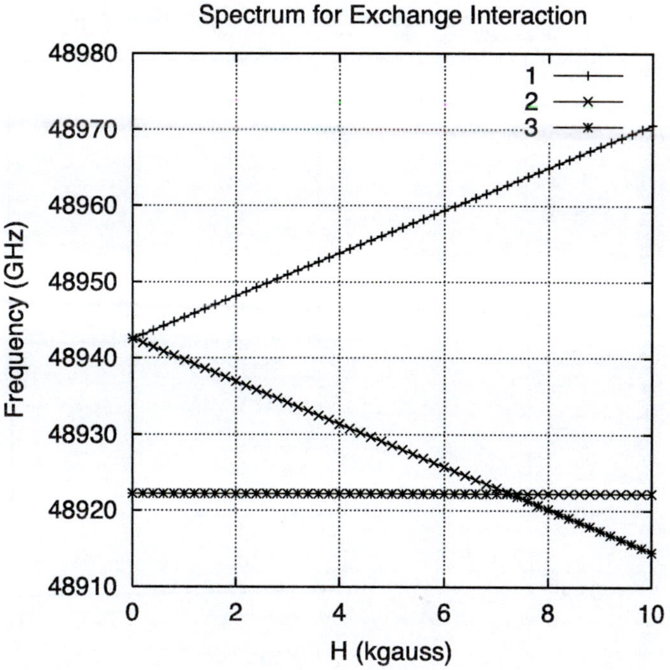

Fig. 11.14 Modification of the E-section spectrum of Fig. 11.9 due to the D-term of the crystal field spin-Hamiltonian of TiO_2 shown in (11.6)

$\frac{\mu_0}{4\pi r^3}\left[\mathbf{m}_1 \cdot \mathbf{m}_2 - 3\frac{(\mathbf{m}_1 \cdot \mathbf{r})(\mathbf{m}_2 \cdot \mathbf{r})}{r^2}\right]$, where $\mathbf{m}_1$ and $\mathbf{m}_2$ are the magnetic moments of the dipoles, and $\mathbf{r}$ is the vector separation between the two dipoles. This energy term manifests itself in the following spin-Hamiltonian for three interacting electrons:

$$\mathscr{H}_{dd} = \frac{\mu_0}{4\pi}4\beta_0^2\left[\frac{\mathbf{S}_1 \cdot \mathbf{S}_2}{r_{12}^3} + \frac{\mathbf{S}_1 \cdot \mathbf{S}_3}{r_{13}^3} + \frac{\mathbf{S}_2 \cdot \mathbf{S}_3}{r_{23}^3}\right.$$

$$\left. -3\frac{(\mathbf{r}_{12} \cdot \mathbf{S}_1)(\mathbf{r}_{12} \cdot \mathbf{S}_2)}{r_{12}^5} - 3\frac{(\mathbf{r}_{13} \cdot \mathbf{S}_1)(\mathbf{r}_{13} \cdot \mathbf{S}_3)}{r_{13}^5} - 3\frac{(\mathbf{r}_{23} \cdot \mathbf{S}_2)(\mathbf{r}_{23} \cdot \mathbf{S}_3)}{r_{23}^5}\right], \quad (11.37)$$

where $\beta_0 = 9.2731 \times 10^{-24}$ amp − meters^2 is the Bohr magneton, and the various vector spin-matrices have been defined earlier. If we assume that the spins lie at the vertices of an equilateral triangle of side 6×10^{-10}m, as in Fig. 11.19, then we can expand (11.37) to get

$$\mathscr{H}_{dd} = 0.24\left[\mathbf{S}_1 \cdot \mathbf{S}_2 + \mathbf{S}_1 \cdot \mathbf{S}_3 + \mathbf{S}_2 \cdot \mathbf{S}_3 - 3S_x^{(1)}S_x^{(2)} - 0.75\left(S_x^{(1)}S_x^{(3)} + S_x^{(2)}S_x^{(3)}\right)\right.$$

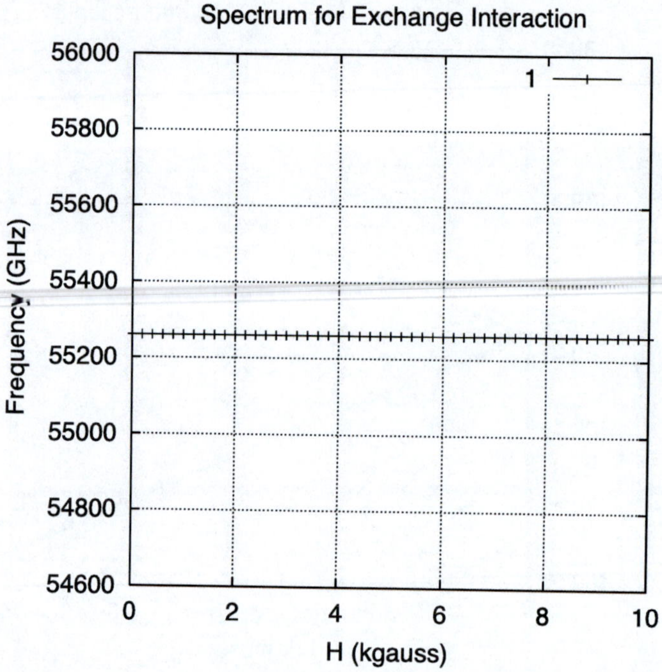

Fig. 11.15 Modification of the F-section spectrum of Fig. 11.9 due to the *D*-term of the crystal field spin-Hamiltonian of TiO_2 shown in (11.6)

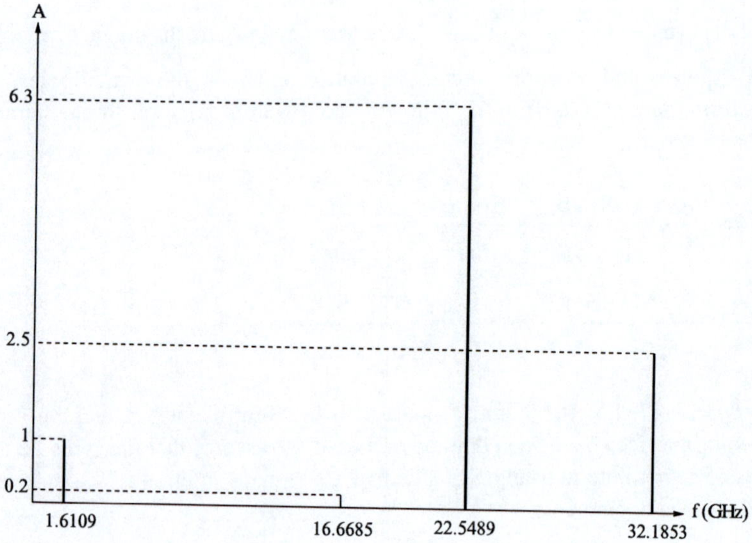

Fig. 11.16 Relative absorption peaks versus frequency

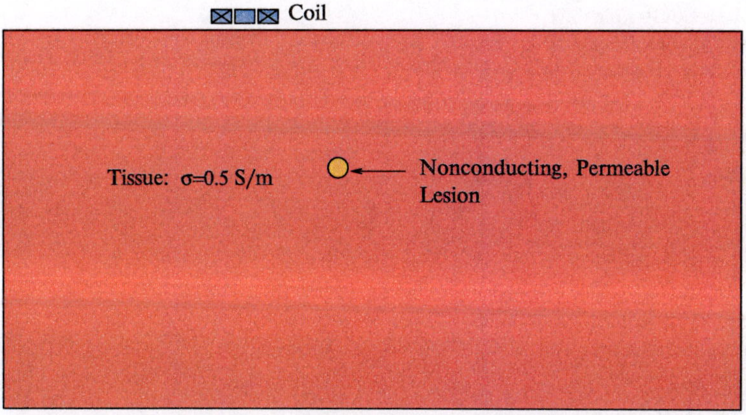

Fig. 11.17 Model of an eddy-current probe scanned past a nonconducting, magnetically permeable lesion embedded in tissue of an averaged conductivity

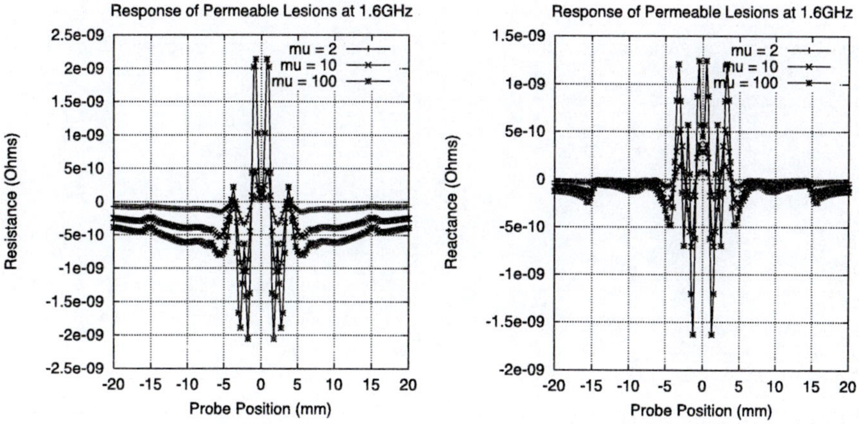

Fig. 11.18 Response of probe to a lesion whose permeability (at 1.6 GHz) is $\mu = 2,\ 10,$ and 100

$$
\begin{aligned}
&+1.299\left(S_y^{(1)}S_x^{(3)} + S_x^{(1)}S_y^{(3)} - S_y^{(2)}S_x^{(3)} - S_x^{(2)}S_y^{(3)}\right)\\
&-2.25\left(S_y^{(1)}S_y^{(3)} + S_y^{(2)}S_y^{(3)}\right)\Big] \,,
\end{aligned}
\tag{11.38}
$$

where the units of $\mathscr{H}_{dd}$ in (11.38) are in GHz. This matrix operator is added to that in (11.27) to get the overall spin-Hamiltonian for Zeeman + Exchange + Dipolar effects.

Fig. 11.19 Illustrating three magnetic dipoles, labeled as spin operators, **S**, situated on the vertices of an equilateral triangle, for the purpose of computing the dipole-dipole interaction

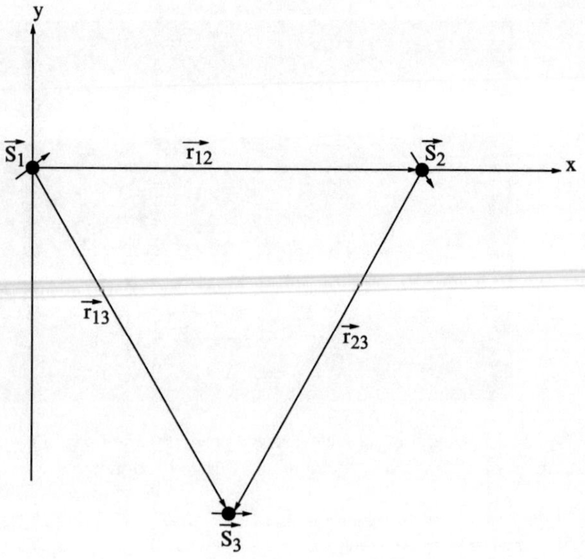

Chapter 12
Carbon-Nanotube Reinforced Polymers

12.1 Introduction

Our goal in this chapter is to develop electromagnetic models for the inspection of carbon-nanotube reinforced polymers (CNRPs), along the lines of our work with CFRPs. Electromagnetic models for carbon nanotube structures are of considerable current interest. In [134] terahertz wave reflection and transmission from carbon nanotubes are investigated. The interesting feature is that the dielectric properties are dispersive, i.e., the permittivity and conductivity are frequency dependent. We will see that shortly in the next section. In the THz regime, the frequency response is affected by the number of tube walls, thickness, aspect ratio, filling factor, and geometrical factor during growth. An extended Drude-Lorentzian model, [31], [66], is used in [134] to simulate the relative dispersive permittivity of carbon nanotubes at THz frequencies:

$$\epsilon_r = \epsilon_c - \frac{\omega_p^2}{\omega(\omega - j\Gamma)} + \frac{\omega_{p1}^2}{-\omega^2 + j\omega\Gamma_1 + \omega_1^2}, \tag{12.1}$$

where ϵ_c is the dielectric constant at infinite frequency, ω_p, ω_1, and ω_{p1} are the electron plasma, phonon, and oscillator frequency, respectively. Γ and Γ_1 are the relaxation rate and spectral width, respectively. If we write $\epsilon_r = \epsilon + \sigma/j\omega$, where ϵ and σ are real, then it is easy to show that

$$\begin{aligned}
\epsilon &= \epsilon_c - \frac{\omega_p^2}{\omega^2 + \Gamma^2} + \frac{\omega_{p1}^2(-\omega^2 + \omega_1^2)}{(-\omega^2 + \omega_1^2)^2 + \omega^2\Gamma_1^2} \\
\frac{\sigma}{\epsilon_0} &= \frac{\omega_p^2\Gamma}{(\omega^2 + \Gamma^2)} + \frac{\Gamma_1^2\omega_{p1}^2\omega^2}{(-\omega^2 + \omega_1^2)^2 + \omega^2\Gamma_1^2}.
\end{aligned} \tag{12.2}$$

H. A. Sabbagh et al., *Advanced Electromagnetic Models for Materials Characterization and Nondestructive Evaluation*, Scientific Computation,
https://doi.org/10.1007/978-3-030-67956-9_12

Table 12.1 Drude-Lorentz parameters for various CNT structures [134]

Structure	ϵ_c	ω_p	ω_{p1}	ω_1	Γ	Γ_1
SWCNT	8.41	$2\pi \times 23$ THz	$2\pi \times 38.9$ THz	$2\pi \times 5.9$ THz	$2\pi \times 24.5$ THz	$2\pi \times 29.6$ THz
DWCNT	5.76	$2\pi \times 10.5$ THz	$2\pi \times 32.2$ THz	$2\pi \times 5.5$ THz	$2\pi \times 24.3$ THz	$2\pi \times 23.3$ THz
H-doped CNT	6.25	$2\pi \times 7.42$ THz	$2\pi \times 4.69$ THz	$2\pi \times 1.53$ THz	$2\pi \times 34.29$ THz	$2\pi \times 3.27$ THz

Both parts are even functions of ω, and $\sigma > 0$, thereby satisfying the Kramers-Kronig relations for a passive linear system. From (12.2) we get the DC conductivity to be $\sigma(0) = \epsilon_0\omega_p^2/\Gamma$.

Examples of these parameters are tabulated in Table 12.1.

The quantum origin of these parameters is not given in [134], but will be needed in our work, if we are not given these data in the literature. We'll say more about this in a later section when we discuss nanographene (see Eq. (12.3) and Fig. 12.2).

12.2 Modeling Piezoresistive Effects in Carbon Nanotubes

12.2.1 The Structure of CNTs

A nanotube is constructed by rolling a graphene sheet along the direction of the chiral vector, $\mathbf{C}_h$, as in Fig. 12.1 [114]. The 'chirality,' determines the electromechanical properties of the tube, and is defined in terms of the indices, (m, n), associated with the nanotube unit vectors, $\mathbf{a}_1$, $\mathbf{a}_2$, of Fig. 12.1. In particular, it determines the electronic band structure, and, therefore, the conductivity. For example, $n = m$ tubes have a zero band gap, and are therefore metallic, while $n \neq m$ have some band gap and are semiconducting, though the subset, $n - m = 3q$, with q an integer, has only a small gap induced by the curvature of the graphene sheet. This makes this subset semimetallic, quasi-metallic, or small-gap semiconducting (SGS). Within the semiconducting and SGS groups, the band gap of the specific tube varies inversely with the diameter or the square of the diameter, respectively [47]. The dependence of the electronic properties on the structure implies that mechanical deformations can alter the band structure, which results in, among other things, piezoresistivity [47].

12.3 Electromagnetic Features of CNTs

The distinguishing electromagnetic features of carbon nanotube (CNT) structures are superparamagnetism and piezoresistivity. It appears that each requires quantum mechanical calculations to generate parametric values: spin Hamiltonians for

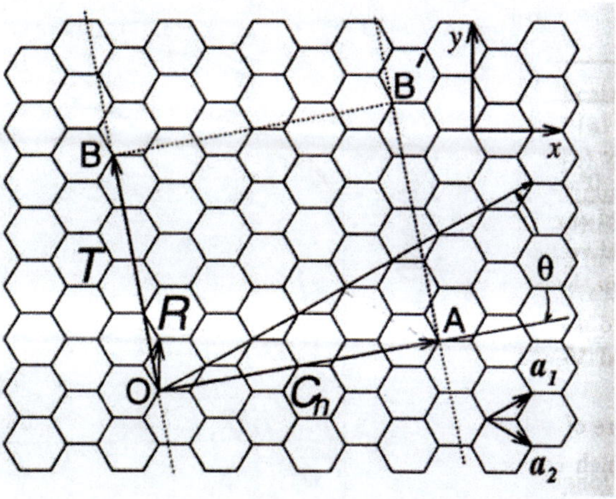

Fig. 12.1 The unrolled honeycomb lattice of a carbon nanotube. A nanotube is constructed when site O is connected to A, and B is connected to B'. **OA** defines the chiral vector, $\mathbf{C}_h$, and **OB** defines the translational vector, **T**, which lies along the axis of the tube. The rectangle, $OAB'B$, defines the unit cell of the nanotube, and **R** denotes a symmetry vector. The figure corresponds to $\mathbf{C}_h = (4, 2)$, $\mathbf{T} = (4, -5)$, and $\mathbf{R} = (1, -1)$, where the components refer to the basis vectors $\mathbf{a}_1$, $\mathbf{a}_2$ (Taken from [114])

superparamagnetism [16] and linear combination of atomic orbitals (lcao) for piezoresistivity [70]. We have already done considerable modeling of paramagnetism, as will be shown shortly, as well as in [96–98]. In this section, we will concentrate on developing computational models for piezoresistive effects.

12.4 Quantum-Mechanical Model for Conductivity

The formula for electrical conductivity of nanographene is [70]

$$\sigma(\omega) = \left(\frac{(2\pi e/h)^2}{3Vf}\right)\sum_{i,f}(E_f - E_i)^2| < f|\mathbf{r}|i > |^2\delta(E_f - E_i - hf)\,, \qquad (12.3)$$

where E_i and E_f are the energies of the initial valence eigenstates, $|i >$, and the final unoccupied conduction eigenstates, $|f >$, respectively. This is akin to 'pumping' from one energy level to a higher one in masers, and accounts for the loss of energy. It can be shown ([70]) that only the diagonal elements of the position matrix elements survive, so that $< f|\mathbf{r}|i >= \sum_{j,l} c_{j,l}^{(f)*} c_{j,l}^{(i)} < l|\mathbf{r}|l >= \sum_{j,l} c_{j,l}^{(f)*} c_{j,l}^{(i)}\mathbf{r}$, where the index, l, runs over both initial and final states. The $c_{j,l}^{(f),(i)}$ are expansion coefficients for the LCAO expansion of the $|i >$ and $|f >$ eigenstates into atomic states (orbitals). The DC conductivity is obtained by taking the limit ω

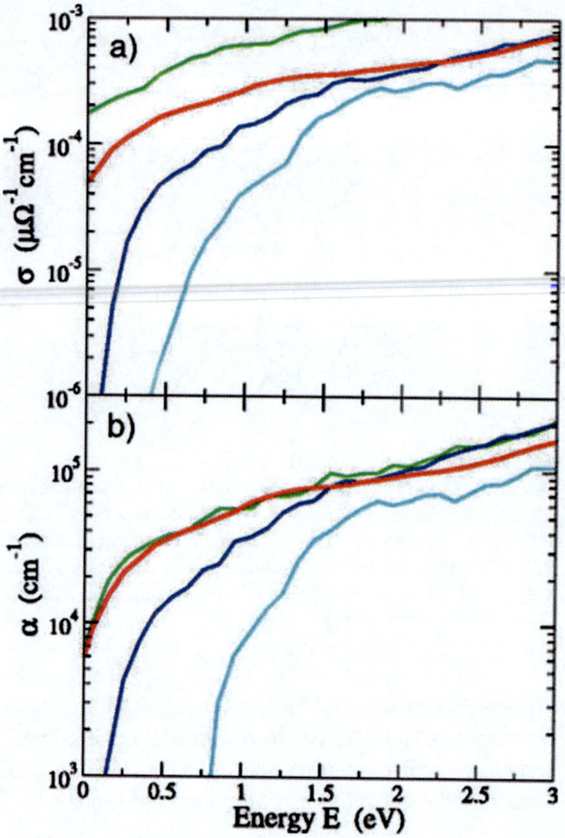

Fig. 12.2 (**a**). Conductivity of 3D-NG (*red curve*) compared to that of a graphene layer (*green curve*), of a low-density a-C cell (*blue curve*), and of a tetrahedral a-C cell (*light-blue curve*). (**b**). Optical absorption of the above materials. The frequency range is [0, 7.25×10^{14} Hz] (From [70])

(or E)$\rightarrow$ 0 in (12.3). The result, (12.3), is obtained through the use of tight-binding molecular dynamics (TBMD) simulations.

The model that leads to (12.3) was executed using the Slater-Koster tight-binding framework that was developed at the Naval Research Laboratory [84].

Figure 12.2 illustrates the conductivity as a function of energy (or frequency) as calculated from (12.3). The frequency range corresponding to these energies is $[0, 7.25 \times 10^{14}]$ Hz. The spiral coils that are defined and analyzed in [111, Chapter 7] have been accurately modeled out to 1.5×10^{12} Hz, which, according to Fig. 12.2 is still virtually 'DC' (Fig. 12.3).

12.5 What Are We Looking At?

There is a clear distinction in the network-like structure of the nanographene sheet, Figs. 12.4 and 12.5, and the yarn-like structure of the CNT sheets of Fig. 12.3. The current version of **VIC-3D®** allows us to model anisotropic and random

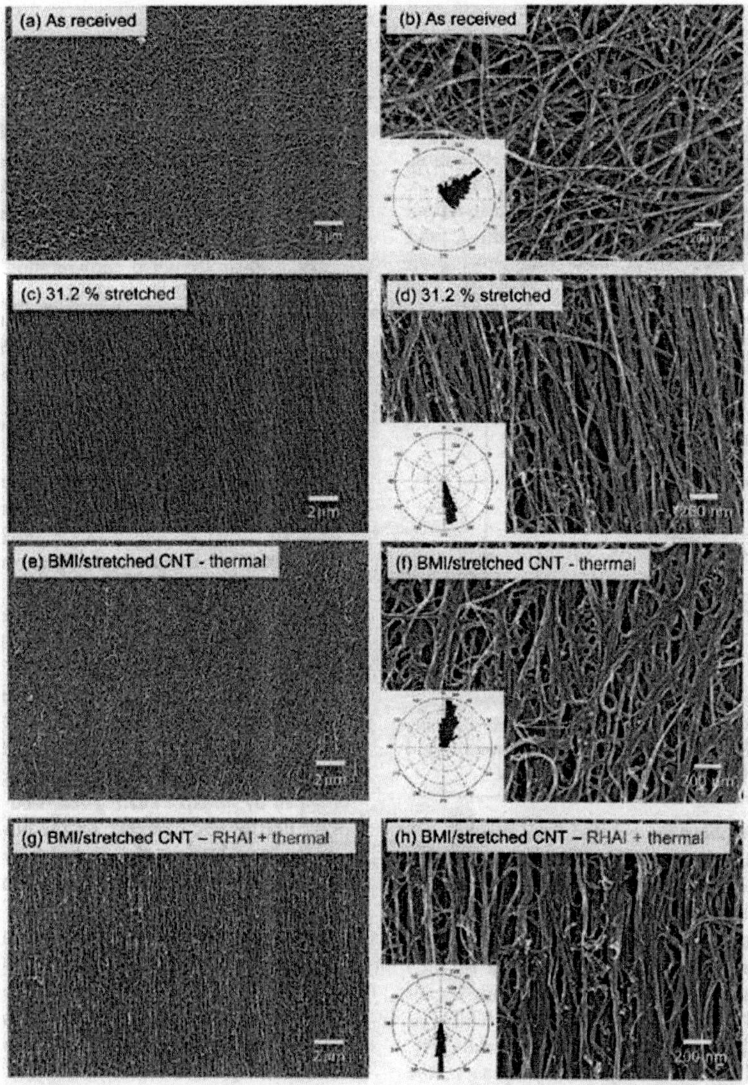

Fig. 12.3 FE-SEM images of (**a** and **b**) as-received, (**c** and **d**) 31.2% stretched pristine CNT sheets, and BMI/stretched CNT sheet nanocomposites cured by (**e** and **f**) thermal and (**g** and **h**) RHAI followed by post-thermal cure. Inset: rose plots of the histogram of angular orientation obtained from each FE-SEM image (From [55])

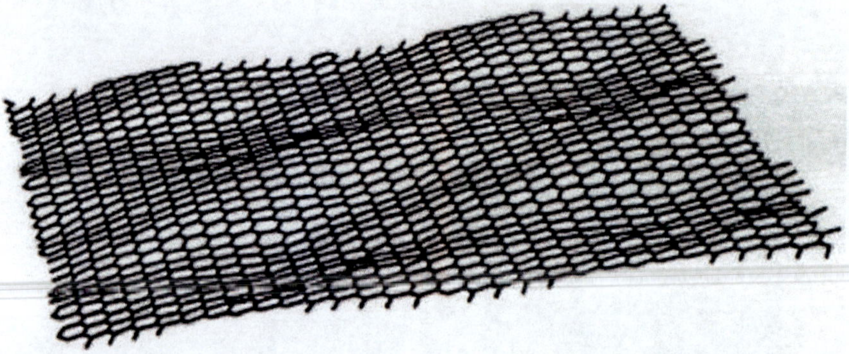

Fig. 12.4 Undulations and rippling of the graphene layer at 300°K, as reproduced by TBMD simulations using the NRL Hamiltonian. The height of the ripples is ~ 1Å. (From [70])

anomalies, given the complex conductivity and permeability, of the anomaly. We will investigate the enhancements needed to analyze problems of the type shown in these figures. The analyses described in [35, 109, 110, 112] will be appropriate here.

12.6 An Example of a Bianisotropic System

In [137, 138] and Fig. 12.6 we have an example of a bianisotropic system in which the electrical characteristics of the system are partially controlled by a magnetic field operating through magnetic spins. To be specific, this is a four-terminal carbon nanotube sensor for magnetic field measurements that relies upon spin-coherent electron transport in single-wall carbon nanotubes.

The Landauer formula for the conductance of the FM-SWCNT-FM magnetic tunnel junction system is [137]

$$G = \frac{2e^2}{h} \sum_{\sigma} Tr \left[\mathrm{Im}(\Sigma_L^r) \mathbf{G}^r \mathrm{Im}(\Sigma_R^r) \mathbf{G}^a \right]_{\sigma\sigma} , \qquad (12.4)$$

where $\sigma(\uparrow\downarrow)$ is the spin index. The retarded (advanced) Green's functions, $\mathbf{G}^{r(a)}$, are given by

$$\mathbf{G}^{r(a)}(E) = \frac{1}{E - H_{tube} - \Sigma_L^{r(a)} - \Sigma_R^{r(a)}} , \qquad (12.5)$$

where the self-energies are

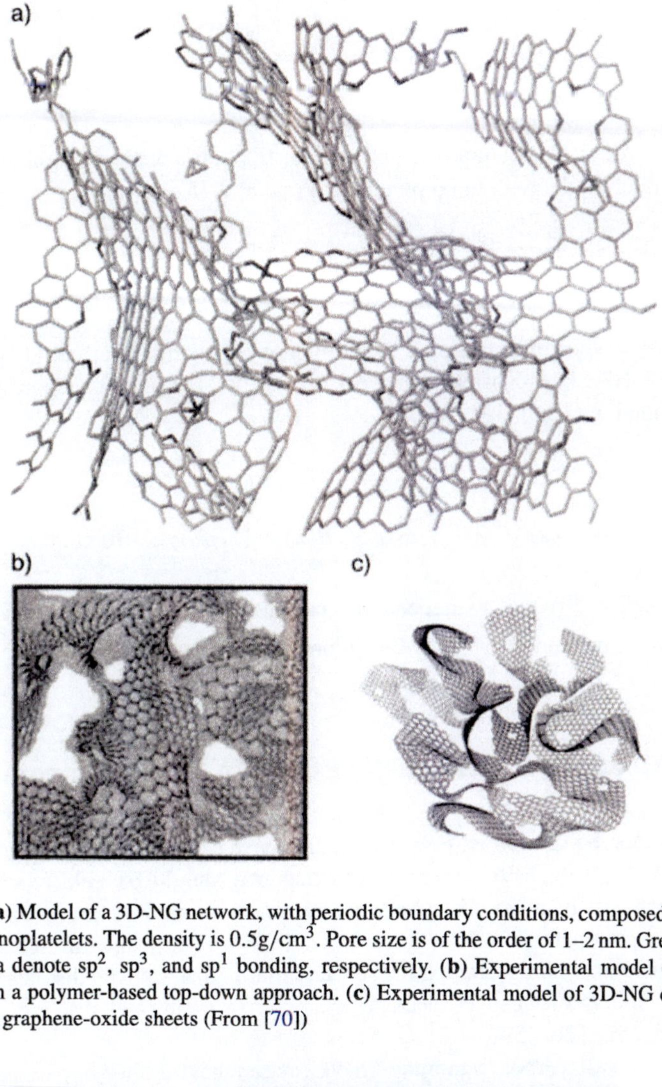

Fig. 12.5 (**a**) Model of a 3D-NG network, with periodic boundary conditions, composed of curved graphene nanoplatelets. The density is $0.5\text{g}/\text{cm}^3$. Pore size is of the order of 1–2 nm. Grey, orange, and magenta denote sp^2, sp^3, and sp^1 bonding, respectively. (**b**) Experimental model of 3D-NG derived from a polymer-based top-down approach. (**c**) Experimental model of 3D-NG derived by assembly of graphene-oxide sheets (From [70])

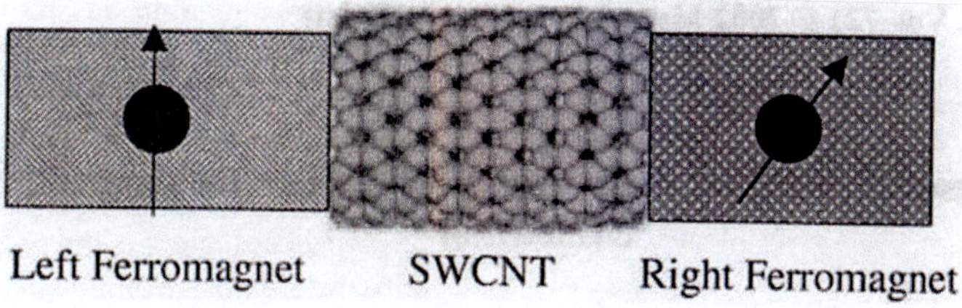

Fig. 12.6 Carbon nanotube magnetic junction model system (From [137])

$$\Sigma_{\alpha}^{r}(E) = \hat{R}_{\alpha} \begin{bmatrix} \Sigma_{\alpha\uparrow}^{r} & 0 \\ 0 & \Sigma_{\alpha\downarrow}^{r} \end{bmatrix} \hat{R}_{\alpha}^{\dagger}, \quad \hat{R}_{\alpha} = \begin{bmatrix} \cos\frac{\theta_{\alpha}}{2} & \sin\frac{\theta_{\alpha}}{2} \\ -\sin\frac{\theta_{\alpha}}{2} & \cos\frac{\theta_{\alpha}}{2} \end{bmatrix}, \tag{12.6}$$

where $\alpha = L, R$. The nanotube Hamiltonian, H_{tube}, is a nearest neighbor π−orbital tight-binding model with bond potential, $V_{pp\pi} = 2.75$ eV:

$$H_{tube} = -V_{pp\pi} \sum_{<ij>} a_i^{\dagger} a_j + c.c.\,. \tag{12.7}$$

This is another example of a tight-binding calculation that could be executed using the NRL Slater-Koster framework. The general class of bianisotropic materials is defined in (12.8) [135]:

$$\begin{aligned} \mathbf{D(x)} &= \epsilon_0 \underline{\underline{\epsilon}}_{bi} \cdot \mathbf{E(x)} + (\sqrt{\epsilon_0/\mu_0}) \underline{\underline{\alpha}}_{bi} \cdot \mathbf{B(x)} \\ \mathbf{H(x)} &= (\sqrt{\epsilon_0/\mu_0}) \underline{\underline{\beta}}_{bi} \cdot \mathbf{E(x)} + (1/\mu_0) \underline{\underline{\chi}}_{bi} \cdot \mathbf{B(x)}\,. \end{aligned} \tag{12.8}$$

The dispersive (frequency dependent) parameters, $\underline{\underline{\alpha}}_{bi}$ and $\underline{\underline{\beta}}_{bi}$, in addition to $\underline{\underline{\epsilon}}_{bi}$ and $\underline{\underline{\chi}}_{bi}$, are needed in **VIC-3D®** in order to work with bianisotropic materials.

12.7 Modeling Paramagnetic Effects in Carbon Nanotubes

Carbon nanotube reinforced polymers (cnrp) have excellent electromagnetic properties [55, 56, 126, 139], which means that one should be able to characterize and evaluate them nondestructively in much the same way as for carbon fiber reinforced polymers [30, 54, 70, 117]. There is a significant difference between the two advanced composites, however, and that is that CNRPs display significant magnetic effects that can be used to characterize them, and that are missing in CFRPs [55, 56, 126, 139].

A single-wall carbon nanotube (SWCN) can have either a paramagnetic or diamagnetic response to an applied magnetic field depending upon the tube's diameter, chirality, Fermi energy level, and the direction of the magnetic field relative to the tube axis [126].

The more interesting magnetic effect occurs because nanotubes, and their constituents can form with sizes smaller than the smallest ferromagnetic domain that can occur in these materials, as suggested in Fig. 12.7 [126]. Hence, the result is that the material behaves paramagnetically as a 'single-domain particle', and because there will be a number of atoms with unpaired spins in the lattice structure, we say that superparamagnetism results [6, 11, 16, 23, 25, 28, 30, 49, 55, 56, 67, 70–72, 74–76, 97, 113, 116–119, 124, 126, 136, 139, pp.410–418]. The magnetic particles

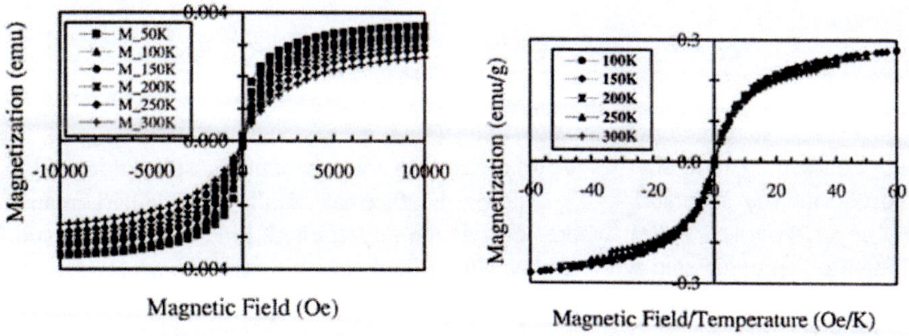

Fig. 12.7 Magnetization of a single-wall CNT sample as a function of magnetic field (left) and magnetic field/temperature (right) at constant temperatures. The fact that no coercivity is observed at temperatures above 100 K (left) and that there is overlapping of curves (right) suggest that we are seeing the effects of single magnetic domains in the magnetic particles that were used as catalysts for the growth of the CNTs, and remained with the CNTs during processing and synthesis of the nanocomposites (From [126])

that we are observing were used as catalysts for the growth of the CNTs, and remained with the CNTs during processing and synthesis of the nanocomposites [126]. The right-hand curve in Fig. 12.7 follows closely the familiar Langevin function, $L(x) = \coth x - 1/x$, where $x = \mu H/kT$. For low fields this function is approximately $\mu H/3kT$, whereas for high fields it gives $1 - kT/\mu H$ for the form of the approach to saturation [16].

We have studied paramagnetic effects for modeling masers [3, 29] and for possible applications to biomedical imaging for atherosclerosis (unpublished). We give an example of how paramagnetic and superparamagnetic effects can be modeled and included in **VIC-3D®**, or even used as a 'quantum sensor', in the next section.

12.7.1 Paramagnetic Spin Dynamics and the Spin Hamiltonian

In order to fully understand the possibilities of using paramagnetic phenomena to characterize CNT structures noninvasively, we must review a bit of electron-spin physics. Our interest is in the dynamic response of spins to time-varying fields. These fields are either applied electromagnetic fields or fluctuating fields due to random vibrations of the crystalline surroundings of the spin system. The discussion in this subsection and the next follows [96], which deals with spin dynamics in the crystalline field of a solid-state maser. Later we will discuss the changes that occur when the spin system is in a noncrystalline environment, such as biological tissue.

The system of equations used to describe spin dynamics is derived from Schrödinger's wave equation of quantum mechanics, and is given by

$$\frac{d\rho_{mn}}{dt} = \frac{j}{\hbar}\sum_{k}\left(\rho_{mk}\mathscr{H}_{kn} - \mathscr{H}_{mk}\rho_{kn}\right) + \sum_{pq} R_{mn,pq}\left(\rho_{pq} - \rho_{pq}^{(T)}\right), \tag{12.9}$$

where ρ_{mn} is the density matrix connecting energy states u_m and u_n of the unperturbed system, $R_{mn,pq}$ are real numbers that account for spin-lattice relaxation, and the superscript, T, denotes the thermal equilibrium density matrix. $\mathscr{H}_{jk} = \mathscr{H}_{0jk} + \mathscr{H}_{1jk}(t)$, where $\mathscr{H}_{0jk}$ is the unperturbed, time-independent spin-Hamiltonian associated with the crystalline field, and $\mathscr{H}_{1jk}(t) = gh\beta\,[\mathbf{H}(t)\cdot\mathbf{S}]_{jk}$ is the time-dependent perturbation. Here g is a constant, $h\beta$ the Bohr magneton, h Planck's constant, $\hbar = h/2\pi$, $\mathbf{H}(t)$ the time-dependent (rf) magnetic field, and $\mathbf{S} = S_x\mathbf{a}_x + S_y\mathbf{a}_y + S_z\mathbf{a}_z$ is the vector spin operator.

Because $\{u_m\}$ is an orthonormal system of eigenstates of $\mathscr{H}_0$, it follows immediately that $\mathscr{H}_{0mm} = E_m$, and all off-diagonal elements of $\mathscr{H}_{0mn}$ vanish. Furthermore, in order to get a linear (i.e., first-order in $\mathbf{H}(t)$) response for the overall system, we must set the diagonal terms of (12.9) to their thermal equilibrium values, $\rho_{mn}(t) = \rho_{mn}^{(T)}$, and solve the off-diagonal terms to first-order in $\mathbf{H}(t)$:

$$\frac{d\rho_{mn}}{dt} = \left(j\omega_{0mn} - \frac{1}{\tau_{mn}}\right)\rho_{mn} + \frac{j}{\hbar}\left(\rho_{mm}^{(T)} - \rho_{nn}^{(T)}\right)\mathscr{H}_{1mn}(t), \tag{12.10}$$

where $\omega_{0mn} = \dfrac{E_n - E_m}{\hbar}$, and the relaxation times, τ_{mn}, replace the $R_{mn,pq}$ of (12.9).

For a sinusoidally time-varying field, we have $\mathscr{H}_{1mn}(t) = \dfrac{g\beta h}{2}\left(\mathbf{H}e^{j\omega t} + \mathbf{H}^* e^{-j\omega t}\right)\cdot \mathbf{S}_{mn}$. If we assume solutions of (12.10) of the form $\rho_{mn} = A_{mn}e^{j\omega t} + B_{mn}e^{-j\omega t}$, then the coefficients of the positive-frequency terms, A_{mn}, and negative-frequency terms, B_{mn}, are given by

$$A_{mn} = \frac{(j/\hbar)\left(\rho_{mm}^{(T)} - \rho_{nn}^{(T)}\right)\tau_{mn} g\beta h/2}{1 - j\left(-\omega + \omega_{0mn}\right)\tau_{mn}}\mathbf{S}_{mn}\cdot\mathbf{H}$$

$$B_{mn} = \frac{(j/\hbar)\left(\rho_{mm}^{(T)} - \rho_{nn}^{(T)}\right)\tau_{mn} g\beta h/2}{1 - j\left(\omega + \omega_{0mn}\right)\tau_{mn}}\mathbf{S}_{mn}\cdot\mathbf{H}. \tag{12.11}$$

The magnetic dipole-moment operator for each spin is $g\beta h\mathbf{S}$, which means that the average dipole-moment for each spin is $\mathbf{m} = \mathrm{Tr}\,[\rho g\beta h\mathbf{S}]$, where Tr is the trace of an operator (sum of the diagonal elements of its matrix representation). The macroscopic dipole-moment per unit volume, $\mathbf{M}$, is obtained by multiplying $\mathbf{m}$ by the number density, N, of spins. Upon evaluating the trace, we find

$$\mathbf{M} = \gamma^2 \sum_{j<k} \mathbf{S}_{kj}\mathbf{S}_{jk}\left(N_j^{(T)} - N_k^{(T)}\right)\tau_{jk}$$

$$\left[\left(\frac{j/\hbar}{1-j(\omega_{0jk}-\omega)\tau_{jk}}-\frac{j/\hbar}{1+j(\omega_{0jk}+\omega)\tau_{jk}}\right)\mathbf{H}e^{j\omega t}\right.$$
$$\left.+\left(\frac{-j/\hbar}{1+j(\omega_{0jk}-\omega)\tau_{jk}}-\frac{-j/\hbar}{1-j(\omega_{0jk}+\omega)\tau_{jk}}\right)\mathbf{H}^{*}e^{-j\omega t}\right], \qquad (12.12)$$

where we have discarded the time-independent static dipole terms, $\mathbf{S}_{mm}\rho_{mm}^{(T)}$, and have set $\gamma^2 = g^2h^2\beta^2$. $N_j^{(T)}$ is the number of spins per-unit-volume occupying the jth energy level when the system is in thermal equilibrium at temperature T. If N is the total number of spins (or systems) in the crystal, then $N_j^{(T)} = \dfrac{N}{Z}\exp(-E_j/kT)$, where $Z = \sum_{j=1}^{J}\exp(-E_j/kT)$ and J is the total number of energy states. Thus, at thermal equilibrium (at positive temperatures), the lower energy states are more densely populated than the higher energy states.

The absorption spectrum, $A(\omega)$, is given by μ_0 times the imaginary part of the generalized magnetic susceptibility, which is the coefficient of $\mathbf{H}e^{j\omega t}$ in (12.12). In the vicinity of the resonant frequency, ω_{0jk}, the absorption spectrum is

$$A(\omega) \approx \mu_0\frac{\gamma^2}{2}\sum_{j<k}|S_{kj}|^2\left(N_j^{(T)}-N_k^{(T)}\right)\frac{\tau_{jk}/\hbar}{1+(\omega_{0jk}-\omega)^2\tau_{jk}^2}$$
$$= \mu_0\frac{\gamma^2}{2}\frac{N}{Z}\sum_{j<k}|S_{kj}|^2\left(e^{-E_j/kT}-e^{-E_k/kT}\right)\frac{\tau_{jk}/\hbar}{1+(\omega_{0jk}-\omega)^2\tau_{jk}^2}. \qquad (12.13)$$

This spectrum consists of 'lorentzian' curves (resonant curves) centered at the frequencies ω_{0jk}, with line-width $1/\tau_{jk}$. The peak of each resonance is proportional to τ_{jk}, and this gives us the familiar trade-off between bandwidth and magnitude of absorption (or magnitude of gain). The term $N_j^{(T)} - N_k^{(T)}$ yields the population difference per unit volume of the jth and kth energy levels when the system is in thermal equilibrium at temperature T. This population difference will be small if the energy differential, $E_k - E_j$, is small compared to the thermal energy, kT, as is the usual case for paramagnetic spin systems at normal temperatures. In addition to τ_{jk}, an important parameter is the 'line-strength', $|S_{kj}|^2$, or the transition matrix element connecting the jth and kth states. It determines the ease with which pump power is absorbed by the spins, or it determines the gain at signal frequencies.

12.7.2 Application to Fe^{3+} : TiO_2

The five unpaired electrons in Fe^{3+} are each in the 3d state, meaning that the ion is in an S-state, with a spin, $S = 5/2$. The total number of spin-states, therefore, is $N_s = 2S + 1 = 6$.

The spin-Hamiltonian, $\mathcal{H}_0$, for the Fe^{3+} : TiO_2 complex is [96]

$$\mathcal{H}_0 = g\beta \mathbf{H}_0 \cdot \mathbf{S} + D\left(S_z^2 - 35/12\right) + E\left(S_x^2 - S_y^2\right) + (a/6)\left(S_x^4 + S_y^4 + S_z^4 - 707/16\right)$$
$$+ (7/36)F\left(S_z^4 - (95/14)S_z^2 + 81/16\right) , \tag{12.14}$$

where the nominal values of the derived constants are $g = 2.0$, $D = 20.35\,\text{GHz}$, $E = 2.21\,\text{GHz}$, $a = 1.1\,\text{GHz}$, $F = -0.5\,\text{GHz}$, and H_0 is the dc magnetic field. S_x, S_y, and S_z are 6×6 Pauli spin-matrices. This Hamiltonian gives us frequency directly, rather than energy. The D term has axial symmetry (about the z-axis), and corresponds to the ion having an electric quadrupole moment, that is acted upon by the crystalline electric fields. The E term represents and additional nonaxially symmetric anisotropy in the xy plane, and corresponds to the ion's possessing an electric moment of higher order than quadrupolar. These are the main terms, as the size of D and E would suggest; the remaining terms are due to the fact that $S > 2$ and that the crystal symmetry is complicated. Clearly, these latter terms are less important, but must be included for completeness.

The eigenvalue equation that determines the unperturbed energy levels (or frequencies in this case) is

$$\mathcal{H}_0 u = E u , \tag{12.15}$$

and when this equation is solved as a function of $\mathbf{H}_0 = H\mathbf{a}_z$, we get the six curves shown in Fig. 12.8. The zero-field energies occur in pairs (Kramers' doublets), as is typical of a system with an odd number of electrons in an electric field (the crystalline field).

Consider the system at $H = 1.78$ kilogauss; the eigenvalues of $\mathcal{H}_0$ are

$$\begin{array}{lll} E_1 = -58.20 \times 10^9 h & E_2 = -54.15 \times 10^9 h & E_3 = -19.60 \times 10^9 h \\ E_4 = -5.64 \times 10^9 h & E_5 = 56.14 \times 10^9 h & E_6 = 81.05 \times 10^9 h \end{array} , \tag{12.16}$$

from which we derive the resonant frequencies (in GHz)

$$\begin{array}{lllll} \omega_{012} = 4.05 & \omega_{023} = 34.55 & \omega_{034} = 13.96 & \omega_{045} = 61.78 & \omega_{056} = 24.91 \\ \omega_{013} = 38.60 & \omega_{024} = 48.51 & \omega_{035} = 75.74 & \omega_{046} = 86.69 & \\ \omega_{014} = 52.56 & \omega_{025} = 110.29 & \omega_{036} = 100.65 & & \\ \omega_{015} = 114.34 & \omega_{026} = 135.20 & & & \\ \omega_{016} = 139.25 & & & & \end{array} . \tag{12.17}$$

The width of the absorption curve for the 1–2 (4.05 GHz) transition of $Fe^{3+} : TiO_2$ is 60MHz. Hence, the spin-lattice relaxation (or simply the *transverse relaxation*) time for the off-diagonal element, ρ_{12}, is $\tau_{12} = 1/2\pi \times 30 \times 10^6 = 5.305 \times 10^{-9}$ s. Figure 12.9 shows the absorption spectrum in the vicinity of 4.05 GHz with this value of τ_{12}.

This example illustrates the utility of the eigenstates in determining the frequency response of a maser. It relies, as we have noted, on knowledge of the crystalline-field

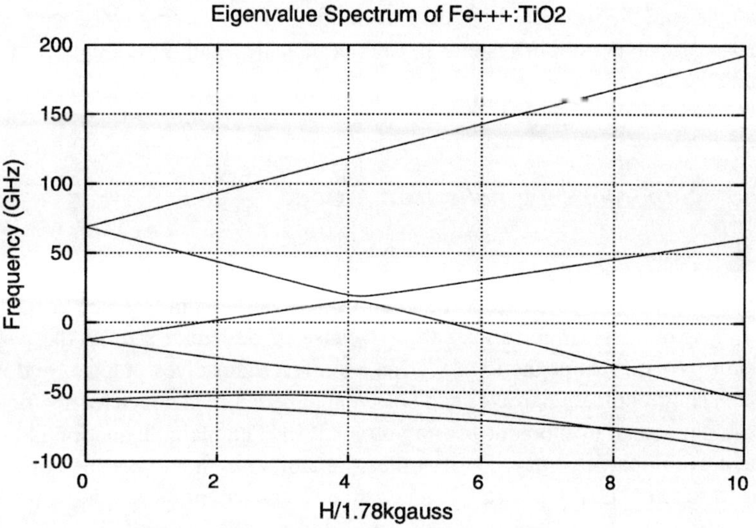

Fig. 12.8 Six-fold energy levels (in frequency units) for Fe^{3+} : TiO_2, as a function of the z-directed magnetic field, H

Fig. 12.9 Absorption spectrum in vicinity of 4.05 GHz, with $\tau_{12} = 5.305 \times 10^{-9}$ s. The half-power width is 60 MHz

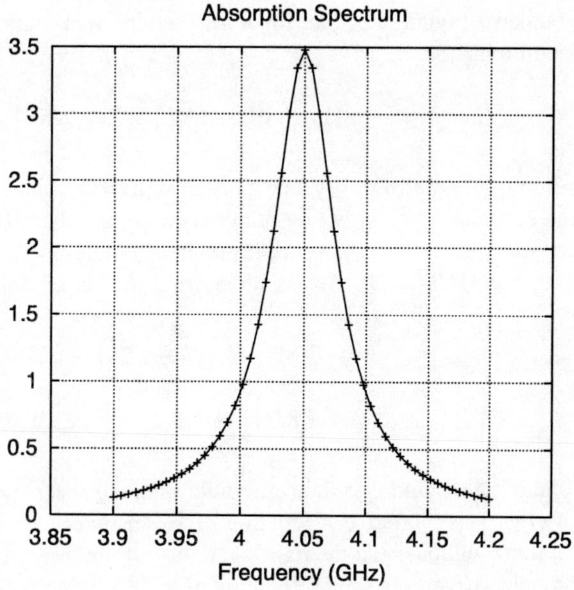

environment of the iron ion. It is this information that is lacking when we consider electron-paramagnetic spin systems in CNT composites, and is the subject of current research.

12.7.3 Superparamagnetic Iron Oxide

This is what started this discussion of paramagnetic spin-systems. Iron oxide, whether it is $\gamma - Fe_2O_3$, called 'maghemite,' or, perhaps magnetite, Fe_3O_4 ([79] is not clear on this), is ferromagnetic. Because of the small size of the particles ($\sim$ 10nm), their ferromagnetic properties manifest themselves in a single domain, and such single-domain particles can behave magnetically in a manner analogous to the paramagnetism of moment-bearing atoms [80]. The main distinction is that the moment of the particle may be 10^5 times the atomic moment, because of the 10^5 atoms ferromagnetically coupled by exchange forces within the single domain.[1]

Two Spins

We'll make a simple quantum-mechanical calculation of a system of two electrons coupled through the exchange interaction in a static magnetic field, $\mathbf{H}_0$. The Hamiltonian is

$$\mathscr{H} = -g\beta\mathbf{H}_0 \cdot \left(\mathbf{S}^{(1)} + \mathbf{S}^{(2)}\right) - 2J_{exch}\mathbf{S}^{(1)} \cdot \mathbf{S}^{(2)} , \quad (12.18)$$

where $g\beta = g \times 0.0014\,\text{GHz/gauss} = 2.8\,\text{GHz/kgauss}$, if we take $g = 2$. J_{exch} is the exchange energy, with a nominal value of 2.1×10^{-21} J. Dividing by Plancks constant, h, gives us the result in frequency units: $J_{exch}/h = 2.1 \times 10^{-21}/6.626 \times 10^{-34} = 3169.3\,\text{GHz}$. Hence, the normalized Hamiltonian for the system becomes

$$\begin{aligned}\mathscr{H} &= -2.8\mathbf{H}_0 \cdot \left(\mathbf{S}^{(1)} + \mathbf{S}^{(2)}\right) - 6338.7\mathbf{S}^{(1)} \cdot \mathbf{S}^{(2)} \\ &= -2.8H_0\left(S_z^{(1)} + S_z^{(2)}\right) - 6338.7\mathbf{S}^{(1)} \cdot \mathbf{S}^{(2)} ,\end{aligned} \quad (12.19)$$

where we assume that the static field is along the z-direction. The eigenspectrum of (12.19) is plotted as a function of H_0 in Fig. 12.10. The left-hand figure shows all four solutions, and the right-hand the bottom three eigenvalues. The two parallel branches have a constant separation of 6338.7, which is exactly $2J_{exch}$, where J_{exch} is the exchange energy. It is important to note that the transition (resonant) frequency

[1]A 'single-domain particle' is a particle that is in a state of uniform magnetization at any magnetic field [80].

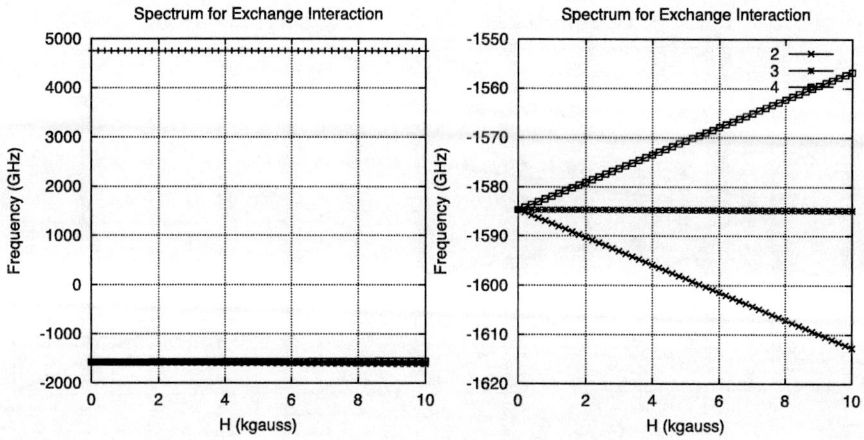

Fig. 12.10 Eigenspectrum of spin-Hamiltonian with exchange interaction. Left: complete spectrum. Right: expanded version of bottom three eigenvalues

between states 2 and 3 is the same between as between 3 and 4, for all values of H_0: $\omega_{0_{23}} = \omega_{0_{34}}$.

The absorption coefficient for the coupled two-spin system is

$$A(\omega) = 2 \times \mu_0 \frac{\gamma^2}{4} \frac{1}{Z} \left(e^{-E_2/kT} - e^{-E_4/kT} \right) \frac{\tau/\hbar}{1 + (\omega_0 - \omega)^2 \tau^2} . \tag{12.20}$$

The response is as if the two coupled spins behave as a single spin-system transiting from 'spin-up' (state 2) to 'spin-down' (state 4), which is what we would expect of a two-level (spin-1/2) system.

For comparison, we write down the result for two non-interacting spin-1/2 particles:

$$A(\omega) = 2 \times \mu_0 \frac{\gamma^2}{4} \frac{1}{Z} \left(e^{-E_2/kT} - e^{-E_3/kT} \right) \frac{\tau/\hbar}{1 + (\omega_0 - \omega)^2 \tau^2} . \tag{12.21}$$

Hence, the effect of the exchange interaction is to increase the density of spins in the thermal term by eliminating the middle energy term, $\exp(-E_3/kT)$. Because there is a greater differential in the energies than there was before, we have effectively a greater population difference between the two energy states 2 and 3 that are separated by $\hbar\omega_0$. Clearly, the more interacting spins we have, the greater this population difference becomes, and the greater the absorption spectrum becomes. Because of these two effects, with something of the order of 10^5 spins interacting through the exchange integral, the spectrum becomes significantly larger than in the simple paramagnetic case, giving rise to the name 'superparamagnetism.' We'll give a further example of this next.

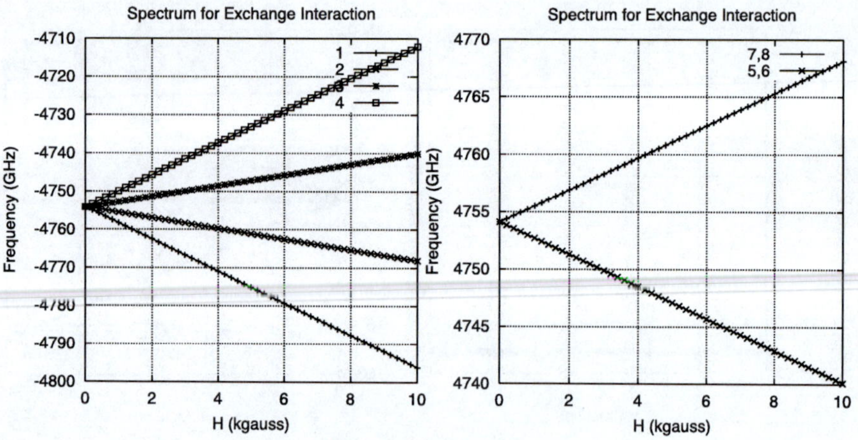

Fig. 12.11 Eigenspectrum of spin-Hamiltonian with exchange interaction for three electrons. Left: Spectrum of bottom four eigenvalues. Right: Spectrum of two largest (degenerate) eigenvalues. The separation of the average value of each spectral cluster is $9508.2 = 3 \times J_{exch}$ for all H_0

Three Spins

We'll extend the previous model to include three electrons interacting through the exchange integral. The Hamiltonian now becomes

$$\mathscr{H} = -2.8H_0\left(S_z^{(1)} + S_z^{(2)} + S_z^{(3)}\right) - 6338.7\left(\mathbf{S}^{(1)} \cdot \mathbf{S}^{(2)} + \mathbf{S}^{(1)} \cdot \mathbf{S}^{(3)} + \mathbf{S}^{(2)} \cdot \mathbf{S}^{(3)}\right), \tag{12.22}$$

where the various three-particle spin-matrices are obtained by taking three-fold left- and right-direct products of the Pauli spin matrices, **s**, with the two-dimensional identity matrix, I_2:

$$\mathbf{S}^{(1)} = \mathbf{s} \otimes I_2 \otimes I_2 \; \mathbf{S}^{(2)} = I_2 \otimes \mathbf{s} \otimes I_2 \; \mathbf{S}^{(3)} = I_2 \otimes I_2 \otimes \mathbf{s}\,. \tag{12.23}$$

The eigenspectrum of (12.22), plotted as a function of the static magnetic field, is shown in Fig. 12.11. As is the case with the two-electron problem, the separation between the lowest energy levels is constant and equal to $2.8H_0$ GHz where H_0 is in kGauss. This is identical to the result for a single electron with a spin of 1/2.

The absorption coefficient for this system is:

$$A(\omega) = \frac{\mu_0\gamma^2}{4Z}\left[3\left(e^{-E_1/kT} - e^{-E_4/kT}\right) + \left(e^{-E_2/kT} - e^{-E_3/kT}\right)\right]\frac{\tau/\hbar}{1+(\omega_0-\omega)^2\tau^2}\,. \tag{12.24}$$

Consider the left-parenthetical term, $3\left(e^{-E_1/kT} - e^{-E_4/kT}\right) = 3e^{-E_1/kT}\left(1 - e^{-(E_4-E_1)/kT}\right)$, of (12.24), where $E_4 - E_1 = 3\hbar\omega_0$. Under the usual conditions of room (or body) temperature, and a magnetic field of a few kGauss, the exponent, $(E_4 - E_1)/kT$ is of the order of 10^{-3}, which means that the term in parenthesis is approximately equal to $3 \times \hbar\omega_0/kT$, so that the absorption coefficient in (12.24) is approximately equal to

$$
\begin{aligned}
A(\omega) &\approx \frac{9\mu_0\gamma^2}{4Z} e^{-E_1/kT} \frac{\hbar\omega_0}{kT} \frac{\tau/\hbar}{1+(\omega_0-\omega)^2\tau^2} \\
&= \frac{\mu_0 g^2 h^2 (3\beta)^2}{4Z} \frac{e^{-E_1/kT}}{kT} \frac{\omega_0\tau}{1+(\omega_0-\omega)^2\tau^2}, \qquad (12.25)
\end{aligned}
$$

where β is the Bohr magneton (the magnetic dipole of a single spin). Therefore, we can conclude from (12.25) that the exchange interaction causes individual spins to align themselves parallel to each other, thereby producing an atomic system of spin-1/2, but with an equivalent dipole three times that of a single spin-1/2 particle. As indicated in (12.24), the effect of the AC field is to cause this single macrospin system to transit from spin-up to spin-down, with all individual spins remaining parallel to each other. In contrast, three noninteracting spins would have an equivalent absorption coefficient of

$$
\begin{aligned}
A(\omega) &= \frac{3\mu_0\gamma^2}{4Z} \left(e^{-E_1/kT} - e^{-E_2/kT}\right) \frac{\tau/\hbar}{1+(\omega_0-\omega)^2\tau^2} \\
&\approx \frac{3\mu_0\gamma^2}{4Z} \frac{e^{-E_1/kT}}{kT} \frac{\omega_0\tau}{1+(\omega_0-\omega)^2\tau^2}, \qquad (12.26)
\end{aligned}
$$

which is much smaller than (12.25). We can imagine what happens when 10^5 particles interact under exchange effects. This confirms, once again, that the exchange interaction associated with ferromagnetic single-domain particles gives rise to the notion of 'superparamagnetism.'

This model of superparamagnetism results from the large value of J_{exch}, because that isolates the upper energy levels of Figs. 12.10 and 12.11 from the lower levels for all (reasonable) values of H_0. The upper energy levels may be degenerate, as in Fig. 12.11 for the three-spin problem, but the lower levels are always nondegenerate, and have the same number of equal intervals as the number of spins. Furthermore, the large value of the exchange energy ensures that the upper levels will be virtually unpopulated compared to the lower levels. These facts are crucial to the theory.

12.8 Inverse Problems

12.8.1 Inverse Problem No. 1

Piezoresisivity manifests itself quantum-mechanically through the change, due to material strain, in the band-gap between the valence and conduction bands. If, as we assume, the atomic-orbital wave functions that go into the conductivity expression, (12.3), depend upon the band structure, then it follows that the conductivity can be directly computed as a function of strain from (12.3). This means that we can determine the state of strain throughout a region of the structure by determining the conductivity throughout that region. This is a classical problem of 'eddy-current' nondestructive evaluation (NDE) by either model-based inversion or voxel-based inversion [111]. This is what **VIC-3D®** was designed to solve. The inclusion of quantum models such as (12.3)into **VIC-3D®** is under development.

The inversion is a two-step process. First, we generate a family of red curves as in Fig. 12.2, each parameterized by a known value of band-gap. Depending upon the frequency of excitation of the 'eddy-current' probe, we may use much of the frequency range of Fig. 12.2, or more likely frequencies that are essentially DC. We create an interpolation table with given values of band-gap, and generate 'forward responses' for the measured impedance using the given $\sigma(\omega)$ corresponding to the nodal values of band-gap in the interpolation table. The forward response table, which we will call the 'surrogate model', is computed using **VIC-3D®** with the appropriate $\sigma(\omega)$. If the frequency range is large, then $\sigma(\omega)$ will not be constant, and we will develop a new data type for **VIC-3D®** to accomodate frequency-dependent electromagnetic parameters.

Once we have inverted the measured impedance data to determine the band-gap, we must still go through a second step to infer the strain. The procedure is reasonably straightforward, and we paraphrase the theoretical model described in [29], which is also based on a tight-binding model. The change in the band-gap for small strains is given by

$$\Delta E_{gap} = \text{sgn}(2p+1)3t_0[(1+\nu)\epsilon\cos 3\theta + \gamma\sin 3\theta] , \tag{12.27}$$

where $p = -1,\ 0,$ or 1, depending upon the value of $\mod(n-m, 3)$, and $n,\ m$ are the chiral indices of the CNT. t_0 is the tight-binding transfer integral, ν is Poisson's ratio, ϵ is the axial strain, γ is the torsional strain, and θ is the chiral angle.

Equation (12.27) works well for semiconducting and armchair CNTs with diameters larger than 1nm. For smaller CNTs, and for primarily metallic CNTs, curvature may play a role in accounting for the band-gap. Using a similar tight-binding approach, it can be shown that for primarily metallic CNTs

$$\Delta E_{gap} = -\text{sgn}\left(\frac{t_0a^2}{4d^2} - \frac{ab\sqrt{3}}{2}\right)\frac{ab\sqrt{3}}{2}\epsilon\cos 3\theta , \tag{12.28}$$

where a is the length of the graphene lattice unit vector, d is the diameter of the CNT, and b is the change in the transfer integral with change in bond length.

The zero-strain band-gap, E^0_{Gap}, is

$$E^0_{Gap} = \frac{2t_0 a}{\sqrt{3}d} \tag{12.29}$$

for a semiconducting CNT, and

$$E^0_{Gap} = 0 \tag{12.30}$$

for metallic CNTs.

To summarize: given the various quantum factors and the solution of (12.3), we can generate a surrogate model for the two state parameters, ϵ and γ, as well as the material parameter, ν, such that we can invert measured impedance data and reconstruct these three parameters using model-based inverse methods. This allows us to describe the distribution of these parameters in space. We can also use voxel-based inverse methods to solve the inverse problem.

12.8.2 A Thermally-Activated Transport Model

Rather than go through a full-blown analysis typified by (12.3), we can simplify the model with no loss of accuracy if we have empirical knowledge of a 'standard cnt.' We use a thermally activated transport model of conductivity to write [29, 80]

$$\sigma(E_g) = \frac{C}{1 + \exp(E_g/kT)}, \tag{12.31}$$

where E_g is the band gap (the Fermi level is assumed to lie at the top of the valence band), C is a factor to be determined, k is Boltzmann's constant, and T the absolute temperature. If we know the conductivity at a given band gap energy, E_0, and temperature, T_0, to be σ_0, then we can determine C to be $\sigma_0\,(1 + \exp(E_0/kT_0))$, and write

$$\sigma(E_g) = \sigma_0 \left[\frac{1 + \exp(E_0/kT_0)}{1 + exp(E_g/kT)}\right]. \tag{12.32}$$

This, then, replaces (12.3), and allows the rest of the inversion process to proceed as before. Alternatively, we can compute σ_0 using (12.3) (assuming we know the E_0 that went into the model), and then use (12.32) to compute the other members of the surrogate interpolation table entries, and continue the inversion process.

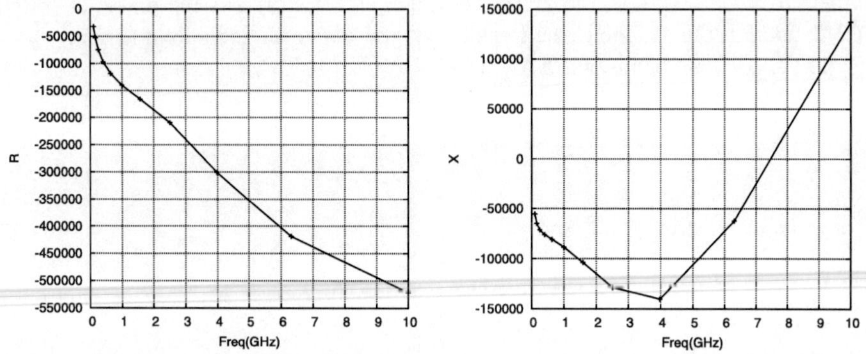

Fig. 12.12 Showing the input model impedance data for $\sigma = 3500$ S/m

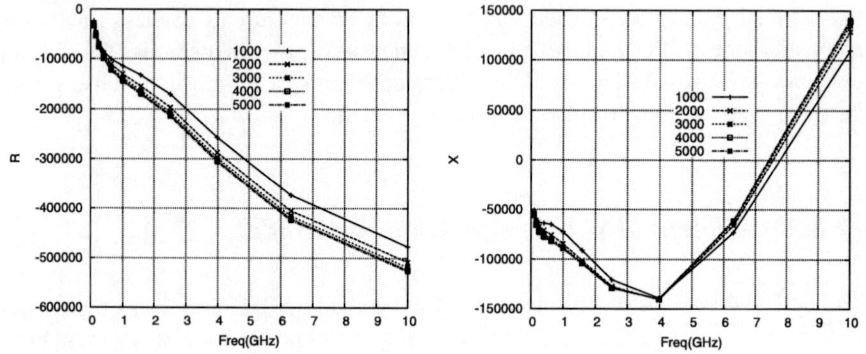

Fig. 12.13 Showing the interpolating impedance functions parameterized by nodal conductivity values listed in the legend

12.8.3 A Simple Inverse Problem

Assume that $\sigma_0 = 5000$ S/m in (12.32), and that $E_0 = 0$. This value of σ_0 is consistent with the 'zero-frequency' result for the red curve in Fig. 12.2. Our objective is to determine E_g from a reconstruction of the conductivity, $\sigma(E_g)$, using the setup suggested in Fig. 12.17. The probe will be excited at 11 frequencies covering the range 0.1 GHz to 10 GHz, and the 'measured' impedance response is shown in Fig. 12.12 for $\sigma(E_g) = 3500$ S/m. These data will be submitted to NLSE, the nonlinear least-squares estimator in **VIC-3D®**, after we have developed the surrogate interpolating system. This is done by modeling the responses at five values of conductivity: $\sigma = 1000,\ 2000,\ 3000,\ 4000,\ 5000$ S/m, which then become the nodal values for the interpolation table. The interpolating impedance responses are shown in Fig. 12.13.

The result of the inversion produces an estimate of $\sigma(E_g)$ to be 3531 S/m, which is quite close to the true value. If we assume that the 'experiment' that produced the measured data in Fig. 12.12 was performed at $T = 300°$ K, then we can solve (12.32) for E_g

$$\begin{aligned} E_g &= kT \ln[2\sigma_0/\sigma(E_g) - 1] \\ &= 0.02586 \ln\left[\frac{10000}{\sigma(E_g)} - 1\right] \\ &= 0.0157\text{eV} . \end{aligned} \tag{12.33}$$

Now, assuming that all other parameters in (12.28) are known, we can solve for ϵ by inserting $\Delta E_{gap} = -0.0157$ into the left-hand side.

This example illustrates the fact that we can apply the methods of eddy-current NDE to carbon nanotubes once we have rigorous models for computing the electromagnetic properties of these tubes. These models require quantum-mechanical calculations. Further examples of more complicated model-based inverse problems, in which a number of variables are jointly determined, are given in [111].

We can scan the system of Fig. 12.17 and apply voxel-based inverse methods to map out E_g throughout the region using the models in Chap. 2.

12.8.4 Voxel-Based Inversion: A Surface-Breaking Checkerboard at 50 MHz

In this experiment we consider the complex checkerboard flaw that is shown in Fig. 12.14, and perform a 16×16 raster scan with the transmitter and receiver probes at 50 MHz. The number in each voxel is the anomalous conductivity, which is the actual conductivity minus the host conductivity, divided by the host conductivity. Thus, a zero indicates that that voxel contains host material, whereas -1 indicates that the actual conductivity is zero.

The checkerboard is a difficult anomaly to reconstruct, because the 'scene' changes so rapidly; i.e., it contains high spatial frequencies and the receiver scan must be fine enough to reconstruct these frequencies. The reconstruction using the LMS-estimator is quite good (corresponding results were obtained using the S-estimator and classical estimator). We show the reconstruction of the middle row and column in Fig. 12.15.

12.8.5 Voxel-Based Inversion: A Buried Checkerboard

When we bury the checkerboard below a layer of host material we get the added complication of increasing the number of unknowns that are to be determined during the reconstruction. This, together with the fact that the checkerboard scene contains

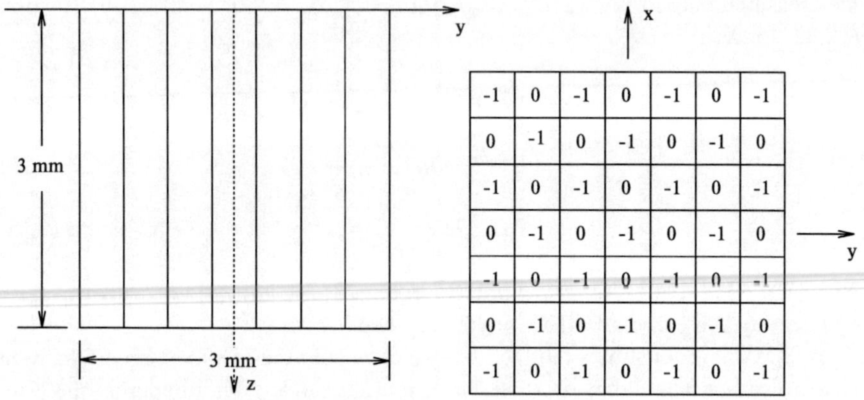

Fig. 12.14 A surface-breaking checkerboard

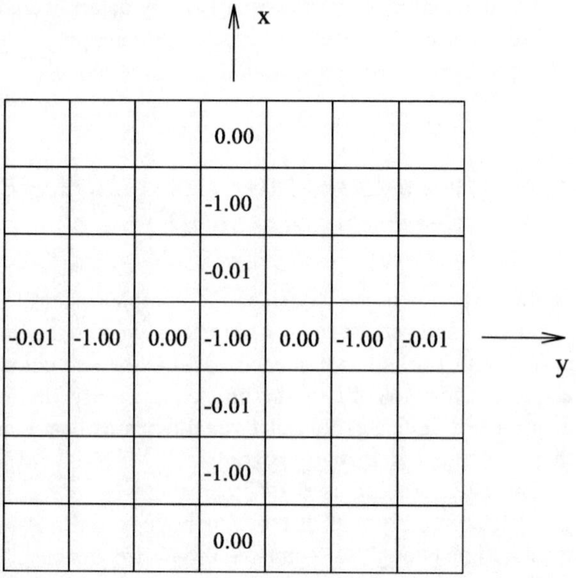

Fig. 12.15 Reconstruction of the middle row and column of the surface-breaking checkerboard

high spatial-frequencies, suggests that we use a 31×31 raster scan for the receiver probe, rather than the 16×16 scan that was used in the preceding example, but covering the same area. The resolution of the receiver scan was thereby improved to 0.2 mm over 0.4 mm.

When we do this we get an excellent reconstruction; the top layer is reconstructed exactly (zero anomalous conductivity), and the reconstructions of the middle row and column of the bottom layer are shown in Fig. 12.16.

Now, we can repeat our earlier calculation to determine the band-gap distribution within the anomalous region. Clearly, where the solution is zero, the band-gap is also

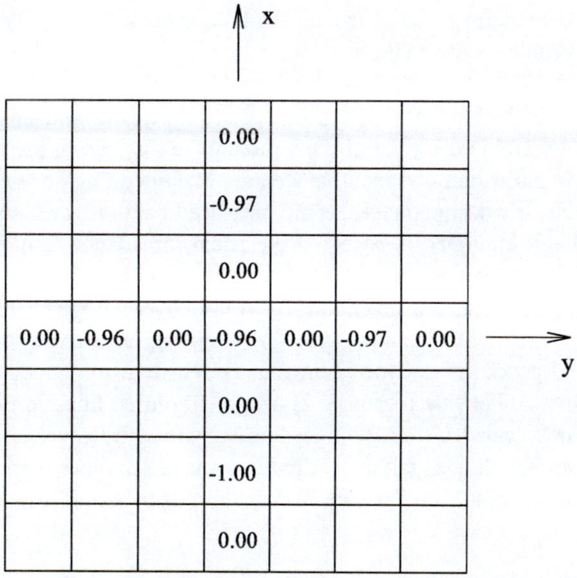

Fig. 12.16 Reconstruction of the middle row and column of the buried checkerboard

zero, as is the strain, ϵ, in (12.28). On the other hand, when the solution is -1, then (12.32) indicates that $E_g >> kT = 0.02586$, and, therefore, that ϵ is quite large.

12.8.6 Spatial Imaging Using Embedded CNT Sensors

These two examples of inverse methods that are available to us via **VIC-3D®** indicate that we can efficiently produce highly accurate numerical results that also include sophisticated stochastic reliability metrics that are not easily achieved using other methods. (See [111] for more on these metrics with model-based inverse methods.)

Our interest is in NDE of CNT structures, in which we exploit the piezoresistivity of the structure itself. We use external inductive sensors typically used in eddy-current NDE. This differs from the use of CNT sensors embedded within a composite structure to measure the damage, as in structural health monitoring (SHM). We believe that our approach yields more reliable estimates of the state of the structure, and would have to be done, even if an embedded sensor indicated an anomaly within the structure.

Wan et al. [133] have developed an approach to damage analysis of 3D braided composite materials using embedded CNT thread sensors. They cast the problem of thread distribution within the braided structure as a combinatorial optimization problem, and use particle swarm optimization (PSO) to solve it. They point out the possibility in some of their experiments that the sensor, itself, may have been damaged, rather than the composite structure. This is a continuing reliability problem with SHM.

Dai et al. [32] use a different approach, in that a separate CNT-array sensor is embedded within the composite structure, and an inverse problem is used to determine the state of the structure. In this sense, their approach is similar to ours in philosophy, except that we are not embedding our sensors. Furthermore, [32] use electrical impedances as the measured data for inversion, which is what we do, but their approach is based on electrical impedance tomography (EIT) to acquire and invert the data. The sensor array used in EIT is conductive and not inductive, as in eddy-current inversion. By conductive, we mean that the sensor is in electrical contact with the structure. The resulting electromagnetic model produces a rather ill-posed inverse problem that is sensitive to modeling errors and measurement noise. For that reason, [32] uses a simplified linear process that uses the maximum *a posteriori* (MAP) algorithm to solve the inverse problem. Furthermore, this approach does not appear to be amenable to more complex (and realistic) problems that include anisotropies and random distributions of anomalies.

12.8.7 Inverse Problem No. 2: Characterizing the CNT via ESR

We can use electron-spin resonance (ESR) as a method of characterizing the CNT, in much the same way that nuclear magnetic resonance (NMR) maps biological structures. That would be a straightforward application of the spintronics theory that we just developed. The method would work at GHz frequencies, rather than MHz, as in MRI, but we expect to work at GHz and higher for NDE of CNT in any case. This would be an interesting application of our spiral coil models in **VIC-3D®**. Information from the absorption curve of Fig. 12.9 could be used for this. It seems likely that if the environment of the spins changes, perhaps because of stresses, then the crystal-field would also change, causing the spin-Hamiltonian to change with it. This would cause the eigenvalues to change to some degree that would have to be determined numerically, thereby shifting the center of the absorption curve.

12.8.8 What Does VIC-3D® Need?

Refer to Fig. 12.17. The anomalous region, which will be the CNT structure, is embedded in a host, which at this time is isotropic and could be air. **VIC-3D®** expects to be given the usual electromagnetic parameters, namely complex conductivity and magnetic permeability, that are to be assigned to each voxel in the anomalous region. The parameters can be (bi)anisotropic and stochastic and even frequency dependent. The results of the quantum-mechanical models are

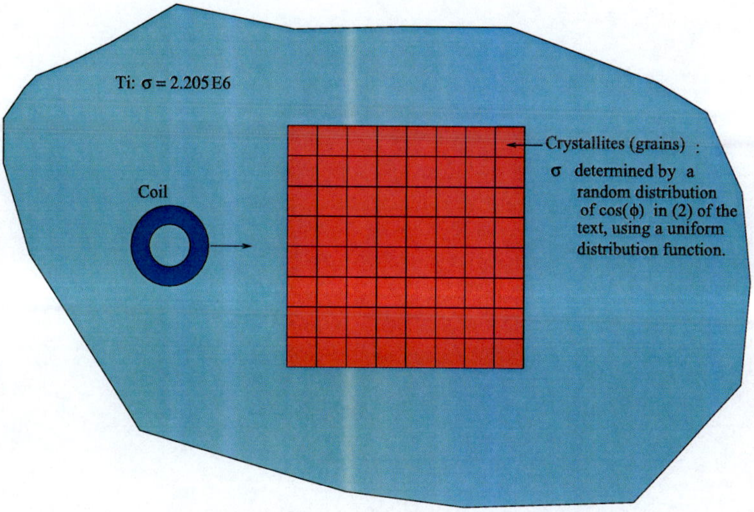

Fig. 12.17 Illustrating the model setup for **VIC-3D®**. There is a second layer identical to the one shown immediately below the one shown. This is due to the fact that **VIC-3D®** requires a minimum of two cells in any direction of the grid

translated into these EM parameters off-line and then read into **VIC-3D®**. The system is excited by one of a number of different EC probes, as stated above, and the computed observable is the change in impedance that the probe sees for the anomalous region relative to the host.

The absorption coefficient developed in the note on paramagnetic effects (Eq. (12.13), Fig. 12.9) is the imaginary part of the magnetic permeability, and is called 'magnetic loss' in the current version of **VIC-3D®**. It is frequency-dependent. Similarly, the σ described in the discussion of nanographene (Eq. (12.3)) is frequency dependent, and is modeled in **VIC-3D®**.

We can simulate the spin-coherent transport feature of the magnetic tunnel junction system of Fig. 12.6 by letting adjacent voxels of Fig. 12.17 carry spin systems that are tied together stochastically by a correlation function with a relatively large correlation length. The problem can then be treated stochastically as is done in our treatment of random surfaces [110].

References

1. A. Abubakar, T.M. Habashy, Near well-bore imaging of the triaxial induction logging data using the multiplicative regularized contrast source inversion method, in *23rd Annual Review of Progress in Applied Computational Electromagnetics, March 19–23, Verona* (Applied Computational Electromagnetics Society, 2007), pp. 653–660
2. B.A. Abu-Nabah, P.B. Nagy, Lift-off effect in high-frequency eddy current conductivity spectroscopy. NDT&E Int. **40**, 555–565 (2007)
3. N.H. Alamusi, H. Fukunaga, S. Atobe, Y. Liu, J. Li, Piezoresistive strain sensors made from carbon nanotubes based polymer nanocomposites. Sensors **2011**(11), 10691–10723 (2011)
4. J. Allen, et al., A technology plan for electromagnetic characteristics of advanced composites. Rochester Institute of Technology, Prepared for Rome Air Development Center (1976)
5. J.L. Allen, et al., Electromagnetic properties and effects of advanced composite materials: measurement and modeling. RADC-TR-78-156, Phase Report (ADA 05804) (1978)
6. P.D. Allen, T.G. St. Pierre, W. Chua-anusorn, V. Ström, K.V. Rao, Low-frequency low-field magnetic susceptibility of ferritin and hemosiderin. Biochimica et Biophysica Acta **1500**, 186–196 (2000)
7. C. Altman, A. Schatzberg, Appl. Phys. **B26**, 147–153 (1981)
8. C. Altman, A. Schatzberg, Appl. Phys. **B28**, 327–333 (1982)
9. C. Altman, A. Schatzberg, K. Suchy, IEEE Trans. Antennas Propag. **AP-32**(11), 1450–1450 (1984)
10. G. Angiulli, T. Isernia, S. Tringali, Modeling realistic contrast maps from MRI images for microwave breast cancer detection. IEEE Antennas Propag. Mag. **53**(1), 113–122 (2011)
11. J.-P. Ansermet, Classical description of spin wave excitation by currents in bulk ferromagnets. IEEE Trans. Magn. **40**(2), 358–360 (2004)
12. J.-P. Aubin, *Approximation of Elliptic Boundary-Value Problems* (Wiley, New York, 1972)
13. H.T. Banks, D. Cioranescu, A.K. Criner, W.P. Winfree, Parameter estimation for the heat equation on perforated domains. J. Inverse Ill-posed Problems **19**, 825–857 (2011)
14. S. Barkeshli, D.J. Radecki, H.A. Sabbagh, On a linearized inverse scattering model for a three-dimensional flaw embedded in anisotropic advanced composite materials. IEEE Trans. Geosci. Remote Sens. **30**(1), 71–80 (1992)
15. V. Barthelmann, E. Novak, K. Ritter, High dimensional polynomial interpolation on sparse grids. Adv. Comput. Math. **12**, 273–288 (2000)
16. C.P. Bean, J.D. Livingston, Superparamagnetism. J. Appl. Phys. **30**(4), 120S-129S (1959)
17. D.W. Berreman, J. Opt. Soc. Am. **62**(4), 502–510 (1972)

H. A. Sabbagh et al., *Advanced Electromagnetic Models for Materials Characterization and Nondestructive Evaluation*, Scientific Computation,
https://doi.org/10.1007/978-3-030-67956-9

18. P.R. Bevington, D.K. Robinson, *Data Reduction and Error Analysis for the Physical Sciences* (McGraw-Hill Book Company, New York, 1992)
19. M.P. Blodgett, C.V. Ukpabi, P.B. Nagy, Surface roughness influence on eddy current electrical conductivity measurements. Mater. Eval. **61**, 765–772 (2003)
20. S.K. Burke, L.R.F. Rose, Proc. Roy. Soc. **418A**, 229–246 (1988); also Appl. Comput. Electromagn. Soc. Newsl. **6**(1), 30–34 (1991)
21. R.H. Byrd, C.L. Dert, A.H.G. Rinnooy Kan, R.B. Schnabel, Concurrent stochastic methods for global optimization. Math. Program. **46**, 1–29 (1990)
22. C. Cai, M. Lambert, Sparse grid-nested sampling for model selection in eddy-current testing, in *Conference Paper, 20th International Workshop on Electromagnetic Nondestructive Evaluation (ENDE 2015), Sendai* (2015)
23. R. Casanas, H. Scharfetter, A. Altes, A. Remacha, P. Sarda, J. Sierra, J. Rosell, In-vitro measurement of iron concentration in human hepatic tissue by magnetic induction methods, in *2001 Proceedings of the 23rd Annual EMBS International Conference, October 25–28, Istanbul, Turkey* (2001), pp. 2971–2974
24. Y. Censor, Row-action methods for huge and sparse systems and their applications. SIAM Rev. **23**, 444–446 (1981)
25. S.B. Chaves, L.M. Lacava, Z.G.M. Lacava, O. Silva, F. Pelegrini, N. Buske, C. Gansau, P.C. Morais, R.B. Azevedo, Light microscopy and magnetic resonance characterization of a DMSA-coated magnetic fluid in mice. IEEE Trans. Magn. **38**(5), 3231–3233 (2002)
26. R.E. Collin, *Field Theory of Guided Waves* (McGraw-Hill, New York, 1960)
27. P.L. Combettes, The foundations of set theoretic estimation. Proc. IEEE **81**(2), 182–208 (1993)
28. B. Coqblin, *The Electronic Structure of Rare-Earth Metals and Alloys: The Magnetic Heavy Rare-Earths* (Academic Press, London, 1977)
29. M.A. Cullinan, M.L. Culpepper, Carbon nanotubes as iezoresistive microelectromechanical sensors: theory and experiment. Phys. Rev. B **82**, 115428 (2010)
30. B.D. Cullity, *Introduction to Magnetic Materials* (Addison-Wesley, Reading, 1972)
31. E. Dadrasnia, S. Puthukodan, H. Lamela, Terahertz electrical conductivity and optical characterization of composite nonaligned single- and multiwalled carbon nanotubes. J. Nanophoton. **8**, 1–10 (2014)
32. H. Dai, G.J. Gallo, T. Schumacher, E.T. Thostenson, A novel methodology for spatial damage detection and imaging using a distributed carbon nanotube-based composite sensor combined with electrical impedance tomography. J. Nondestructive Eval. **35**, 26 (2016)
33. B.J. Debusschere, H.N. Najm, P.P. Pébay, O. Knio, R.G. Ghanem, O.P. LeMaítre, Numerical challenges in the use of polynomial chaos representations for stochastic processes. SIAM J. Sci. Comput. **26**, 698–719 (2006)
34. T.M.R. Ellis, I.R. Phillips, T.M. Lahey, *Fortran 90 Programming*. (Addison-Wesley, Harlow, 1994)
35. T. Fast, A.E. Scott, H.A. Bale, B.N. Cox, Topological and euclidean metrics reveal spatically nonuniform structure in the entanglement of stochastic fiber bundles. J. Mater Sci. **50**, 2370–2398 (2015)
36. R. Fletcher, *Practical Methods of Optimization*, 2nd edn. (Wiley, Hoboken, 1987)
37. J. Foo, G. Karniadakis, Multi-element probabilistic collocation method in high dimensions. J. Comput. Phys. **229**, 1536–1557 (2010)
38. K.O. Friedrichs, H.N. Shapiro, et al., *Integration of Functionals*. Institute of Mathematical Sciences (New York University, New York, 1957)
39. C. Gabriel, S. Gabriel, E. Corthout, The dielectric properties of biological tissues: I. Literature survey. Phys. Med. Biol. **41**, 2231–2249 (1966)
40. S. Gabriel, R.W. Lau, C. Gabriel, The dielectric properties of biological tissues: II. Measurements in the frequency range 10 Hz to 20 GHz. Phys. Med. Biol. **41**, 2251–2269 (1966)
41. S. Gabriel, R.W. Lau, C. Gabriel, The dielectric properties of biological tissues: III. Parametric models for the dielectric spectrum of tissues. Phys. Med. Biol. **41**, 2271–2293 (1996)

42. Z. Gao, J. Hesthaven, On ANOVA expansions and strategies for choosing the anchor point. Appl. Math. Comput. **217**(7), 3274–3285 (2010)
43. J. Garcke, Sparse grid tutorial. http://en.wikipedia.org/wiki/Sparse-grid
44. J. Garcke, M. Griebel, Data mining with sparse grids using simplicial basis functions. http://en.wikipedia.org/wiki/Sparse-grid
45. W. Graf, J. Hamm, J.E. Nanevicz, D.E. Tremain, Engineering effects of advanced composite materials on avionics. SRI International, prepared for U.S. Army Avionics R&D Command (ADA 104015) (1981)
46. M. Griebel, Sparse grids and related approximation schemes for higher dimensional problems, in *Proceedings Foundations of Computational Mathematics 2005, (FoCM05), Santander*, ed. by L. Pardo, A. Pinkus, E. Suli, M. Todd (Cambridge University Press, Cambridge, 2006), pp. 106–161
47. R. Grow, in *Carbon Nanotubes: Properties and Applications*, ed. by M. O'Connell (Taylor & Frances, New York, 2006)
48. J.C. Hancock, P.A. Wintz, *Signal Detection Theory* (McGraw-Hill Book Company, New York, 1966)
49. R. Hergt, W. Andrä, C.G. d'Ambly, I. Hilger, W.A. Kaiser, U. Richter, H.-G. Schmidt, Physical limits of hyperthermia using magnetite fine particles. IEEE Trans. Magn. **34**(5), 3745–3754 (1998)
50. G.T. Herman, L.B. Meyer, Algebraic reconstruction techniques can be made computationally efficient. IEEE Trans. Med. Imag. **12**(3), 600–609 (1993)
51. G.T. Herman, A. Lent, H. Hurwitz, A storage-efficient algorithm for finding the regularized solution of a large, inconsistent system of equations. J. Inst. Math. Appl. **25**, 361–366 (1980)
52. M. Hestenes, *Conjugate Direction Methods in Optimization*. (Springer, New York, 1980)
53. K. Kalyanasundaram, P.B. Nagy, A simple numerical model of the apparent loss of eddy current conductivity due to surface roughness. NDT&E Int. **37**, 47–56 (2004)
54. A.C. Katageri, B.G. Sheeparamatti, Carbon nanotube based piezoresistive pressure sensor for wide range pressure sensing applications - a review. **4**(08), 665–671 (2015). ISSN: 2278-0181
55. J.-W. Kim, G. Sauti, E.J. Siochi, J.G. Smith, R.A. Wincheski, R.J. Cano, J.W. Connel, K.E. Wise, Toward high performance thermoset/carbon nanotube sheet nanocomposites via resistive heating assisted infiltration and cure. Appl. Mater. Interfaces **6**, 18832–18843 (2014)
56. J.-W. Kim, G. Sauti, R.J. Cano, R.A. Wincheski, J.G. Racliffe, M. Czabaj, E.J. Siochi, Structural CNT composites part II: assessment of CNT yarns as reinforcement for comosite overwrapped pressure vessels. Paper Number 1500, (2015)
57. A. Klimke, Sparse grid interpolation toolbox user's guide, V. 5.1 (2008). http://www.ians.uni-stuttgart.de/spinterp
58. A. Klimke, B. Wolhmuth, Algorithm 847: spinterp: piecewise multilinear hierarchical sparse grid interpolation in matlab. ACM Trans. Math. Softw. **31**(4), 561–579 (2005)
59. C.M. Krowne, *IEEE Antennas Propagation Symposium Digest, Boston, MA, June 25–29* (1984), pp. 569–572
60. C.M. Krowne, IEEE Trans. Microwave Theory Tech. **MTT-32**(12), 1617–1625 (1984)
61. C.M. Krowne, IEEE Trans. Antennas Propag. **AP-32**(11), 1224–1230 (1984)
62. C.L. Lawson, R.J. Hanson, *Solving Least Squares Problems* (Prentice-Hall, Englewood Cliffs, 1974)
63. M. Loève, *Probability Theory* (D. Van Nostrand, New York, 1955)
64. D.L. Luenberger, *Introduction to Linear and Nonlinear Programming* (Addison-Wesley, Boston, 1973)
65. X. Ma, N. Zabaras, An adaptive high-dimensional stochastic model representation technique for the solution of stochastic partial differential equations. J. Comput. Phys. **229**, 3884–3915 (2010)
66. I. Maeng, C. Kang, S.J. Oh, J.-H. Son, K.H. An, Y.H. Lee, Terahertz electrical and optical characteristics of double-walled carbon nanotubes and their comparison with single-walled carbon nanotubes. Appl. Phys. Lett. **90**, 1–3 (2007)

67. S. Maenosono, S. Saita, Theoretical assessment of FePt nanoparticles as heating elements for magnetic hyperthermia. IEEE Trans. Magn. **42**(6), 1638–1642 (2006)
68. C.-S. Man, R. Paroni, Y. Xiang, E.A. Kenik, On the geometric autocorrelation function of polycrystalline materials. J. Comput. Appl. Math. **190**, 200–210 (2006)
69. Y. Marzouk, D. Xiu, A stochastic collocation approach to Bayesian inference in inverse problems. Commun. Comput. Phys. **6**(4), 826–847 (2009)
70. C. Mathioudakis, P.C. Kelires, Modelling of three-dimensional nanographene. Nanoscale Res. Lett. **11**, 151 (2016)
71. R.D. McMichael, P. Krivosik, Classical model of extrinsic ferromagnetic resonance linewidth in ultrathin films. IEEE Trans. Magn. **40**(1), 2–11 (2004)
72. E. Merzbacher, *Quantum Mechanics* (Wiley, New York, 1961)
73. E.K. Miller, Adaptive sparse sampling to estimate radiation and scattering patterns to a specified uncertainty with model-based parameter estimation. IEEE Antennas Propag. Mag. 103–113 (2015)
74. G. Mone, E. Svoboda, Precision-Guided Tumor Killers. Popular Science (2006), p. 56
75. P.C. Morais, E.C.D. Lima, D. Rabelo, A.C. Reis, F. Pelegrini, Magnetic resonance of magnetite nanoparticles dispersed in mesoporous copolymer matrix. IEEE Trans. Magn. **36**(5), 3038–3040 (2000)
76. P.C. Morais, G.R.R. Gonçalves, K.S. Neto, F. Pelegrini, N. Buske, Study of particle-particle interaction in magnetic fluids using magnetic resonance. IEEE Trans. Magn. **38**(5), 3225–3227 (2002)
77. J.J. Moré, B.S. Garbow, K.E. Hillstrom, User Guide for Minpack-1, ANL-80-74, Argonne National Laboratory (1980)
78. M. Nakhkash, Y. Huang, M.T.C. Fang, Application of the multilevel single-linkage method to one-dimensional electromagnetic inverse scattering problem. IEEE Trans. Antennas Propag. **47**(11), 1658–1668 (1999)
79. Nanoparticles beat back atherosclerosis, Science News, June 11, 2016
80. W. Obitayo, T. Liu, A review: carbon nanotube-based piezoresistive strain sensors. J. Sens. **2012**, 652438 (2012)
81. F.W.J. Olver, D.W. Lozier, R.F. Boisvert, C.W. Clark (eds.), *NIST Handbook of Mathematical Functions* (Cambridge University Press, Cambridge, 2010)
82. C.N. Owston, Carbon fibre reinforced polymers and non-destructive testing. Br. J. NDT **15**(6), 2–11 (1973)
83. C.N. Owston, Eddy current methods for the examination of carbon fibre reinforced epoxy resins. Mater. Eval. **34**, 237–250 (1976)
84. D.A. Papaconstantopoulos, M.J. Mehl, The Slater-Koster tight-binding method: a computationally efficient and accurate approach. J. Phys. Condens. Matter **15**, R413–R440 (2003)
85. J.L. Phelps, In-service inspection methods for graphite-epoxy structures on commercial transport aircraft (NASA-CR-165746). Final Report, Boeing Commercial Airplane Co., Seattle (1981)
86. R. Prakash, Non-destructive testing of composites. Composites **11**, 217–224 (1980)
87. R. Prakash, C.N. Owston, Eddy-current method for the determination of lay-up order in cross-plied Crfp laminates. Composites **7**(2), 88–92 (1976)
88. W.H. Press, S.A. Teukolsky, W.T. Vetterling, B.P. Flannery, *Numerical Recipes in C*, 2nd edn. (Cambridge University Press, Cambridge, 1992). Reprinted 1997
89. A.H.G. Rinnooy Kan, G.T. Timmer, Stochastic global optimization methods part II: Multi level methods. Math. Program. **39**, 57–78 (1987)
90. T.M. Roberts, Explicit eigenmodes for anisotropic media. IEEE Trans. Magn. **26**(6), 3064–3071 (1990)
91. T.M. Roberts, H.A. Sabbagh, A model for eddy-current interactions with advanced composites, in *Conference Proceedings: Review of Progress in Quantitative Nondestructive Evaluation* ed. by D.O. Thompson, D.E. Chimenti, vol. 5B (Plenum Press, New York, 1986), pp. 1105–1111

92. T.M. Roberts, H.A. Sabbagh, L.D. Sabbagh, Electromagnetic interactions with an anisotropic slab. IEEE Trans. Magn. **24**(6), 3193–3200 (1988)
93. T.M. Roberts, H.A. Sabbagh, L.D. Sabbagh, Electromagnetic scattering for a class of anisotropic layered media. J. Math. Phys. **29**, 2675–2681 (1988)
94. A.H.G. Rinnooy Kan, G.T. Timmer, Stochastic global optimization methods part i: Clustering methods. Math. Program. **39**, 27–56 (1987)
95. S.G. Ruehm, C. Corot, P. Vogt, S. Kolb, J.F. Debatin, Magnetic resonance imaging of atherosclerotic plaque with ultrasmall superparamagnetic particles of iron oxide in hyperlipidemic rabbits. *Circulation* **103**, 416–422 (2001)
96. H.A. Sabbagh, Maser spin dynamics, Goddard Space Flight Center, Greenbelt, X-523-66-448 (1966)
97. H.A. Sabbagh, Notes on the Spin-Hamiltonian, Goddard Space Flight, Greenbelt, X-520-66-3 (1966)
98. H.A. Sabbagh, Thermal noise in spin-phonon systems. IEEE Trans. Sonics Ultrasonics **SU-16**(3), 147–156 (1969)
99. H.A. Sabbagh, A model of eddy-current probes with ferrite cores. IEEE Trans. Magn. **MAG-23**(3), 1888–1904 (1987)
100. L.D. Sabbagh, et al., A computational model for electromagnetic interactions with advanced composites, in *Proceedings, 4th Annual Review of Progress in Applied Computational Electromagnetics, Monterey, CA, March 22–24* (1988)
101. H.A. Sabbagh, Splines and their reciprocal-bases in volume-integral equations. IEEE Trans. Magn. **29**(6), 4142–4152 (1993)
102. H.A. Sabbagh, R.G. Lautzenheiser, Inverse problems in electromagnetic nondestructive evaluation, in *Nonlinear Phenomena in Electromagnetic Fields*, ed. by T. Furuhashi, Y. Uchikawa (Elsevier, Amsterdam, 1992), pp. 177–180
103. H.A. Sabbagh, R.G. Lautzenheiser, Inverse problems in electromagnetic nondestructive evaluation. Int. J. Appl. Electromagn. Mater. **3**, 253–261 (1993)
104. H.A. Sabbagh, R.G. Lautzenheiser, Inverse problems in electromagnetic nondestructive evaluation, in Review of Progress in Quantitative Nondestructive Evaluation, ed. by D.O. Thompson, D.E. Chimenti, vol. 13 (Plenum Press, New York, 1994), pp. 911–918
105. H.A. Sabbagh, L.D. Sabbagh, S.N. Vernon, Verification of an eddy-current flaw inversion algorithm. IEEE Trans. Magn. **22**(6), 1881–1886 (1986)
106. H.A. Sabbagh, T.M. Roberts, L.D. Sabbagh, A computational model for electromagnetic interactions with advanced composites, in *Proceedings, 2nd Annual Review of Progress in Applied Computational Electromagnetics, Monterey, CA, March 18–20* (1986)
107. H.A. Sabbagh, D.J. Radecki, S. Barkeshli, B. Shamee, J.C. Treece, S.A. Jenkins, Inversion of eddy-current data and the reconstruction of three-dimensional flaws. IEEE Trans. Magn. **26**(2), 626–629 (1990)
108. H.A. Sabbagh, E.H. Sabbagh, R.K. Murphy, Recent advances in modeling eddy-current probes, in *Review of Quantitative Nondestructive Evaluation*, ed. by D.O. Thompson, D.E. Chimenti, vol. 21 (American Institute of Physics, College Park, 2002), pp. 423–429
109. H.A. Sabbagh, R.K. Murphy, E.H. Sabbagh, J.C. Aldrin, J.S. Knopp, M.P. Blodgett, Stochastic-integral models for propagation-of-uncertainty problems in nondestructive evaluation, in *39th Annual Review of Progress in QNDE, Denver, Colorado, July 15–20* (2012)
110. E.H. Sabbagh, R.K. Murphy, H.A. Sabbagh, M. Cherry, A. Pilchak, J.C. Aldrin, C. Annis, Stochastic-integral models for characterizing random grain noise in titanium alloys, in *40th Annual Review of Progress in QNDE, Baltimore, July 21–26* (2013)
111. H.A. Sabbagh, R. Kim Murphy, E.H. Sabbagh, J.C. Aldrin, J.S. Knopp, *Computational Electromagnetics and Model-Based Inversion: A Modern Paradigm for Eddy-Current Nondestructive Evaluation* (Springer, New York, 2013)
112. H.A. Sabbagh, R.K. Murphy, E.H. Sabbagh, J.C. Aldrin, C. Annis, J.S. Knopp, Stochastic inverse problems: models and metrics, in *41st Annual Review of Progress in QNDE, Boise, July 20–25* (2014)

113. V.L. Safonov, H.N. Bertram, Impurity relaxation mechanism for dynamic magnetization reversal in a single domain grain. Phys. Rev. B **61**(22), R14893–R14896 (2000)
114. R. Saito, G. Dresselhaus, M. Dresselhaus, *Physical Properties of Carbon Nanotubes* (Imperial College, London, 2005)
115. A. Schatzberg, C. Altman, J. Plasma Phys. **26**(part 2), 333–344 (1981)
116. M.P. Sharrock, Measurement and interpretation of magnetic time effects in recording media. IEEE Trans. Magn. **35**(6), 4414–4422 (1999)
117. A.E. Siegman, *Microwave Solid-State Masers* (McGraw-Hill Book Company, New York, 1964)
118. T.J. Silva, C.S. Lee, T.M. Crawford, C.T. Rogers, Inductive measurement of ultrafast magnetization dynamics in thin-film permalloy. J. App. Phys. **85**(11), 7849–7862 (1999)
119. L.B. Silveira, J.G. Santos, F. Pelegrini, C. Gansau, N. Buske, P.C. Morais, Magnetic resonance study of zero-field-frozen magnetite-based biocompatible magnetic fluid. IEEE Trans. Magn. **39**(5), 2642–2647 (2003)
120. C.D. Skouby, Electromagnetic effects of advanced composites. McDonnell Aircraft Company, prepared for Office of Naval Research (ADA 010882) (1975)
121. S.A. Smolyak, Quadrature and interpolation formulas for tensor products of certain classes of functions. Soviet Math. Dokl. **4**, 240–243 (1963)
122. D.K. Stiles, B. Oakley, Simulated characterization of atherosclerotic lesions in the coronary arteries by measurement of bioimpedance. IEEE Trans. Biomed. Eng. **50**(7), 916–921 (2003)
123. M. Stoyanov, User manual: TASMANIAN sparse grids. Oak Ridge National Laboratory (2013)
124. R. Street, D.C. Crew, Fluctuation aftereffect in magnetic materials. IEEE Trans. Magn. **35**(6), 4407–4413 (1999)
125. K. Suchy, C. Altman, J. Plasma Phys. **13**(part 3), 437–449 (1975)
126. K.J. Sun, R.A. Wincheski, C. Park, Magnetic property measurements on single wall carbon nanotube polyimide composites. J. Appl. Phys. **103**(2), (2008)
127. A. Tarantola, *Inverse Problem Theory and Methods for Model Parameter Estimation* (Society for Industrial and Applied Mathematics, Philadelphia, 2005)
128. J.C. Treece, T.M. Roberts, S.D. Schunk, Electromagnetic imaging for reconstruction of flaws in advanced composites, in *Conference Proceedings: Review of Progress in Quantitative Nondestructive Evaluation*, ed. by D.O. Thompson, D.E. Chimenti, vol. 7A (Plenum Press, New York, 1988), pp. 349–356
129. J.C. Treece, T.M. Roberts, D.J. Radecki, S.D. Schunk, Detecting micro-structures and flaws in composites using eddy-current instrumentation, in *Conference Proceedings: Review of Progress in Quantitative Nondestructive Evaluation*, ed. by D.O. Thompson, D.E. Chimenti, vol. 8B (Plenum Press, New York, 1989), pp. 1519–1526
130. J.C. Treece, H.A. Sabbagh, B.F. Shamee, Final report: eddy-current detection of prepreg FAWT. Sabbagh Associates, SA/TR-1/90 (1990)
131. P.M. van den Berg, A. Abubakar, Contrast source inversion method: state of art. Prog. Electromagn. Res. PIER **34**, 189–218 (2001)
132. V. Volpe, Conductivity and electromagnetic shielding characteristics of graphite/epoxy laminates. J. Compos. Mater. **14**, 189–197 (1980)
133. L. Wan, Y. Ma, J. Guo, Damage analysis of 3D braided composite material using embedded carbon nanotube thread sensors. Mater. Eval. **74**(6), 919–928 (2016)
134. Y. Wang, M.N. Afsar, R. Grignon, Complex permittivity and permeability of carbonyl iron powders at microwave frequencies. IEEE Antennas Propag. Soc. Int. Symp. **4**, 619–622 (2003)
135. W.S. Weiglhofer, A. Lakhtakia, Waves and fields: from uniaxial to biaxial mediums, in between and beyond, in *8th International Conference on Electromagnetics of Complex Media, Lisbon, 27–29 September 2000*. Avaiable through DTIC, ADP011588 (2000)
136. D. Weller, A. Moser, Thermal effect limits in ultrahigh-density magnetic recording. IEEE Trans. Magn. **35**(6), 4423–4439 (1999)

137. R.A. Wincheski, M. Namkung, S.M. Paik, J. Smits, Carbon nanotube based magnetic tunnel junctions for electromagnetic nondestructive evaluation. Mater. Res. Soc. Symp. Proc. **721**, E6.10.1–E.6.10.5 (2002)
138. R.A. Wincheski, M. Namkung, P. Williams, J. Smits, Four terminal carbon nanotube sensor for magnetic field measurement, in *Presented at 2004 MRS Spring Meeting, April 14, San Francisco* (2004)
139. R.A. Wincheski, J.-W. Kim, G. Sauti, E. Wainwright, P. Williams, E.J. Siochi, Nondestructive evaluation techniques for development and characterization of carbon nanotube based superstructures, in *Presented at the Annual Review of Progress in Quantitative Nondestructive Evaluation* (2014)
140. M.-F. Wong, J. Carette, A. Hadjem, J. Wiart, Stochastic electromagnetic modeling with uncertain dielectric properties using FDTD, in *24th Annual Review of Progress in Applied Computational Electromagnetics, ACES, Niagara Falls* (2008), pp. 450–455
141. J.M. Wozencraft, I.M. Jacobs, *Principles of Communication Engineering* (Wiley, New York, 1965)
142. D. Xiu, Efficient collocational approach for parametric uncertainty analysis. Commun. Comput. Phys. **2**(2), 293–309 (2007)
143. D. Xiu, *Numerical Methods for Stochastic Computations: A Spectral Method Approach* (Princeton University Press, Princeton, 2010)
144. X. Xu, J. Qiu, H. Ji, T. Takagi, Detection of delamination in laminated CFRP composites using eddy current testing: simultion and experimental study. Int. J. Appl. Electromagn. Mech. **57**, 177–192 (2018)
145. F. Yu, P.B. Nagy, Numerical method for calculating the apparent eddy current conductivity loss on randomly rough surfaces. J. App. Phys. **95**, 8340–8351 (2004)
146. N. Zabaras, Solving stochastic inverse problems: a sparse grid collocation approach, in *Computational Methods for Large-Scale Inverse Problems and Quantification of Uncertainty*, ed. by People on Earth (Wiley, Hoboken, 2001)
147. R. Zorgati, et al., IEEE Trans. Magn. **27**(6), 4416–4431 (1991)

Index

H. A. Sabbagh et al., *Advanced Electromagnetic Models for Materials Characterization and Nondestructive Evaluation*, Scientific Computation,
https://doi.org/10.1007/978-3-030-67956-9